Keiji Sano
Takao Asano
Akira Tamura

Acute Aneurysm Surgery

Pathophysiology and Management

Springer-Verlag Wien GmbH

Keiji Sano, M.D., D.M.Sc., F.A.C.S. (Hon.), Professor of Neurosurgery and Chairman, Department of Neurosurgery, Teikyo University School of Medicine, and Emeritus Professor of Neurosurgery, University of Tokyo, Japan

Takao Asano, M.D., D.M.Sc., Professor of Neurosurgery, Saitama Medical Center, Saitama Medical School, Kawagoe, Japan

Akira Tamura, M.D., D.M.Sc., Professor of Neurosurgery, Teikyo University School of Medicine, Tokyo, Japan

© 1987 by Springer-Verlag Wien
Originally published by Springer-Verlag Wien New York in 1987
Softcover reprint of the hardcover 1st edition 1987

The use of registered names, trademarks, etc. in the publication does not imply, even in the absence of a specific statement, that such names are exempt from the relevant protective laws and regulations and therefore free for general use.

Product Liability: The publisher can give no guarantee for information about drug dosage and application thereof contained in this book. In every individual case the respective user must check its accuracy by consulting pharmaceutical literature.

With 133 partly colored Figures

Library of Congress Cataloging-in-Publication Data. Sano, Keiji. Acute aneurysm surgery. 1. Intracranial aneurysms—Surgery. 2. Intracranial aneurysms. I. Asano, Takao, 1943– . II. Tamura, Akira, 1943– III. Title. RD594.2.S26 1987. 616.8′1. 87-9428

ISBN 978-3-7091-3986-8   ISBN 978-3-7091-3984-4 (eBook)
DOI 10.1007/978-3-7091-3984-4

PREFACE

Research into the Western literature reveals an incidence of 10–12 and of 10–11 per 100 000 of the population of the U.S.A. of subarachnoid hemorrhage (SAH) and of ruptured intracranial aneurysms respectively. In Europe the corresponding figures vary from 6 to 24 and 3 to 10, whereas in Japan they are higher and are found to be 21 to 24 and 15. Without doubt the management of subarachnoid hemorrhage remains of the most important clinical problems in neurosurgical practice world-wide.

Some years ago, the novel by James Clavell called "Shogun" and its television and movie counterparts created a sensation. The word Shogun has become well known. The first Shogun of the type referred to in this novel was Yoritomo (1147–1199) of the Minamoto family. After he won a decisive victory over the Taira family and put an end to the longtime struggle between these families, Yoritomo was appointed "Shogun", the head of all samurais, by the emperor in 1192. He then went on to create the Shogunate Government in Kamakura, the first such government in Japanese history.

On the last day of December, 1198, on a cold windy day when he was attending the opening ceremony of a new bridge across a river nearby, all of a sudden he groaned in agony, lost consciousness and fell from his horse. His contemporaries believed that he had seen the dreadful ghosts of his defeated enemies, but the details of this incident were not fully recorded. Apparently he recovered from this ictus and assumed a normal life style again. But then suddenly, on January 13, 1199, he died. (The same year that his contemporary, King Richard, Cœur de Lion, of England died.) No autopsy was done. From this brief account of his death, however, one may surmise that he suffered initially from a subarachnoid hemorrhage, survived the first attack, and then died of rebleeding.

His case reminds one of the death of Carl August, the Crown Prince of Sweden (Stefan Zweig, Marie Antoinette, 1932; Bengt Ljunggren et al., Surgical Neurology, 1984). He too fell from his horse while inspecting two Hussar regiments on May 31, 1810, and died within an hour. First it was thought that he had been poisoned. This rumor led a mob to carry out the notorious lynching of the innocent Marshal, Count Hans Axel von Fersen, who once had been Marie Antoinette's (1755–1793) lover. The necropsy, however, revealed he had died of a spontaneous subarachnoid hemorrhage. The death of Crown Prince Carl August caused a change in the Swedish dynasty from the Holstein-Gottorp family to the

Bernadotte family. Incidentally, Yoritomo's death also caused the change of the ruling family from the Minamoto to the Hojo.

Subarachnoid hemorrhage due to ruptured intracranial aneurysm may cause sudden death which, in case of a man of power, may shake the state, as would an assassination, because of its unexpectedness. In order to prevent such bleeding, or at least its recurrence as seen in Yoritomo's case, surgery for ruptured aneurysm should be performed in the acute or early stage. In the acute stage of SAH, however, we inevitably confront many problems we must solve and many difficulties we must overcome.

This book deals with these problems and difficulties in the hope of finding their solutions and of improving the management of patients with aneurysmal SAH. Even so, it is true, that sometimes we feel like asking with Marcellus:

What might be toward, that this sweaty haste
Doth make the night joint-labourer with the day ...

(Hamlet, Act I, Sc. 1)

Tokyo, September 1987   Keiji Sano

CONTENTS

I. INTRODUCTION

During the past half century, major developments in surgical techniques, diagnostic methods, anesthesia, and adjunctive treatments in the care of patients with subarachnoid hemorrhage (SAH) have been achieved. For instance, the microsurgical techniques by the use of an operating microscope has greatly increased the ease and safety of the direct attack to a ruptured aneurysm. The recent invention of computed tomography (CT) scan has greatly contributed to evaluation of amount and distribution of subarachnoid blood, brain damage due to intracranial hematoma, brain edema, infarction, hydrocephalus, and so on, thus enabling an exact judgement as to the feasibility of operative interventions and the prognosis of patients. The use of hypertonic solutions and neuroleptic drugs combined with advanced cardiorespiratory controls during anesthesia markedly decreased the hazards encountered in aneurysm operations.

In 1950, Ask-Upmark and Ingvar stated that without operation only about one case out of five cases with SAH is to be expected to make a good recovery and become able to take up his old occupation; one case out of five remains crippled and the other three cases die sooner or later from SAH.

Since then, several studies on the natural course of SAH due to aneurysmal rupture have been conducted, obtaining similar or only slightly better results. According to estimation of Kassell and Drake (1982), there are approximately 28,000 cases of SAH due to aneurysmal rupture yearly in North America, of which 10,000 will die or be significantly disabled as a result of the initial hemorrhage, leaving 18,000 available for treatment. 8,000 of these patients will die or be disabled as a result of rebleeding (3,000), vasospasm (3,000), and a variety of medical and surgical comtlications (2,000), leaving 10,000 functional survivors (36%), or roughly one out of 3 cases with SAH. There are also numerous reports concerning the result of surgical treatment of ruptured aneurysms. Most of these surgical series were based on the principle of delayed operation, while in more recent ones, results of early operation in selected patients were also included. Inasmuch as the superiority of the surgical over conservative treatments for ruptured intracranial aneurysms has been amply proven (Alvord and Thorn 1976, Sundt and Whisnant 1978, Drake 1981, Sano 1983), the surgical treatment for ruptured cerebral aneurysms cannot yet be regarded as satisfactory, so far as the overall outcome of the whole group of SAH patients is considered.

Recent clinical studies (Adams *et al.* 1981 a, Ljunggren *et al.* 1981, Ropper and Zervas 1984, Kassell 1984, Nishimoto *et al.* 1985) showed that the management mortality with all the patients including all clinical grades ranged between 27 and 50%, regardless of the operative principle as to the timing of surgery. Obviously, this high mortality is due to the fact that SAH patients being initially in poor condition occupy a considerable percentage and they succumb in the acute stage not responding to any current mode of therapy, or without ever having the chance to receive hospital care. With patients in good condition, much better results have been obtained. Nevertheless, even in this group of patients, the mortality ranged from a few percent to 30%. Further, the percentage of patients with unfavorable outcome, namely those who are unable to resume prior occupation due to minor or major neurological deficits, was found to be as much as 50%.

For the purpose of improving surgical results, the use of a clinical grading system such as devised by Hunt (1974) has proved valuable because the system predicts the outcome of SAH patients with fair certainity and helps to select those SAH patients who are amenable to surgical treatment. In essence, however, such grading systems tell no more than that patients in good condition generally carry good prognosis and those in bad condition bad prognosis. The problem most germane to the surgical treatment of SAH is the timing of operation. It seems clear that acute surgery is theoretically superior to delayed surgery because prevention of rebleeding is earlier accomplished and vigorous medical treatments can be instituted without fear of rebleeding. In this respect, pioneering neurosurgeons found that the acute surgery was dangerous because of technical difficulties in handling angry, swollen brains and of the generally poor results of such operations. Therefore, the principle of intentional delayed operation has been supported by the majority of succeeding neurosurgeons. At least, however, these technical problems of acute surgery have been overcome by the recent development in microsurgical techniques and ancillary measures. Since the feasibility of acute surgery has come to be widely recognized, an international cooperative study has been carried out to compare the surgical result of conventional, delayed operation with that of acute surgery. The study showed that the acute surgery is not so much fraught with danger as had previously been thought. But, the superiority of early surgery over delayed surgery was not statistically proven (Kassell 1986).

As briefly reviewed above, it has increasingly been appreciated that the surgical treatment *per se* affords a rather limited benefit in the overall management of SAH due to rupture of intracranial aneurysms. Neurosurgeons may be satisfied with this share, considering that the purpose of surgical intervention is merely to prevent rebleeding. Needless to say, however, we are daily confronted with all the problems that an SAH patient carries. So far as complications such as acute or chronic hydrocephalus, electrolyte imbalances, cardiopulmonary disorders, and so on, are concerned, they have been well studied and effective therapies against them have been established. What is left behind awaiting urgent solution is the problem concerning ischemic brain damage.

In the course of SAH due to aneurysmal rupture, the brain is successively exposed to major ischemic insult at least on two occasions; the first is at the time of aneurysmal rupture, which causes acute ischemic neurological deficits (AINDs): the second, namely delayed ischemic neurological deficits (DINDs), will take place concomitant with the occurrence of cerebral vasospasm.

In an interim report of the International Cooperative Study, Kassell and Torner (1984) stated that three major causes of mortality and morbidity in 1,272 patients with SAH due to aneurysm rupture were vasospasm (33.5%), direct effect of SAH (25.5%), and rebleeding (17.3%). This direct effect should be interpreted as AINDs, because, at the time of hemorrhage, increase of intracranial pressure, disturbance of cerebral blood flow and microcirculation inevitably occur to cause AINDs.

The severity of AINDs is of paramount importance as it reflects the degree of brain damage inflicted by aneurysmal rupture and plays a decisive role in the future outcome of a SAH patient. The DIND, which is so often linked with the occurrence of vasospasm, is another major factor affecting the outcome Because these two ischemic events to-

gether exert decisive influences over the fate of SAH patients, no further improvement in the management mortality or morbidity would not be expected without exploring measures to combat with them. Progresses in basic researches are in urgent need to develop effective therapies against those two major complications of SAH.

The following chapters dealing with the basic aspect of ischemic brain damage are intended first to give an overview on hitherto accumulated findings on pathogenetic mechanisms involved in AINDs (Chapter II) and DINDs (Chapter III). Chapter IV deals with the roles of phospholipid and free fatty acid (FFA) metabolism in the pathogenesis of vasospasm and ischemic brain edema. Since this new biochemical field is considered to be rather unfamiliar to most neurosurgeons, the chapter is started with a brief account of the basic knowledge (Chapter IV A). In Chapters IV B and IV C, we present our hypotheses concerning vasospasm and ischemic brain edema, in the hope that they might help to unravel the mysteries pertaining to the underlying pathogenetic mechanisms. Chapter V deals with various grading systems for SAH patients. In Chapter VI we discuss surgical indications and decision making, in Chapter VII perioperative care, and in Chapter VIII surgical techniques for various aneurysms are described.

In this book, the term "acute" or "early" (*i.e.*, acute surgery, early surgery, acute stage or early stage) is put to denote that the one in question is "within one week or so after SAH".

II. ACUTE ISCHEMIC NEUROLOGICAL DEFICITS (AINDs)

Introduction

SAH due to rupture of intracranial aneurysms is characterized by the suddenness of the appearance of symptoms. The most frequent symptom is headache which may be accompanied by nausea and vomiting. Often, the patient rapidly lapses into coma of variable severity and duration. Although an exact figure has not been known, it has been estimated that a considerable percentage of SAH patients die immediately after ictus which accounts for 2% of all sudden deaths (Freytag 1966). In the remaining patients, some may recover consciousness or be freed from headache in hours or days, and some may present more persisting major neurological deficits. In SAH, it is recognized as a common pattern that the symptoms or the neurological deficits are the most severe immediately after its onset (AINDs), which tend to gradually improve later. Then, in about half of the patients, secondary deteriorations in the neurological status take place (DINDs). These characteristics in the evolution of symptoms and signs are telling us that the major brain damage is inflicted at the time or close to the time of aneurysmal rupture, and another harmful event occurs later. In this chapter, the nature of brain damages pertaining to AINDs and pathomechanisms involved in their occurrence will be delineated.

A. Pathological Studies with Autopsied Materials

1. Intracranial Hematomas Associated with Aneurysmal Rupture

In the majority of aneurysmal ruptures, the bleeding occurs into the subarachnoid space. The dissipated blood mingles with the cerebrospinal fluid bringing out the typical bloody appearance of CSF, and frequently forms a clot within the subarachnoid space in the vicinity of ruptured aneurysm (subarachnoid hematomata). Sometimes the bleeding occurs also into the brain, forming an intracerebral clot. The occurrence of subdural clot is also known. As stated in the preceding paragraph, the major brain damage is considered to be inflicted in the very beginning of aneurysmal rupture, hence it may well be a process related to formation of either or both of the subarachnoid and intracerebral hematomas.

It seems clear that the intracerebral hematoma is associated with the destruction of cerebral parenchyma where it is formed. This can result in the occurrence of focal neurological deficits and more diffuse, fatal brain damage, depending on the size and localization of the hematoma. In the early reports dealing with autopsied cases, the intracerebral hematoma with or without intraventricular rupture was considered as the prime cause of major neurological deficits and death (Richardson and Hyland 1941, Robertson 1949, Tomlinson 1959, Bebin and Currier 1957, Cromptom 1962). Bebin and Currier (1957) stated; "Deaths due to hemorrhage from ruptured intracranial aneurysms are thought to result from hemorrhage into areas other than (but usually in addition to) the subarachnoid space. The most common is intracerebral with or without intraventricular hemorrhage". In contrast to their view, Freytag (1966) reported with 250 medicolegal cases dying from ruptured aneurysms that intracerebral bleeding was associated with subarachnoid hemorrhage in only 24% of the cases who died immediately after rupture but was seen in 71% of those who survived for some time. The relative low incidence of intracerebral bleeding in these medicolegal cases as compared to foregoing hospital autopsied cases was readily explained by the fact that the former group mostly consisted of sudden deaths while the latter group consisted of cases surviving for much longer periods. It was concluded that it was a massive subarachnoid hematoma at the base of the brain which was solely responsible for the fatal issue. The sudden shifing of vital brain structures by blood rapidly accumulating in the basal cistern under arterial pressure was considered to be the cause of immediate loss of consciousness and respiratory paralysis. A similar conclusion to the above had already been reached by other authors (Robertson 1949, Crompton 1962). Nowadays, the presence or absence of intracerebral or subarachnoid hematoma is readily detected by CT scans. Our recent experiences with CT scans conform to the Robertson's statement that the prognosis of intracerebral rupture is worse than that of massive subarachnoid hemorrhage, although death tends to occur more rapidly in the latter. Thus, the combinations of immediate death and massive subarachnoid hematoma, and that of major neurological deficits and intracerebral hematoma seem to have been well established, although these two types of hematomas often coexist. However, could all the spectrum of brain damages due to aneurysmal rupture be explained on the basis of the destructive effect of these hematomas?

2. Cerebral Infarction Following SAH—Evidence of Cerebral Ischemia

Robertson (1949) for the first time pointed out that intracerebral rupture is not the only cause of neurological deficits due to SAH. The presence of ischemic changes have been observed in the brain, sometimes in vascular territory remote from that of the vessel bearing the aneurysm. He considered spasm of arteries may be the cause of these ischemic changes. The incidence and significance of cerebral infarction has been confirmed by succeeding autopsy studies.

Tomlinson with the 32 autopsied cases concluded that ischemic lesion would account for most disabilities in those surviving with ruptured aneurysms (Tomlinson 1959). Birse and Tom (1960) reported a high incidence of focal infarction found on microscopic examination of tissue appearing grossly normal, or, at most, edematous. In 7 of 8 cases, softening was present in the areas of supply of the parent vessel of the aneurysm, but remote infarction also was found in 6 cases. The pathological characteristics of the ischemic changes have then been described in detail (Smith 1963, Cromptom 1964a). According to Cromptom (1964a), the infarction accompanying a ruptured aneurysm was different in appearance from that due to a cerebral embolus in that the former is pale and bloodless in most cases, whereas it is hemorrhagic and obvious in the latter. Cortical as well as ganglionic infarctions were usually in the distribution of the artery bearing the aneurysm, showing no predilection for watershed areas. The laminar involvement of cortex was found in a few of Crompton's cases, but not in Smith's. The characteristic pale cortical infarct was of different types (Crompton 1964a). In some cases it consisted of total necrosis with a sharp margin, involving cortex and a large wedge of underlying white matter. In other cases it was largely confined to the cortex with slight involvement of the underlying white matter at some points. And there were cases with much less well-defined cortical necrosis, which was patchy and perivascular in distribution. The impression gained was that these types of infarct represented differing grades or degrees of ischemia.

Upon reviewing the autopsy reports of the earlier era, it is evident that brain damage in the acute stage causing AINDs and that in the more chronic stage causing DINDs are intermingled in them, because they mostly consist of hospital autopsied cases, with survival periods encompassing the occurrence of cerebral vasospasm. As later described again in more detail, the association between cerebral vasospasm, DINDs, and small or large cerebral infarcts localized in the vascular territory which underwent vasospasm has been more clearly revealed since the advent of CT scan. Here, it needs to be pointed out that the exact nature of the brain damage afflicted in the acute stage of SAH has never been well demarcated in those preceding autopsy reports using conventional methods of brain fixation. This is partly because histological evidence of ischemic cerebral lesions is not reliable within a short period of time after SAH, especially concerning their exact extent and severity (Crompton 1964a); and partly because in the modern practice of neurosurgery, the brain pathology is severely distorted by supportive measures to maintain life such as the prolonged mechanical ventilation.

Nevertheless, as described by many authors (Tomlinson 1959, Falconer 1954, Birse and Tom 1960, Smith 1963, Crompton 1964), small, patchy foci of cortical necrosis are frequently present and even if they may not show focal neurological lesions, they may result in the disturbance of mental functions, which is one of the most distressing expressions of the morbidity of ruptured cerebral aneurysms (Crompton 1964 a). Although the pathogenetic mechanism underlying this patchy widespread ischemic lesion has not been known, it may be caused by diffuse vasospasm or generalized circulatory disturbances associated with aneurysmal rupture. Until now, this patchy cortical infarctions have not been well delineated in the CT scans. This is not to wonder, considering that cortical lesions may be too minute to be detected by CT scans.

In the acute stage of SAH, the presenting neurological deficits often cannot be explained by such lesions as intracerebral clots or acute hydrocephalus, that will be readily detected by CT scan. Thus the patchy widespread cortical necrosis remains as a possible brain lesion underlying AINDs. AINDs determine the clinical grade of the patient, which is known to exert a most decisive influence on his prognosis. Therefore, it seems of utmost importance to elucidate the pathogenetic mechanism of AINDs, the pathological expression of which is presumably the widespread cortical necrosis. In the next chapter, the factors involved will be analyzed on the basis of experimental as well as clinical studies.

B. Factors Involved in the Pathogenesis of AINDs

1. Immediate Changes in ICP and CBF Following Aneurysmal Rupture: Clinical Studies

Since subarachnoid hemorrhage (SAH) is a bleeding from a major cerebral artery which in severe cases forms a large intracerebral or cisternal clot in a short period of time, it is duely anticipated that the aneurysmal rupture is accompanied by drastic changes in the intracranial pressure (ICP) and cerebral blood flow (CBF). Preceding CBF studies in SAH patients have invariably shown that there is a close relationship between the severity of clinical grade and the lowering of CBF (Heilbrun *et al.* 1972, Sakurai *et al.* 1975) in the acute stage of SAH.

However, from obvious reasons, these CBF studies have been carried out days after the onset of SAH, and they are considered to reflect either the late CBF changes associated with AINDs or those associated with concomitant DINDs. Therefore, results of clinical CBF studies will be discussed in the next chapter in relation to DINDs and cerebral vasospasm.

The ICP changes associated with aneurysmal rupture has been incidentally recorded at the time of rebleeding in SAH patients under continuous ICP monitoring (Nornes 1972). A sudden rise of epidural pressure (EDP) up to 2,000 mm H₂O has been recorded simultaneous to a clinical deterioration suggestive of aneurysmal rebleeding. This acute ICP elevation was clearly different from plateau waves observed in association with intracranial lesions other than SAH. Two different pressure patterns were found in patients who had verified recurrent hemorrhages. One was associated with massive hematoma while the other occurred with edema but only minimal hematoma;

the terms "hemorrhagic-compressive lesion: SAH type II" and "ischemic-edematous lesion: SAH type I" have been used for these two conditions.

In SAH type I, a majority of rebleedings were arrested at EDP levels of approximately the diastolic blood pressure. The EDP then returned to considerably lower levels within minutes. In SAH type II, the ICP approached the systolic blood pressure and both aneurysm leakage and CBF were arrested, usually leading to death within hours. The term "brain tamponade" was thought to be descriptive of this picture. In addition to SAH types I and II, a short-lasting abrupt ICP elevation which was not associated with a definite rebleeding was observed. It was suggested that those pressure patterns are determined by factors such as the volume of extravasated blood, the vasomotor reaction, and the intracranial spatial buffering capacity (Nornes 1973).

2. Immediate Changes in ICP and CBF Following Aneurysmal Bleeding: Experimental Studies

The ICP elevation following aneurysmal rupture was shown to be of such an extent as to temporarily exceed the diastolic blood pressure (Nornes and Magnes 1972, Nornes 1973). CBF is determined by the formula:

$$CBF = k \times PP/CVR = k \times (SAP - ICP)/CVR$$

PP: perfusion pressure; CVR: cerebrovascular resistance; k: constant; SAP: systemic arterial pressure (Lundberg et al. 1974).

The above relationship indicates that the brain is exposed to severe global ischemia when the ICP is extremely elevated following aneurysmal rupture. Although this global ischemia appears to play an important role in the occurrence of AINDs, its exact evaluation as to the severity and duration has been difficult to perform in clinical situations. For the purpose of investigating the initial ischemic event following SAH, we undertook an experiment using an animal SAH model (Asano and Sano 1977).

Although various SAH models have been devised in the past, it is open to question whether or not in any of them an exact reproduction of human SAH due to aneurysmal rupture has been accomplished. Nevertheless, the model we employed was considered to mimic human SAH as much as possible, as SAH was induced by extraction of a

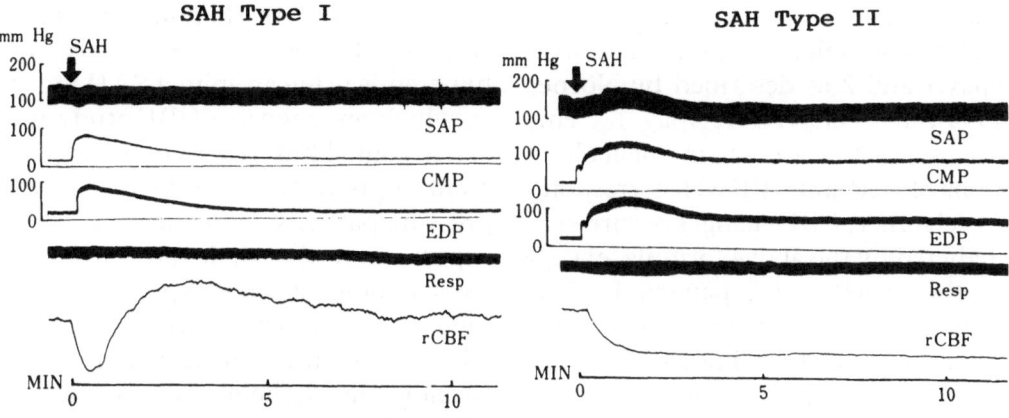

Fig. II-1. ICP patterns in the canine SAH model

kneedle (4–0 atraumatic kneedle with thread) previously inserted into the intracranial portion of the internal carotid artery, and CSF leakage during the experiment was strictly avoided. The animal (adult mongrel dogs) was initially anesthetized with halothane inhalation during the operative procedure. Then the dog was immobilized with gallamine infusion and mechanical ventilation was instituted. The regional cortical blood flow (rCBF) measured by the use of a thermocouple, epidural pressure (EDP), cisterna magna pressure (CMP), systemic arterial pressure (SAP), and EEG were continuously monitored during the experiment. Three hours after induction of SAH, the brain was perfused with a mixed solution of formalin and carbon black solution to examine the state of cerebral microcirculation.

With this SAH model, ICP patterns very similar to those in human SAH were reproduced (**Fig.II-1**). Further, the concomittant changes in other parameters such as rCBF, SAP and EEG were recorded. The ICP patterns obtained could be divided into two types as shown in the figure. In pattern I, ICP rose abruptly near to the arterial diastolic pressure, stayed there for several minutes, and then gradually fell to a level slightly higher than normal. The time-course of rCBF was almost a mirror image of that of ICP, except that there was an overshoot before returning to the original level. The magnitude and duration of this rCBF overshoot tended to be more pronounced as the ICP elevation was greater. EEG activity was invariably suppressed while ICP was elevated, but it gradually improved as rCBF showed an overshoot and then came back to a normal level. However, at three hours after SAH, EEG still showed some slowing. In pattern II, ICP elevation was so severe and prolonged that rCBF remained depressed during the experiment and there was no recovery of EEG activity. In most cases, a marked elevation of SAP (Cushing reflex) occurred. This caused a further elevation of ICP, but did not help to improve the depressed rCBF.

The ICP patterns I and II of our SAH model quite well corresponded to SAH types 1 and 2 as described by Nornes (1973), respectively. It may be emphasized that moment to moment alterations of ICP and CBF well correspond to each other. CBF changes occurred as predicted by the above formula, except for the overshoot in pattern I. This overshoot of rCBF represents reactive hyperemia which is known to occur following a temporary increase of the CSF pressure (Häggendahl *et al.* 1970 a, b). Jakubowski *et al.* (1982) also showed the occurrence of reactive hyperemia using an animal SAH model in which an extensive CBF study was carried out. Thus, results of our experiment support Nornes's view that the pressure patterns following aneurysmal rupture show the whole range from full spatial compensation (type 1 SAH) to total decompensation (type 2 SAH). This acute intracranial hypertension heralds the ischemic events which follow the aneurysmal rupture.

3. Intracranial Hypertension Following SAH

The effect of intracranial hypertension on cerebral blood flow and metabolism has been the object of numerous clinical and experimental studies. Among hitherto established concepts regarding the relationship between ICP and CBF, those pertinent to SAH are briefly reviewed in this section. Although these basic concepts are indispensable to understand the pathomechanism operating in SAH, they would still leave some aspects unexplained as they have been mostly derived from experimental models much simpler than SAH. The complexity of SAH seen as a pressure loading phenomenon will also be depicted.

a) The Pressure Buffering Capacity and Volume/Pressure Relationship

The cranial cavity which contains CSF, vessels, and the brain is surrounded by the thick meninges and bone. Although the meninges are slightly distensible (Löfgren *et al.* 1973) and the skull can expand in infants (Weed 1935), they essentially behave as a rigid container in most situations. In other words, the craniospinal intradural space is nearly constant in volume. Since its contents, namely the brain, blood, and CSF are nearly noncompressible, a change in volume of one of the components of the intracranial cavity necessitates a corresponding change of the volume of one or more of the other components. This concept well known as Monro-Kellie doctrine is valid in the occurrence of intracranial space-occupying lesions, and may be expressed by the formula below (Lundberg *et al.* 1974).

$$V_{brain} + V_{blood} + V_{CSF} + V_{expansive\ lesion} = V_{intracran}.$$

Since the mass of brain cannot be rapidly decreased, an acute increase in the volume of an intracranial lesion must be accomodated by loss of other components, namely blood and CSF. The readiness with which this accomodation takes place is expressed as the pressure buffering capacity. When this capacity is exceeded, ICP becomes elevated. Such a relation is well illustrated by the pressure/volume curve (Langfitt 1969). In this experiment, Langfitt used a slowly expanding (1 ml/hour) supratentorial balloon. The horizontal portion of the curve represents the pressure buffering capacity of the animal, which is a sum of several compensatory mechanisms, namely shift of brain tissue, shift of CSF into the spinal subarachnoid space, increased absorption of CSF, and squeezing out of blood from the cerebral vascular bed.

It is well known that the pressure/volume relationship is greatly affected by the rapidity of the volume increase. Löfgren (1973) showed that during rapid elevation of the CSF pressure (0.08–1.45 ml/sec), the compensation is due to expansion of the spinal dural sack (70%) and compression of the cerebral venous bed (30%). At slower elevation of the CSF pressure (4.8–153 µl/min), however, the absorption of CSF via arachnoid villi appears to be the principal pressure buffering mechanism (Mann *et al.* 1978). Also it was shown that compared to other mammalian species, man has a high rate of CSF formation of about 400 µl/min (Bering *et al.* 1963, Cutler *et al.* 1968) and lowest resistance of arachnoid villi to CSF outflow (Mann *et al.* 1978). Thus, at least 2 ml/min of CSF can be absorbed in the presence of elevated pressure in man (Nelson *et al.* 1971). The exhaustion of pressure buffering capacity is represented by the steep portion of the curve (spatial decompensation). At this stage, the brain is "tight" and even small increments of

the intracranial contents may be accompanied by critical rises in ICP.

Thus, the ICP changes in SAH appear to conform to the general principle of the pressure/volume relationship, the SAH types I and II representing the spatial compensation and decompensation, respectively. However, compared to the usual pressure/volume relationship in space occupying lesions, SAH is unique in the point that a very high ICP level is reached in the beginning of aneurysmal rupture. This is probably ascribed to the nondistensibility of the intracranial cavity, into which the intraarterial pressure of the aneurysm bearing artery is instantaneously transmitted. As the pressure buffering system starts to function, ICP is kept lower than it otherwise would have been. This again increases the pressure differential between the intraarterial pressure and ICP, resulting in further bleeding and elevation of ICP. Therefore, to some extent the pressure buffering (mainly CSF outflow) may have the effect of delaying hemostasis and permitting a larger accumulation of blood intracranially (Löfgren and Zwetnow 1972). Usually, however, this ICP rise acts as a tamponade, permitting formation of an effective clot within minutes.

The subsequent course of ICP following hemostasis is primarily determined by the efficiency of CSF outflow. When it is effective, there is an exponential fall in ICP as seen in SAH type I. By the same token, the prolonged ICP elevation seen in SAH type II would largely represent the malfunction of CSF outflow system. The impaction of arachnoid villi by red blood cells will increase the CSF outflow resistance (Steiner et al. 1975). Also, SAH frequently causes obliteration of the basal cistern by formation of cisternal clot. This condition leads to obstruction not only to communication of pressure between the supra- and infratentorial spaces, but also to the pathway of CSF outflow. It has been shown that the transtentorial obstruction markedly affects the pressure/volume relationship (Löfgren et al. 1973), and when it is complete, elevation of the ICP above the systolic blood pressure has no effect on the spinal fluid pressure (Langfitt et al. 1964). Thus it may be stated that SAH is a condition which is especially liable to cause spatial decompensation. When further space-occupying events such as acute hydrocephalus, edema, rebleeding, and vascular congestion occur, ICP will steeply rise as the brain is already "tight". Then, transtenorial herniation will follow because of the preexisting pressure difference between the supra- and infratentorial spaces.

As shown above, ICP control is considered to be of prime importance in the management of acute SAH. For this purpose, continuous CSF drainage from the lateral ventricle is clearly the most effective method, although it may increase the danger of rebleeding when used before aneurysmal clipping.

b) Effects of Increased ICP on CBF: Autoregulation

It is well known that a generalized increase in ICP per se is not harmful to the brain. It affects the brain through cerebral ischemia due to diminution of the effective perfusion pressure which

is the difference between the systemic arterial pressure and the ICP (Zwetnow 1968, Zwetnow et al. 1970, Häggendahl et al. 1970a, Johnston and Rowan 1974). In patients suffering from brain tumors, the rise in ICP above a certain level, was associated with a definite decrease in CBF (Kety et al. 1948). The acute increase in ICP by an intrathecal injection of Ringer's solution caused a significant CBF (internal carotid flow) reduction at 380 mm H_2O, and at a CSF pressure of 920 mm H_2O mean flow averaged 25% less than the control value (Greenfield and Tindall 1965). Such ICP elevation in humans was not accompanied by deterioration of neurological conditions (Evans et al. 1951, Wolff 1963). In animal experiments, however, a very high ICP exceeding the mean arterial pressure, was associated with an immediate cessations of CBF, electrical activity and spontaneous respiration, together with a marked rise in the SAP (Cushing 1901, Neely and Youmans 1963, Kramer and Tuynman 1967).

Since the existence of autoregulation of CBF was demonstrated in man (Lassen 1959), numerous studies have been conducted on this subject. It has been established that in the normal brain, CBF is maintained constant despite rather wide variations in cerebral perfusion pressure (Lassen 1974). The autoregulation has a lower limit as well as an upper limit, which are influenced by arterial PCO_2 (Harper 1966, Ekström-Jodal et al. 1972). As well as by the activity of sympathetic system (Edvinsson and MacKenzie 1977). As to the underlying mechanism, a myogenic responses of the smooth muscle cells of the arteriolar wall to stretch induced by variations in transmural pressure (Bayliss effect: Bayliss 1902) has been considered as the most direct one (Reivich 1968, Ekström-Jodal 1970, Lassen 1974). In addition to the myogenic mechanism, operations of metabolic as well as neural factors have been suspected, although their roles remain controversial (Purves 1972, 1978).

Insomuch as the CBF autoregulation has an obvious physiological importance in the maintenance of normal CBF and brain function in the face of varying cerebral perfusion pressure, its vulnerablity to various insults such as lactacidosis, hypercarbia, and any kind of intracranial lesions has also been shown (Harper 1966, Rapela and Green 1967, Lassen 1974). If autoregulation is impaired, CBF becomes totally dependent on the cerebral perfusion pressure. Therefore, the intracranial hypertension, an ominous sign of grave intracranial lesions, can cause a significant CBF diminution.

To study the effect of intracranial hypertension on CBF and its autoregulation, either one or both of two different methods for ICP elevation, i.e., the intrathecal infusion of artificial CSF under pressure and the inflation of a balloon placed in the epidural space, have been employed. In the former type of experiment, it was a common finding that the CBF did not change appreciably if the ICP was varied from about –15 to about 100 mm Hg, i.e. if the cerebral perfusion pressure varied between about 140 and 30–50 mm Hg (Häggendahl et al. 1970, Zwetnow 1970). Or, even a marked increase in

CBF, associated with the development of systemic hypertension at ICP levels between 50 and 96 was observed (Johnston et al. 1972). On the other hand, an acute increase in ICP produced by expansion of the extradural ballon or injection of saline into the CSF spaces invariably caused a fall in cerebral blood flow. However, when the balloon was gradually inflated by multiple small injections and the resting ICP between injections increased slowly, normal CBF was frequently maintained despite marked intracranial hypertension ranging from 35 to 50 mm Hg (Langfitt et al. 1965). Thus, the ICP level at which CBF was reduced, i.e., the lower limit of autoregulation was found to be similar in both types of experiments, although it was markedly influenced by the rapidity of ICP elevation. Johnston et al. (1973) further examined the changes in CBF during expansion of a subdural balloon in one of two different sites. With an infratentorial balloon, CBF was linearly related to cerebral perfusion pressure, and autoregulation appeared to be lost from the outset. Whereas with a supratentorial balloon, CBF remained constant as ICP was increased to levels around 60 mm Hg, and autoregulation appeared to be effective during this phase. Thus, autoregulation is affected by not only the rate of development but also the method of production of the intracranial hypertension.

SAH is therefore regarded as a condition in which autoregulation of CBF is especially liable to be lost, because the ICP elevation is sudden and the brain stem is usually compressed by cisternal clot. In fact, global impairment of CBF autoregulation was a regular finding in SAH patients (Heilbrun *et al.* 1972, Sakurai *et al.* 1975, Ishii 1979) and in an experimental model of SAH (Jakubowski *et al.* 1982). Preservation or recovery of autoregulation was found to be a good prognostic sign.

c) Effects of Intracranial Hypertension on the Systemic Arterial Pressure: Cushing Response

The fact that cerebral compression, *i.e.*, increased ICP causes a rise in blood pressure has long been recognized. Using the technique of fluid infusion into the subarachnoid space, Cushing (1901) found that an increase of intracranial tension occasions a rise of blood pressure which tends to find a level slightly above that of the pressure exerted against the medulla. He suggested the existence of a regulatory mechanism on the part of the vasomotor center, which enables the blood pressure to remain at a point just sufficient to prevent the persistence of an anemic condition of the brain stem. Since then, whether or not the systemic hypertensive response (SHR) is triggered by an ischemia of the vasomotor center of the medulla, and it represents a means of preserving CBF in the presence of increased ICP, has become the subject of numerous studies. Regarding the first issue, *i.e.*, the triggering mechanism of SHR, various hypotheses have been suggested such as ischemia or hypoxia (either generalized or localized to the brain stem) (Kramer and Tuynman 1967, Hoff and Reis 1969, Zwetnow 1970, Goodman *et al.*

1972, Fitch *et al.* 1977), local distortion of the brain stem (Thompson and Malina 1959, Weinstein *et al.* 1964), direct stimulation of the brain stem or spinal cord (Alexander and Kerr 1964, Dickenson and McCubbin 1963, Johnston 1972a), and alteration in cerebral perfusion pressure acting through unidentified intracranial baroreceptors (Rodbard and Saiki 1952). Regarding the second issue, Langfitt *et al.* (1965) observed that a minimal difference between the rising arterial and intracranial pressures is sufficient to produce a steady improvement in blood flow. Johnston (1972) also reported the rather early occurrence of SHR during a graded increase in ICP (50–90 mm Hg), which caused a marked increase in CBF. However, in the majority of reports, exhaustion of the autoregulatory mechanism seemed to be a prerequisite for SHR to appear (Cushing 1901, Kramer and Tuynman 1967, Zwetnow 1970, Häggendahl *et al.* 1970). In this condition, SHR was not able to restore CBF to normal, only diminishing the degree of flow reduction (Kramer and Tuynman 1967, Häggendahl *et al.* 197a, Fitch *et al.* 1977). Of particular importance is the fact that in the case of supratentorial ballon inflation, the appearance of systemic hypertension is preceded by other responses such as bradycardia, arrhythmia, constriction of both pupils, with unilateral pupillary dilatation, and widening of the pulse pressure. The finally occurring maximal arterial hypertension always ensued in progressive heart failure and death (Kramer and Tuynman 1967, Fitch *et al.* 1977). This heart failure was the result of strain of the left ventricle, followed by dilatation and ischemic lesion of the myocardium, leading to forward failure, as shown by rise of the venous pressure and pulmonary edema (Brown 1956, Kramer and Tuynman 1967, Ducker *et al.* 1968). Thus the systemic hypertension of SHR was regarded as a preterminal event (Fitch *et al.* 1977).

Because of the unanimity of views on the mechanism and role of SHR, the principle regarding the clinical handling of systemic hypertension due to acute intracranial hypertension has remained ambiguous. Needless to say, decompressive measures and supportive therapies such as mechanical ventilation and administration of hypertonic solutions should be immediately instituted. In case an overt systemic hypertension is present, however, would it be better to leave it alone or treat it with infusion of ganglion blocking agents such as Arfonad (Kramer 1970)? So far as SHR represents an adaptive response to acute intracranial hypertension, artificial reduction of SAP will exert an adverse influence on the cerebral circulation, although it may salvage the heart and the life of patient. In such a case, the resulting clinical picture might at best be that of a vegetative state or brain death. If this is to be avoided, the use of ganglion blockers or other antihypertensive drugs should be withheld unless there is an ample margin of cerebral perfusion pressure as judged by ICP monitoring.

This line of thinking also applies to the management of acute SAH, in which a vigorous antihypertensive therapy is occasionally carried out for the purpose of preventing aneurysmal rebleeding. As this is a complicated issue, it is difficult to reach a clear-cut conclusion. Obviously, acute surgery eliminates the danger of systemic hypertension, no matter whether it is spontaneous or artificially induced.

d) Events Following Cessation of Aneurysmal Bleeding: Hyperemia, Vasoparalysis, and Delayed Hypoperfusion

As soon as the aneurysmal bleeding stops, the function of pressure-buffering mechanism, being reflected as a gradual fall of ICP, becomes apparent. We have already seen that SAH can be divided into two types, namely the compensated (SAH type 1) and the decompensated (SAH type 2) ones. Since autoregulation is quickly lost in the beginning of aneurysmal bleeding, the subsequent time course of CBF is primarily governed by the cerebral perfusion pressure, the components of which are SAP and ICP. However, the denominator of perfusion pressure, *i.e.*, cerebral vascular resistance (CVR) may also change. The present section focuses on the changes in CVR during and following aneurysmal rupture.

The fact that cerebral vessels dilate in response to a raised ICP has been well known since the first report of Wolff and Forbes (1928). This vasodilatation, *i.e.*, the decrease in CVR, represents a form of autoregulation, so long as the normal level of CBF is maintained in the presence of raised ICP. When the lower limit of autoregulation is surpassed, or autoregulation itself is impaired, CBF decreases. In the event of further fall of CBF, the maximal vasodilation by means of metabolic, myogenic, and neurogenic mechanisms will be maintained. This is due to trains of metabolic changes, among which acidosis of the extracellular fluid (ECF) caused by the accumulation of CO_2 and lactic acid presumably is the major factor (Häggendahl *et al.* 1970 a, Zwetnow 1970). As to the chemical link between CBF and metabolism, however, involvement of factors other than H^+, such as K^+, adenosine, and Ca^{++} has been suggested (Siesjö *et al.* 1970, Kushinsky *et al.* 1972, Astrup *et al.* 1978, Heuser 1978, Rubio *et al.* 1978, Winn *et al.* 1981). Presently, therefore, more than one metabolic factor are considered to be involved in CBF regulation. Further, their relative contributions to vasodilation may vary temporally within the duration of the stimulus (Winn *et al.* 1981).

The effects of acute decompression on ICP and CBF have been extensively studied in experiments using the intrathecal infusion of artificial CSF or expansion of the extradural balloon (Häggendahl *et al.* 1970 a, Langfitt *et al.* 1965, 1968, Siesjö and Zwetnow 1970). The acute decompression, in the presence of vasodilation due to preceding cerebral ischemia, provoked a rapid increase in CBF, which usually exceeded the normal level. Although the magnitude of this reactive hyperemia were quite variable from animal even in the same experimental condition, it roughly reflected the severity of preceding ischemic insult. Thus, like any other organs, the reactive hyperemia is regarded as a flow repay to the preceding flow debt so that the brain may resume its normal activity. The duration, rather than the height of the peak flow of reactive hyperemia was considered to give an important information on the severity of the disturbance of tissue metabolism (Häggendahl *et al.* 1970 b). While it is assumed that the reactive hyperemia is partly caused by the metabolic consequence of the tissue hypoxia, the CBF can return to normal in spite of a remaining or even progressive extracellular (CSF) lactacidosis (Zwetnow *et al.* 1967, Kjällquist *et al.* 1970). This is suggestive of the involvement of factors other than H^+ in the control of CBF (Siesjö and Zwetnow 1970).

Langfitt *et al.* (1965) suggested that cerebrovascular dilation is a consistent response to an increased ICP, and that continued vascular engorgement, causing a further increase in pressure and ultimate ischemic vasomotor paralysis, is the primary cause of the brain swelling which occurs secondary to an increased intracranial volume irrespective of its cause. This phenomenon of cerebral "vasoparalysis" is also manifested by disappearance of the CBF response to hypercapnia. In this regard, it may worth mentioning that in experiments where CSF pressure was raised in hypercapnic animals, CO_2 could dilate the vessels further in a situation when maximal vasodilatation was assumed to be attained due to the markedly reduced perfusion pressure (20–30 mm Hg) (Häggendahl *et al.* 1970 a, Ekström-Jodal *et al.* 1969). Thus, "vasoparalysis" represents a terminal event of cerebral vessels, which are maximally dilated and has lost the capacity to react to any kind of stimulus. This concept has gained so much popularity that it has frequently been adopted to explain the cause of brain swelling due to any causes. In this regard, however, it should be recalled that whereas an increase in the cerebral blood volume may well occur following intracranial balloon compression or trauma (Langfitt et al. 1968), cerebral edema may also be developing. Subsequently, it has been shown that the increased ICP following trauma is not the result of an increase in the cerebral blood volume (CBV) (Lowell et al. 1971), and that the increase in the resting ICP between balloon inflations is probably due to developing cerebral edema (Miller et al. 1973). Therefore, it may be safely stated that the concept of vasoparalysis in its original implication (Langfitt et al. 1965) is valid only when the pressure-buffering capacity has been exhausted due to the preexisting space-occupying lesion or developing brain edema. The role of vasoparalysis may reside in the aggravation of brain edema due to an increased filtration pressure in terminal vascular trees (hydrostatic edema) (Langfitt et al. 1968). Another issue pertinent to the CBF changes following SAH is delayed hypoperfusion. As discussed above, transient global ischemia either due to arterial occlusion or raised ICP, is followed by reactive hyperemia. As reactive hyperemia ceases, CBF further decreases and stabilizes at a level slightly below the normal value (Hossmann et al. 1973, Levy et al. 1979, Kofke et al. 1979). This postischemic hypoperfusion may be coupled with an increased cerebral metabolic activity (Hossmann et al. 1976, Levy and Duffy 1977, Nemoto et al. 1981, Ginsberg et al. 1985), leading to an imbalance between oxygen demands of the tissue and the oxygen availability, hence to secondary postischemic relative hypoxia (Hossmann 1982). The reasons for the occurrence of postischemic hypoperfusion and hypermetabolism have been ascribed to loss of CBF regulation (loss of CO_2 reactivity in the presence of autoregulation), and partial uncoupling of oxidative metabolism, respectively (Hossmann 1982). Consequently, lactacidosis ensues, which may be more excessive in incomplete ischemia than in complete ischemia. Since excessive lactacidosis was shown to be toxic to the brain (Myers 1979, Welsh et al. 1980) and ischemia due to SAH is in most cases incomplete, these events appear particularly relevant to the pathomechanism underlying AINDs.

Thus, reactive hyperemia, vasoparalysis, and postischemic hypoperfusion coupled with hypermetabolism comprise the major hemodynamic and metabolic changes which follow the acute decompression or recirculation in experimental conditions. In SAH, cerebral circulation is affected not merely by these events but also by a number of other factors, such as the cardiorespiratory condition, chemical effects of extravasated blood, brain damage due to intracerebral or subarachnoid clot, effects of surgery, and so on. None the less, the knowledge about basic patterns of hemodynamic changes following transient global ischemia is indispensable for the understanding of pathomechanisms and proper clinical handling of SAH.

4. Pathological Sequelae of Aneurysmal Rupture: Microcirculatory Disturbance and Selective Vulnerability

Since the hemodynamic event following aneurysmal rupture can be regarded as transient global ischemia due to intracranial hypertension, the resultant brain damage, *i.e.*, AINDs, may represent a type of selective vulnerability. The pathomechanism of selective vulnerability has long been a controversial issue and from time to time, it has swung between two extremes, namely the vascular theory (Spielmeyer 1925) and the theory of pathoclisis (Vogt and Vogt 1922). The vascular theory had a revival through discovery of no-reflow phenomenon following transient global ischemia (Ames *et al.* 1968). The theory of pathoclisis, which ascribes the occurrence of selective vulnerability to a topographical difference in physico-chemical characteristics and metabolic activity of a particular site of the brain, has been a main target of basic researches related to ischemic brain damage. These two concepts which directly pertain to the relationship between cerebral blood flow and metabolism, do not oppose each other but are mutually dependent. How these two mechanisms are involved is the central issue concerning the pathogenesis of AINDs.

a) Microcirculatory Disturbances Following Global Ischemia

The no-reflow phenomenon (NRP) was for the first time demonstrated in the rabbit brain by Ames *et al.* (1968). In this study, complete global ischemia was induced by a cuff applied around the neck inflated to the pressure of 350 mm Hg. On recirculation after ischemia of variable durations, the status of cerebral reperfusion was examined by intravascular injection of carbon black solution. Even after an arrest of cerebral circulation for 5 minutes, areas of poor filling with predilection for locations in the thalamus, basal ganglia, and arterial boundary zones of the cerebral cortex were observed. The extent and severity of impairment of reperfusion increased as the duration of complete cerebral ischemia was lengthened. They were also dependent on the period at which carbon black infusion was started and on the height of reperfusion pressure (Cantu and Ames 1969). This impairment of reperfusion was termed "no-reflow phenomenon" (NRP). Based on microscopical and electron microscopical findings, the cause of NRP was initially attributed to microvascular obstruction due to structural changes of the endothelium (swelling and bleb formation) and perivascular glial swelling (Chiang *et al.* 1968). However, later studies (Kuypers and Matakas 1974, Wade *et al.* 1975, Ames 1975, Fisher *et al.* 1979) indicate that NRP is better explained by mechanisms other than pericapillary glial swelling or the structural changes of the endothelium, such as precapillary shunting, increased blood viscosity due to red cell aggregation or vascular constriction.

Regardless of the underlying mechanism, NRP, whenever it occurs, will obviously become an important factor influencing the brain recovery after ischemia. In various models of transient global ischemia, the occurrence of NRP has been confirmed (Ginsberg and Myers 1972, Cuypers and Matakas 1974, Wade *et al.* 1975, Hallenbeck and Furlow 1979, Kagström *et al.* 1983 a, b). We also examined the occurrence of no-reflow phenomenon following transient global ischemia due to artificially raised intracranial pressure in dogs (Asano *et al.* 1976). Consistent with the report of Marshall *et al.* (1975), NRP was not observed following a transient CSF compression at a pressure of 200 mm Hg for 15 minutes. However, when the concomittant Cushing response was inhibited by intravenous infusion of Arfonad, prominent NRP was observed in areas the distributions of which were in good correspondence with those reported by Ames *et al.* (1968) **(Fig. II-2-A)**. The occurrence of no-reflow was accompanied by remarkable changes in the shapes of erythrocytes and platelets in the superior sagittal sinus (Kim and Sano 1977: **Fig. II-2-B**). Contrarily, occurrence of NRP was not observed following transient global ischemia due to bilateral carotid occlusion in gerbils (Levy *et al.* 1975) or four-vessel occlusion in rats (Pulsinelli *et al.* 1982 b). Such a discrepancy as to the occurrence of NRP following transient global ischemia has recently been resolved by the studies of Kagström *et al.* (1983 a, b). They carefully examined the occurrence of NRP by the use of different models of global ischemia, *i.e.*, the CSF compression ischemia and the ischemia induced by four-vessel occlusion combined with arterial hypotension. The

influences of the duration of ischemia and the time interval between the onset of recirculation and CBF measurements on the extent of NRP were also examined. Further, the result obtained with complete cerebral ischemia was compared with that of incomplete ischemia. In the four-vessel occlusion model, a true NRP occurred following 15 minutes of ischemia, in spite of the fact that a normal perfusion pressure was quickly restored at the termination of ischemia. Also in the CSF compression model, the occurrence of NRP was confirmed in spite of the hemodilution due to seeping of artificial CSF into the general circulation. The NRP observed were similar in distribution to those described by Cantu and Ames (1969). Thus, it was concluded that complete ischemia of whatever cause carries the potential of inducing NRP. Of interest is the finding that whereas perfusion impairments following 15 minutes of compression ischemia did not persist, unequivocal residues of no-reflow areas were observed after 90 minutes of recirculation following 30 minutes of compression ischemia. From this result, it was suggested that with extended periods of ischemia, "delayed hypoperfusion" might not be a secondary event that follows upon an initial hyperemia, but might represent a lingering perfusion defect that is already present in the beginning of the recirculation period. In incomplete ischemia of similar duration, on the other hand, NRP was not at all observed, even if ischemic flow rates fell toward zero. Therefore it was considered likely that the NRP is confined to situations with complete cessation of

A

Carbon black
perfusion
10 min after
recirculation

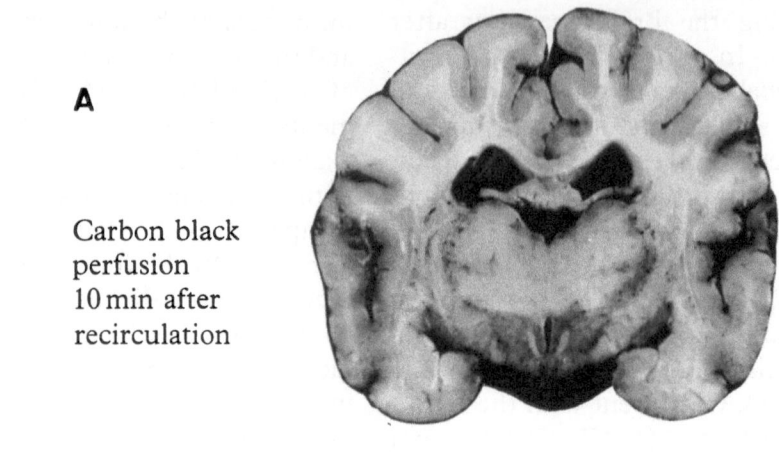

B   Platelet   RBC

control

30 min after
recirculation

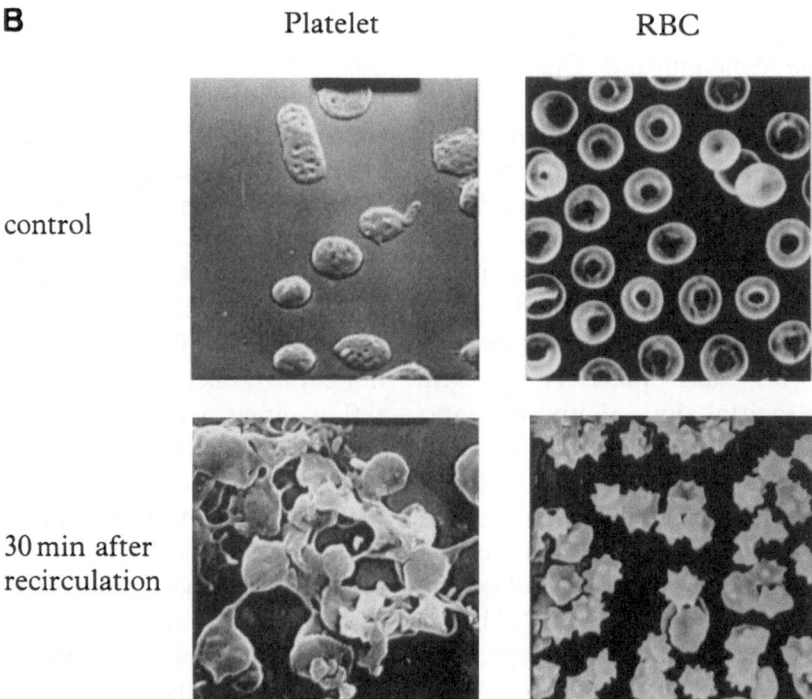

Fig. II-2. Above: An extensive no-reflow following transient complete ischemia due to CSF compression combined with administration or Arfonad (from Asano *et al.* 1976). Below: Pronounced morphological changes of red cells and platelets in the superior saggital sinus blood. The blood sample was obtained from the dog showing the no-reflow phenomenon as shown above. Most of the platelets were in activated forms and massive aggregates were frequently found. The discocytic shapes of normal erythrocytes (RBCs) were transformed into echinocyte, keratocyte, or schizocyte, but no RBC aggregates were found (from Kim and Sano 1977)

blood flow through capillary and pre-capillary vessels, possibly because blood during recirculation is "shunted" through vessels in which opening pressure is rapidly reached (Fischer and Ames 1972).

The above findings related to the occurrence of NRP are of particular relevance to the pathomechanism underlying AINDs. Although it may not be complete, global ischemia is induced due to acute intracranial hypertension in the beginning of aneurysmal rupture. In the canine SAH model (Asano and Sano 1977), the duration of this initial global ischemia extended as long as 10 minutes in SAH type 1 and was much more prolonged in SAH type 2 (**Fig. II-1**). Therefore, it appears likely that NRP occurs following aneurysmal rupture. As literature concerning this subject is scanty, we examined whether or not perfusion defects are found in the canine SAH model (Asano and Sano 1977).

Three hours after induction of SAH, the brain was perfused with carbon black solution and perfusion defects were examined in coronally cut brain slices (**Fig. II-3-A**). To our surprise, perfusion defects of variable extents were found in a high percentage of SAH type 1 animals ($^{20}/_{26}$). Their distribution was always symmetrical and showed a predilection for localization in the thalamus, basal ganglia, and cortical arterial boundary zones, conforming with the pattern of NRP originally reported by Cantu and Ames (1969). In the majority of the SAH type 2 animals, the area of nonfilling in cerebral hemispheres was quite extensive, sparing a small area in the region of hypothal-amus. The infratentorial structures showed good filling even in this condition (**Fig. II-3-B**). These results suggest that compared to other types of transient global ischemia, SAH is particularly liable to develop NRP.

However, it is now open to question whether or not the perfusion impairment observed in our SAH model represented the true NRP, because the time course of CBF in those areas was not followed and the brain perfusion with carbon black was carried out as late as 3 hours after SAH. In this regard, Kagström et al. (1983a) pointed out that the perfusion impairment demonstrated hours after the onset of reperfusion may represent either delayed hypoperfusion or lingering NRP following an extended complete ischemia. Since the ischemia observed in SAH type 1 was not much extended (within 10 minutes), being followed by an immediate reactive hyperemia in the cortical area, it seems possible that what we observed was delayed hypoperfusion. In an SAH model similar to ours, significant edema developed in the brain regions where reperfusion impairment was observed (Shigeno et al. 1983). Jakubowski et al. (1983) used a SAH model somewhat different from ours and showed a prolongation of the central conduction time (CCT) following SAH. The prolonged CCT, frequently found in patients in persistent vegetative state, has been attributed to a selective synaptic delay mainly in certain vulnerable regions, such as the thalamus (Hansotia 1985). These studies together with ours indicate that certain brain regions are selectively vulnerable to the initial global ischemia

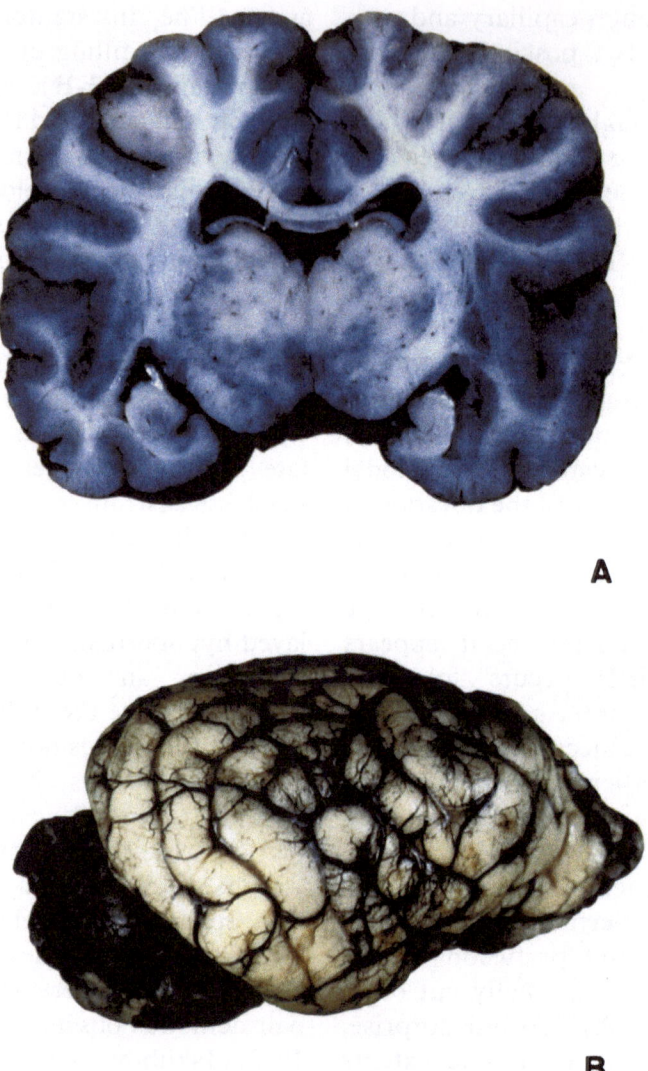

A

B

Fig. II-3. A) Distribution of no-reflow areas in SAH type I. B) A pronounced no-reflow in the cerebral hemispheres in SAH type II. The infratentorial structures were well perfused

following SAH, in regard with microcirculation, electrical activity, and structural integrity.
Compared to other types of global ischemia such as CSF compression, proximal vessel occlusion, and cardiac arrest, SAH is enriched with conditions which hinder the maintenance of normal cerebral circulation: first, loss of autoregulation occurs immediately after aneurysmal rupture; second, the reperfusion pressure is adversely affected when the raised ICP lingers; third, the dissipated blood may exert a

deleterious influence on the cerebral microcirculation due to its vaso-constrictive effect (Meinen *et al.* 1975, Boullin *et al.* 1981, Brandt *et al.* 1981, Sasaki *et al.* 1984), fourth, aneurysmal rupture accompanies some injury of the vessel wall of the parent artery (Stehbens 1984) which may cause luminal narrowing due to vasoconstriction or thrombus formation; fifth, the obliteration of the basal cisterns by hematoma leads to development of transtentorial pressure difference, which aggravates the supratentorial pressure-buffering capacity (Langfitt *et al.* 1964); sixth, the cisternal clot may cause kinking of surface arterioles or their compression in the Virchow–Robin space (Crompton 1964b). In consideration of these particular features of SAH, it is tempting to speculate that the occurrence of NRP is facilitated in SAH and it may bear a causal relationship with the development of AINDs. The implication of the above conjecture to clinical handling of acute SAH is obvious. Nevertheless, it needs to be emphasized that the true nature of perfusion defects observed in experimental SAH models has not yet been clarified. The therapeutic principle would greatly differ depending on whether the perfusion defect is the NRP or delayed hypoperfusion. Since prevention or treatment of AINDs is an essential part of acute aneurysm surgery, further exploration of the problem is warranted.

b) Selective Vulnerability of the Brain to Global Ischemic Insults

In the previous sections, various factors operating in the acute stage of SAH were discussed in relation to the pathogenetic mechanism underlying the occurrence of AINDs. It seems clear that global ischemia due to raised ICP is one of the principal causes of AINDs. However, the link between each of those factors such as the alterations in the ICP, CBF, or the cerebral microcirculation after SAH, and the distribution of the final brain damage, remains elusive. To this missing link would pertain the issue concerning the selective vulnerability of the brain to hypoxia or ischemia.

Brierley *et al.* (Brierley and Excell 1966, Brierley *et al.* 1969, Brierley 1979) have shown that arterial boundary (watershed) zones of the brain are particularly vulnerable to ischemia induced by prolonged systemic hypotension in monkeys. Neuronal alterations were consistently seen in the arterial boundary zones of the neocortex, and less frequently in the hippocampus, basal ganglia and cerebellum (Brierley *et al.* 1971). The vulnerability of the boundary zones to systemic hypotension not complicated with hypoxemia or acidosis, was attributed to the unique anatomy of cortical arterial supply, in which flow is first reduced to a critical level in the boundary zones where perfusion is less than in the major arteries (the borderzone hypothesis) (Brierley 1979). Although comparable pattern of ischemic neuropathology was produced in different conditions such as hypoxia combined with occlusion of the common carotid arteries, prolonged decompression and renovascular hypertension (Brierley 1973, 1979), a different pattern of vulnerability has also been reported.

Miller and Myers (1972) showed that in monkeys subjected to systemic circulatory arrest (more than 12–14 minutes), the damage produced centered in the brain stem, the spinal cord, and the Purkinje cell layer of the cerebellum (the brain-stem injury pattern). On the other hand, when the postarrest period was complicated by prolonged arterial hypotension, a second injury pattern also appeared leading to a significant damage to the cerebral cortex and the basal ganglia (the cerebral cortical injury pattern). Thus, the presence or absence of postarrest hypotension emerged as a major variable affecting both the general clinical outcome and the patterning of lesion distribution.

The brains of patients who had been in a vegetative state due to nontraumatic lesions invariably showed widespread ischemic neuronal damage in the cerebral hemispheres (multifocal infarction and/or laminar necrosis), sparing the brain stem (Brierley et al. 1971, Ingvar et al. 1978, Caronna 1979, Dougherty et al., 1981). The neocortical damage was maximal in the occipital and parietal lobes, and showed no predilection for the arterial boundary zones. In brief, preceding autopsy studies have shown the vulnerability to hypoxia of the neocortical mantle and the hippocampus, and the relative invulnerability of the diencephalon and the brain stem. It is noteworthy that in the report of Dougherty et al. (1981), two SAH patients who bacame vegetative soon after the onset of SAH and remained so until death were included. Neuropathological findings of these two patients were similar to those of others. Assuming that these brains retained the acute brain damage directly incurred by SAH, it seems probable that the brain damage underlying AINDs is of a similar pattern to those of other ischemic insults. This surmise is in harmony with preceding autopsy reports which showed the existence of patchy, widespread foci of cortical and ganglionic necrosis in the brains of patients dying from SAH (Falconer 1954, Tomlinson 1959, Birse and Tom 1960, Smith 1963, Crompton 1964 a). Thus in SAH, like cardiac arrest wherein the hemispheric brain injury was ascribed to the prolonged postarrest systemic hypotension (Miller and Meyers 1972), the lingering effects of intracranial hypertension and arteriolar constriction causing a generalized decrease in CBF even in the absence of systemic hypotension, may account for the occurrence of AINDs better than the primary, global cessation of blood flow.

On the other hand, the study on the pathomechanism underlying selective vulnerability has greatly been facilitated by the precise pathological definition of ischemic cell changes (McGee-Russel et al. 1970, Brown and Brierley 1972, Brierley 1973), the advent of new ischemia models using small animals, and the development of techniques quantitatively to measure the microregional changes in CBF and metabolism. Among recent developments, the studies concerning the "delayed neuronal death" (Kirino 1982, Kirino and Sano 1984) in CA 1 region of the hippocampus are particularly noteworthy.

Although the vulnerability of the hippocampus to ischemia has long been known (Sommer 1880, Vogt and Vogt 1922, Spielmeyer 1925, Lorente de No 1934, Sano and Malamud 1953, Ito et al. 1975, Pulsinelli et al. 1982c), Kirino (1982) for the first time showed that a slow extensive loss of the CA 1 neurons occurs following bilateral carotid occlusion for 5 minutes in Mongolian gerbils. As this change in CA 1 pyramidal cells was very slow, only becoming apparent by light microscopy 2 days following ischemia, it was termed "delayed neuronal death" and considered to represent a hitherto unknown, unique type of ischemic cell change, the regional selectivity of which was hardly explainable by a disturbance in the local vascular bed.

Subsequent studies on the underlying mechanism using similar or somewhat different animal models revealed that: i) during the early recirculation period, an intense 2-deoxyglucose uptake together with hypoperfusion was noted in the hippocampus (Suzuki et al. 1983 a); ii) there was pronounced electrical hyperactivity in the zone CA 1 on the first postischemic day, followed 1 day later by electrical quiescence which coincided with the earliest morphological changes (Suzuki et al. 1983 b); iii) transection of the excitatory perforant path decreased neuronal damage in the CA 1 region, supporting the view that the damage was mediated by excessive release of excitatory transmitters such as glutamate and aspartate (Blomqvist and Wieloch 1985, Wieloch et al. 1985). Thus far, it was beautifully shown that neuronal activation in the limbic system may act as an important mediator and amplifier of the deleteri-

ous reaction leading to neuronal damage (Wieloch 1985). This concept is in harmony with the early postischemic uncoupling wherein metabolism increases disproportionately to flow (Ginsberg et al. 1985). This line of thinking leads to an important clinical implication that cell damage caused by cerebral ischemia can possibly be minimized by specific antagonists to excitatory amino acids (Wieloch et al. 1985, Rothman and Olney 1986).

This therapeutic implication, however, may not directly be applied to clinical situations, because in either of SAH or cardiac arrest, the predominant brain pathology has been shown to be a diffuse, patchy cortical necrosis as described above. This pattern may well conform to the classical principle of selective vulnerability, which implies that the telencephalic cortex and the cerebellar cortex are more vulnerable to anoxia than are the brain stem and spinal cord (Weinberger et al. 1940, Ingvar et al. 1978). Although various hypotheses such as the no-reflow phenomenon, the borderzone hypothesis, the uncoupling between metabolism and blood flow (delayed hypoperfusion coupled with hyper- or preserved metabolism), and delayed neuronal death presumably caused by excitatory amino acid (EAA)-mediated neurotoxicity have been put forward, the nature of brain lesions encountered in clinical situations still remains elusive. Particularly in SAH, the features germane to the pattern of selective vulnerability due to transient global ischemia are subject to distortion because of the occurrence of intracranial complications such as the intracerebral clot, hydrocephalus, and vaso-spasm, or systemic complications affecting the cardiorespiratory functions. Although the problem concerning the pathomechanism underlying AINDs or selective vulnerability would defy immediate solution, the most impor-tant clinical implication of hitherto suggested hypotheses is that "the point of no return", *i.e.*, the irreversible damage, may not be reached within a brief period of ischemia, but it is rather shifted into the period of recirculation.

III. DELAYED ISCHEMIC NEUROLOGICAL DEFICITS (DINDs)

Introduction

A high percentage of SAH patients who survived the first few days following the ictus shows a secondary deterioration of the neurological status such as the drop of the level of consciousness and the appearance of hemiparesis. The neurological deficits which appear delayed to the onset of SAH are called "delayed neurological deficits" and they may be due to any kind of intracranial and systemic disorders occurring after SAH. Thus, delayed neurological deficits may be ascribed to the occurrence of acute hydrocephalus, electrolyte imbalances, rebleeding from the aneurysm, progressive elevation of ICP due to the presence of intracerebral hematoma, brain edema, or cardiorespiratory disorders. In addition to the above complications, there is a group of disorders which is ascribable to a primary impediment of cerebral circulation. A number of autopsy studies have shown the frequent occurrence of focal or wide-spread brain infarction in such cases. In distinction from disorders due to other causes, it is called "delayed ischemic neurological deficits" (DINDs).

Although it has long been conjectured that DINDs are due to the occurrence of cerebral vasospasm, *i.e.*, the angiographical narrowing of the luminal diameter of major intracranial arteries, their causal relationship had been far from clear. It is only recent that it became to be analyzed on the firm basis of multidimensional studies on cerebral blood flow, metabolism, and the brain damage. Further, the true nature of cerebral vasospasm still is an enigma in spite of extensive research efforts. In this chapter, the historical development of the concept of cerebral vasospasm is first described. Then, the current concept of vasospasm in relation to its clinical features and the trend of basic research on its pathogenethic mechanism will be discussed.

A. Historical Consideration

1. The Birth of the Concept of Cerebral Vasospasm

In the beginning of this century, the pathology, symptomatology, and clinical importance of ruptured and unruptured intracranial aneurysms have been well documented, although an accurate diagnosis of the condition during life was thought impossible except for rare occasions (Fearnsides 1916, Symonds 1923). The carotid ligation, which had been widely practiced in a rather indiscriminate fashion during the preceding century, was for the first time carried out for a non-fistulous intracranial aneurysm in 1924 by Wilfred Trotter (Schornstein 1940). In this case, the aneurysm was of traumatic origin and the indication severe epistaxis.

In 1927, Egas Moniz introduced cerebral angiography. Although its diagnostic value was immediately recognized, its use was restricted to selected cases during the following decade. The reports on the result of carotid ligation for cerebral aneurysms published around 1940 comprised of cases in which the diagnosis was mainly made either by operation or localizing symptoms. However, some of those reports are of interest, because into them, the origin of the concept of cerebral vasospasm as well as the origin of later controversies concerning vasospasm can be traced. Jefferson reported on the clinical manifestations and treatment of the saccular aneurysms of the internal carotid artery in the cavernous sinus (Jefferson 1938). Amenability of the lesion to therapy by the carotid ligation was shown, but hazards with the operation were also stressed. As seen in the contemporary papers (Dandy 1939, Schorstein 1940), clear distinction had been made in this period between the immediate and late occurrences of cerebral ischemia following carotid ligation. The former, i.e., immediate

ischemia, was attributed to insufficiency of collateral circulation via the circle of Willis, and the risk of its occurrence was considered to be foreseen by trial digital compressions (Matas test), by test ligatures, and by angiographic studies. Regarding the latter he wrote, "The ill result that cannot be avoided, because it cannot certainly be foreseen, is the late paralysis. Perthes (1920) drew attention to this type of accident, believing it to be due to embolism of the cerebral vessels from clot formed at the site of ligature". Dandy also believed that the deficits which are late in appearing, i.e., develop twelve hours to several days later, are due to cerebral thrombosis and embolism (Dandy 1942).

However, Schorstein (1940), analyzing 60 cases of carotid ligation reported in the literature, could not find evidence of postoperative embolism occluding the cerebral circulation. Therefore he concluded that the above commonly accepted theory had no foundation in fact and he ascribed the delayed appearance of neurological deficits to an anoxemic process. His statement pointed to the heart of the forthcoming problem: "The actual occurrence of neural dysfunction depends on the severity of the anoxemia and on the extent of the area affected; the initiation of the pathological process and its clinical manifestation need not coincide in time. Out methods of neurological examination at the bedside enable us to detect a late stage in the evolution of cerebral anoxemia; an instrument more delicate than the tendon hammer might register its onset. The variability of clinical signs and the delayed appearance of neurological disabilities can be correlated with the conception of an anoxemic process, influenced in its progress by the general health of the patient, particularly his blood-pressure and blood-volume." He also referred to the theory of arterial spasm formulated by Forbes and Wolff (1928) and Penfield (1933).

Nevertheless, he resorted to the finding of Forbes and Cobb (1937) that the narrowing of cerebral vessels in re-

sponse to sympathetic stimulation cannot be a potent cause of cerebral ischemia, and considered that an accumulation of carbondioxide in consequence of a reduced blood-flow should have a vasodilating effect in excess of the hypothetical sympathetic vasoconstriction. The above idea of Schorstein was immediately refuted by Dandy (1942), which was the start of the long-lasting controversy concerning the issue of cerebral vasospasm.

2. Foundation of the Concept of Vasospasm

During the latter half of 1930s, feasibility of the direct attack of intracranial aneurysms either by muscle coating (Tönnis 1936), trapping (Dott 1937), or neck clipping (Dandy 1938) was successively shown, heralding the era of surgical treatment of intracranial aneurysms. The accumulation of surgical cases during the succeeding two decades was naturally accompanied by a remarkable progress in the fields of pathology and radiology. These findings together founded the clinical concept of cerebral vasospasm as described below.

Robertson (Robertson 1949) reported that focal necrosis can occur in SAH at sites remote from the aneurysmal lesion, without an obvious reason like intracerebral hemorrhage or arterial thrombosis. To explain such a finding, he postulated cerebral vasospasm. This hypothesis was soon supported by the angiographical demonstration of cerebral vasospasm by Ecker and Riemenschneider (1951), and succeeding autopsy reports (Tomlinson 1959,

Birse and Tom 1960, Smith 1963, Crompton 1964 a, b). Thus, in most of the successively published reports of surgical series during 1950s, cerebral vasospasm was situated among the main factors influencing the outcome of surgery.

Regardless of the type of operation undertaken, those reports (Norlén and Olivecrona 1953, Falconer 1954, Logue 1956, Botterell et al. 1956, Pool 1959) consistently demonstrated that surgery carried out in the acute stage of SAH generally yielded poor result. The danger of operating in the acute stage of SAH (from one week up to one month after SAH) was found to correspond to the experiences with carotid ligation shortly after SAH (Jefferson 1938, Schorstein 1940). In each of the above series, the occurrence of severe vasospasm was angiographically confirmed in some of the cases. Thus, Norlén and Olivecrona (1951) ascribed the reasons for the unfavorable result of early operation to i) impairment of the cerebral circulation caused by the presence of blood in the subarachnoidal space or clots in the brain substance with consequent rise in the intracranial pressure, and ii) the intense vasoconstriction nearly always present in the cerebral arteries shortly after hemorrhage has occurred and which can be demonstrated in the arteriograms. It may be noticed that the above statement is quite akin to that of Schorstein (1940), differing only in the point that the cause of "anoxemic process" was expressed more lucidly as "cerebral vasospasm". Thus, those early surgical experiences led to formation of a consensus that the occurrence of vasospasm in the acute stage of SAH is a reality, and that the timing of operation is a factor of great importance. To avoid ischemic complications, the neck clipping was considered to be preferable to carotid ligation, even when the aneurysm appeared amenable to the latter type of operation (Norlén and Olivecrona 1951).

The belief held in this period was that vasospasm occurs immediately after aneurysmal rupture. The intense vasoconstriction thus induced was considered as a protective mechanism to prevent further hemorrhage (Ecker and Riemenschneider 1951, Norlén and Olivecrona 1953, Falconer 1954, Logue 1956, Connolly 1961, Fletcher *et al.* 1959, DuBoulay 1963). For instance, Fletcher *et al.* (1959), analyzing one hundred consecutive angiograms, stated: "In two patients vasospasm was found within 24 hours of the time of rupture. It is therefore to be expected that spasm would frequently be found sooner after aneurysmal rupture if angiography were carried out at an earlier date." The immediate arterial constriction following aneurysmal rupture subserving hemostasis appeared to be an adequate response from the teleological standpoint. Further, this view was supported by the pre-existing physiological finding that the larger arteries comprising the circle of Willis have the capacity to constrict in response to mechanical, neuronal, and chemical stimuli (Forbes and Wolff 1928, Echlin 1942, 1965, Pool 1958).

The another belief formed in this period is well represented by the following statement: "Now this spasm, although the crucial factor in stopping hemorrhage and saving life, is harmful in other ways, for it must render the brain ischemic in its territory of supply and on the intensity and duration of this initial spasm will depend to a large degree the extent and severity of the neurological signs, and, although the spasm may relax quite quickly (but usually not completely), recovery from the acute ischemic damage produced in the first few minutes after the bleed may take much longer—hours, days, or weeks—and may, in fact, never be complete (Logue 1956)".

These views, preponderant in this period, are no longer tenable because it has clearly been shown until now that vasospasm does not occur immediately after SAH in humans. None the less, the conviction of neurosurgeon as to the occurrence and role of vasospasm promoted subsequent clinical and basic studies on vasospasm, regardless of "a climate of doubt" surrounding this concept (Denny-Brown 1951, Pool 1958).

3. *Criticism on the Concept of Vasospasm*

The next some fifteen years were started with reports of surgical series which further confirmed the deleterious influence of vasospasm on the outcome of SAH patients. Stornelli and French (1964), analyzing the effect of vasospasm on morbidity in 43 verified SAH patients, showed that vasospasm was the critical factor determining prognosis regardless of the therapy applied. In the entire group of 43 patients, diffuse intracerebral vasospasm, invariably accompanied by elevated spinal fluid pressure, was present in all 4 who died and in only 3 of 28 patients who recovered. Allcock and Drake (1965) analyzed the records of 175 patients. Operation (direct attack) was carried out in 128 of them, and postoperative angiography in 83. When operation was carried out within 10 days following SAH, the incidence of postoperative vasospasm was high (51–61%), compared to that (9%) when operation was carried out later. The incidence of focal signs and stupor in the pre- and postoperative groups was very much higher (65–80%) in those patients showing arterial spasm. Therefore the authors concluded that postoperative spasm was seen more commonly if there was a short period between the hemorrhage and the surgical attack, and recommended to postpone operation for 7–10 days after a bleeding, or longer if spasm was still present.

Wilkins *et al.* (1968) analyzed the charts and angiograms of 259 SAH patients, in which 120 was due to either verified or suspected rupture of an aneurysm. Pre-operative spasm was found in 44 (36.7%) of these patients, in contrast to the very low incidence in other lesions (6.4%) such as craniocerebral trauma, infection, and tumors. No relationship was found between intracranial arterial spasm and either arterial hypertension or the electrocardiographic abnormalities that accompany SAH. It seems noteworthy that the mortality in the no-spasm group ($^{16}/_{56}$: 29%) was a little greater than that in the spasm group ($^9/_{35}$: 26%) in this study. The authors also noted that in a significant number of their patients, the onset of spasm was delayed for several days after SAH.

The clinical data as described above led to formation of an opinion which was lucidly put by Robertson (Robertson 1973): "the leading cause of death and morbidity in the patient with a ruptured intracranial aneurysm who reaches the hospital in other than a moribund state is from the effects of cerebral vasospasm". But, not the all hitherto reported data were in support of this view.

As early as 1959, Potter reported that the occurrence of vasospasm does not necessarily lead to development of corresponding neurological deficits. He concluded that other factors such as the collateral circulation through meningeal arterial anastomoses should be considered before abnormal neurological signs are attributed entirely to spasm.

Crompton (1964 a, b) made a detailed analysis in a consecutive autopsy series of 159 patients dying from ruptured intracranial aneurysms. In the 119 cases in which significant infarction was found, cortical infarction was found in 103 cases and ganglionic infarction in 65 cases. Cortical infarction was usually in the distribution of the artery bearing the aneurysm. Thrombus within the lumen of an artery only contributed to infarction on two occasion. In one, the thrombus originated in the ligated common carotid artery, and in the other in the ruptured aneurysm. Aneurysms lying on or near the midline resulted in bilateral infarction, and ganglionic in-farction, more often than lateralized or peripherally situated aneurysms. Hundred and nine out of the above 119 patients, and 33 out of the 40 patients without cerebral infarct had undergone angiography. The incidences of spasm in patients with and without cerebral infarcts were 37% ($^{40}/_{109}$) and 12% ($^{4}/_{33}$), respectively. Thus there was a considerable difference in the incidence of spasm in the cases with and without cerebral infarcts. However, factors other than vasospasm, such as stenotic atheroma of the cerebral arteries, large subarachnoid hematomata, vascular hypotension, surgery, and etc were also considered to contribute towards the occurrence of cerebral infarction. Combination of these factors was often found in patients with infarction.

In a similar study comparing the angio-graphical and pathological findings, Schneck (1964), and Schneck and Kricheff (1964) found that spasm was present in 62% of patients with infarcts but also in 57% without such lesions. They considered, therefore, that the mere presence of spasm was not in-dicative of the likelihood of cerebral infarction, although narrowing of the vascular lumen by more than 60% and diffuseness of the spasm appeared to be of somewhat greater significance.

Parallel with the angiographical and pathological studies, the influence of vasospasm on rCBF has been inves-tigated. Those reports invariably showed that SAH causes a generalized CBF reduction, which correlates well with the clinical grade of the patient (James 1966, Parkes and James 1971, Ferguson et al. 1972, Heilbrun et al. 1972, Mathew et al. 1974, Ito et al. 1975, Nilsson 1977, Grubb et al. 1977).

However, the correlation between the angiographical spasm and decreased rCBF was generally poor, although in some cases with a severe degree of vasospasm, there was a marked re-duction in flow (Kagström et al. 1966, Zingesser et al. 1968, Heilbrun et al. 1972, Bergvall et al. 1973, Methew et al. 1974, Nilsson 1977, Kelly et al. 1977).

With those controversial data in the background, the preexisting climate of doubt surrounding the concept of va-sospasm culminated in the paper of Milikan (1975). He started with three fundamental questions as to whether there is 1. any clinical picture consis-tently present coincident with known cerebral vasospasm, 2. any relationship between mortality and known vaso-spasm, 3. any relationship between seri-ous brain damage (morbidity) and known vasospasm. As he found no definitive study of patients in the liter-ature, he carried out a survery of cases himself. The charts and angiograms of 198 consecutive patients with proved acute SAH and arteriographically demonstrated intracranial aneurysm were analyzed. Vasospasm was found at the time of angiogram in 41% of all pasients. A larger percentage (52.2%) of those patients without vasospasm had abnormal neurological signs than of those having vasospasm (45.7%).

There was no significant difference in the number and quality of the neurological abnormalities present in pasients with vasospasm and without vasospasm at the time of angiography.

Further in the surgically treated pasients, the mortality was 19% in 63 patients with known vasospasm, whereas it was 19.5% in 82 patients without known vasospasm. In the conservatively treated patients, the mortality was 33% in 18 patients with known cerebral vasospasm and 34% in 23 patients without spasm. Scrutiny of individual cases was added and the author reached the following conclusion: 1. there is no clinical picture consistently present coincident with known cerebral vasospasm; 2. cerebral vasospasm has no effect on the mortality from SAH due to ruptured aneurysm; and 3. there is no relationship between the frequency and severity of the complications from surgical or conservative treatment and the presence or absence of vasospasm.

In as much as this total denial of the role of vasospasm was not congruous to the experience of most neurosurgeons, it obviously demanded a reevaluation of the concept of vasospasm. The first backlash was prepared by the reexamination of the time of appearance of vasospasm. As previously stated, vasospasm had been considered to

occur immediately after aneurysmal rupture. Subsequent angiographical studies revealed, however, that this notion was false. In 1966, Kagström et al. reported that, as a rule, angiography performed during the first few days after the initial bleeding will show no spasm and that spasm is most intense during a period from the sixth to the twelfth day after the bleeding. Wilkins et al. also reported that of the 19 patients who had angiography within 24 hours after SAH, none showed evidence of spasm (Wilkins et al. 1968). This finding has then been supported by succeeding reports (Gurdjian et al. 1969, Bergvall et al. 1973, Odom 1975, Wilkins 1976) and became to be widely accepted. In addition, DuBoulay and Gado (1974) reported that there was no statistical evidence to support the idea that spasm shown at the time of angiography conferred any useful degree of protection.

The above clinical finding that human vasospasm is delayed in its onset unequivocally showed that any study analyzing the correlation between vasospasm and patient's condition should take their temporal relationship into account. This viewpoint had been lacking in previous studies and necessitated a renewed clinical analysis. Simultaneously, great changes were taking place in the field of neurosurgery, due to the introduction of microsurgical techniques, the advent of CT scan, and the development of methods to measure rCBF or brain metabolism. The current concept of vasospasm has been constructed vis-à-vis the recent developments in operative techniques and diagnostic measures.

B. The Current Concept of Cerebral Vasospasm

1. The Time Course of Cerebral Vasospasm and Its Relation to Neurological Deficits

The pioneer observation of Kagström *et al.* (1966) that human vasospasm following SAH due to rupture of intracranial aneurysms is delayed in its onset has been supported by succeeding reports.

Saito *et al.* (1977), by repeated angiograms at intervals of a few days, closely studied the time course of vasospasm and its relationship with neurological deficits in 96 consecutive cases of SAH due to aneurysmal rupture. It was shown that at least 4 days elapsed between SAH and the onset of vasospasm. Vasospasm subsided on an average, 2 weeks after the onset. The delayed onset of human vasospasm has further been confirmed (Weir *et al.* 1978, Kwak *et al.* 1979, Kim *et al.* 1979, Kodama *et al.* 1980), and universally accepted in the International Symposium on Cerebral Arterial Spasm held in Amsterdam in 1980 (Wilkins 1980). Further, Saito *et al.* classifying angiographic vasospasm into three types, *i.e.*, type 1: extensive and diffuse; type 2: multisegmental or multitapering; and type 3: local, noted that type 1 vasospasm was prognostically the most grave. This result was consistent with the preexisting view that only severe vasospasm is capable of causing neurological deficits (Schneck and Kritcheff 1964, Simeone *et al.* 1972, Odom 1973). However, elucidation of the temporal relationship between the occurrence of severe vasospasm and neurological deficits required a new method for the analysis of clinical records.

Fisher *et al.* (1977) set aside all neurological deficits and manifestations occurring on days 1, 2 and 3 after rupture and have included for analysis only events occurring on day 4 or later (the day of the hemorrhage was counted as day 1). Each of the possible causes of the delayed neurological deficit other than vasospasm, such as recurrence of bleeding, acute hydrocephalus, electrolyte disturbance, systemic hypotension, and cerebral embolism was carefully excluded as a factor. Angiography was planned so that it was carried out between the 4th and 13th day after SAH, during which period vasospasm was known to develop. The degree of vasospasm was graded according to the diameter of the residual lumen from 0 (no narrowing) to 4 +. Of 50 cases thus investigated, 25 developed a delayed, new neurological deficit that in each case was ischemic in origin and 25 remained free of new deficit. Of 31 patients with grade 3 + or 4 + vasospasm, 25 (80%) developed a delayed ischemic deficit. Of 19 patients with grade 0, 1 + or 2 + vasospasm, none developed a delayed ischemic deficit. All 25 patients who had a delayed ischemic deficit showed grade 3 + or 4 + vasospasm. In this way, it was shown that without exception, delayed ischemic deficits (DIDs) after day 3 did not occur without severe vasospasm, although vasospasm may occur without a deficit. The DID occurred most often on day 8. Thus the authors concluded that vasospasm accounted for all DIDs and that in the absence of vasospasm DID did not occur.

The above study of Fisher *et al.* for the first time separated the neurological deficits of the first 3 days from the deficits that developed later, and succeeded in demonstrating the temporal relationship between vasospasm and DINDs. The correlation between the two has been more and more reinforced by studies adopting computerized tomographic scanning (CT scan).

2. CT and Vasospasm

Needless to say, the diagnosis of intracranial lesions has greatly advanced since the advent of CT. Especially, clinical management of SAH has been much improved because the nature of underlying lesions such as intracerebral or subarachnoid clot, acute hydrocephalus, brain edema and infarction, and occasionally the aneurysm, are easily discernible on CT scans (Paxton and Ambrose 1974, New and Scott 1975, Pressman *et al.* 1975, Davis *et al.* 1976, Liliequist *et al.* 1977, Lim and Sage 1977, Modesti and Binet 1978, Adams *et al.* 1983). Thus, CT made it possible exactly to differentiate between the brain lesions due to vasospasm and other causes, and contributed to establish the current concept of vasospasm. The relationship between the angiographical spasm, neurological deficit, and CT finding was closely examined by Saito *et al.* (1979). Forty-four consecutive cases of ruptured aneurysm, in which vasospasm was verified by angiogram were selected. Of 35 patients who underwent CT scans, 25 (71%) displayed low-density area (LDA) in the territory of the cerebral arteries involved in vasospasm. There was a tendency that the incidence of LDA was higher as vasospasm was more diffuse and severe.

In 47 cases of verified ruptured aneurysm, Fisher *et al.* (1980) investigated the relationship of the amount and distribution of subarachnoid blood detected by CT to the later development of cerebral vasospasm. The following findings were obtained: when the subarachnoid blood was not detected or was distributed diffusely, severe vasospasm was almost never encountered (1 of 18 cases); in the presence of subarachnoid clots larger than 5 × 3 mm or layers of blood 1 mm or more thick in fissures and vertical cisterns, severe spasm followed almost invariably (23 of 24 cases). There was an almost exact correspondence between the site of the major subarachnoid blood clots and the location of severe vasospasm. Further, every patient with severe vasospasm manifested delayed symptoms and signs, with excellent topographical correlation. Similar results were reported in the succeeding papers (Mizukami *et al.* 1980, Suzuki *et al.* 1980, Kistler *et al.* 1983). Thus it has been established that the extent and location of blood in the subarachnoid space determine the severity and location of vasospasm, and the role of subarachnoid clot in the pathogenesis of vasospasm became evident. Of importance is the fact that by the use of CT scans, the patients in jeopardy of developing symptomatic cerebral vasospasm can be identified (Kistler *et al.* 1983), and administration of any drugs or therapeutic measures can be instituted before the occurrence of vasospasm.

Regarding the nature of human vasospasm, the observation reported by Fox and Ko (1978) is of interest. In 3 SAH patients, apparent leakage of contrast material was observed around the parent vessels and aneurysm. This phenomenon was seen on the CT only within a few days after SAH and was no longer present during the phase of actual vasospasm. Similar contrast enhancement (CE) was confirmed in larger series and its pathogenetic significance as to the occurrence to vasospasm, or subsequent cerebral infarction was suggested (Mizukami *et al.* 1980, Hirata *et al.* 1982, Tazawa *et al.*

1983, Doczi *et al.* 1984). This transient, abnormal leakage of contrast material was ascribed to variable mechanisms such as penetration through the wall of dilated artery (Fox and Ko 1978), increased vascular permeability especially in the venous side (Mizukami *et al* 1980), and meningeal hyperemia (Moran *et al.* 1978). Recently, Doczi *et al.* (1984) suggested that the area of CE is not in the subarachnoid space, but in the gyri which border the subarachnoid space. Those authors believed that the CE was caused by increased blood volume in the small vessels of the cerebral cortex due to impairment of small-vessel autoregulation. Considering the recent finding that the disruption of blood-arterial wall barrier in the major cerebral arteries occurs following experimental SAH (Sasaki *et al.* 1985 a), it seems plausible that the abnormal CE following SAH is in some way related to the occurrence of vasospasm.

3. Recent Findings on Cerebral Circulation and Metabolism Following SAH

As already described, most of the earlier studies showed only a poor correlation between vasospasm and regional changes of the cerebral circulation. Recent studies, however, has shown a more definite relationship between vasospasm and regional derangement of cerebral circulation.

Ishii (1979) analyzed the result of rCBF measurement with 49 patients in whom no intracerebral hematoma was found on CT, and the CBF study was carried

out within 30 days after the onset of SAH. Like the results of other studies (Nilsson 1977, Grubb *et al.* 1977, Gelmers *et al.* 1979, Mickey *et al.* 1984), a definite correlation between the diminution in mean CBF, *i.e.*, global ischemia, and the severity of the clinical grade was found. Also, it was found that most of the patients with vasospasm showed focal ischemia in the regions supplied by the involved artery. All of the 6 patients with diffuse, severe vasospasm, showed focal areas of decreased flow below 30 ml/100 gm/ minute in addition to a reduction in mean CBF. A similar combination of vasospasm and focal ischemia was reported by Gelmers et al. (1979), and Mickey et al. (1984). The latter authors further suggested that the only CBF decreases directly produced by vasospasm was the regional ischemia seen in major arterial distributions.

Further, Grubb et al. (1977), using sracer methods employing radioactive oxygen-15, showed a significant increase in the cerebral blood volume (CBV) (to 58% above normal) in parients with severe neurological deficits associated with severe vasospasm. This large increase in CBV was considered to suggest that cerebral vasospasm consists of constriction of the large, radiographically visible extraparenchymal vessels accompanied by a massive dilatation of intraparenchymal vessels. Their study for the first time demonstrated the importance of intraparenchymal vessels in the pathophysiology of cerebral vasospasm. It sometimes happens that no cerebral infarction occurs in spite of the presence of severe angiographical spasm. This discrepancy has been attributed to operation of factors such as the degree of constriction of the involved artery and the efficacy of collateral circulation. To these factors, the status of intraparenchymal vessels, namely the cerebral microcirculation, must be added. As those authors suggests, it might be that angiographical spasm can only set the stage for events which either further decrease brain perfusion or lead to clotting in the parenchymal microvasculature, which constitute the final step leading to tissue ischemia and infarction.

4. Organic Changes of the Arterial Wall Associated with Vasospasm

Although there are numerous autopsy reports of SAH patients, relatively little attention had been directed to the morphological changes in the cerebral vessels. In this regard, Conway and McDonald (1972) made a study with the intradural arteries obtained from autopsy cases dying from SAH. They found that in all patients surviving 4 weeks or more, the lumina of the intracranial arteries were narrowed by subendothelial granulation tissue which thickened the intima. The presence and the degree of intimal thickening correlated with the distribution and amount of subarachnoid blood or its breakdown products. Although the presence or absence of angiographical vasospasm was not investigated, these authors suggested that early spasm, induced by substances released from blood into the subarachnoid space causes either mechanical or anoxic damage to the vessel wall and subsequent intimal proliferation in the most severely affected arteries.

Hughes and Schianchi (1978) conducted a histological study of cerebral blood vessels in 20 selected cases with known complication of cerebral vasospasm. The large arteries were systematically examined in places known formerly to have been in spasm. In cases dying within 3 weeks after SAH, only slight swelling of the tunica intima was found but there was a conspicuous necrosis of the smooth muscle of the tunica media, With the constant presence of cells, which they called "plump" cells, and considered to be macrophages. In cases dying later than 3 weeks, the above "acute" changes were replaced by the "late" changes of medial fibrosis and medial atrophy with an intimal thickening by subendothelial fibrosis, the latter causing a concentric narrowing of the arterial lumen. This "late" change, being identical to the previous finding of Conway and McDonald (1972), was remarkably similar to that of Heubner's arteritis (endoarteritis obliterans). The early changes were considered to be those of recent mural necrosis, while the late changes were regarded as repair phenomena. As to the link between vasospasm and mural damage, the authors considered it unlikely that both the vasospasm and the histological changes are caused simultaneously by another mechanism, or that the arterial damage causes the vasospasm. Hence they preferred the chronological sequence that SAH first induces vasospasm, which then causes arterial histological changes. Medial necrosis and intimal thickening have consistently been found in succeeding similar autopsy studies, but different stresses were put on the sequence of events.

Mizukami et al. (1976, 1980) described marked inflammatory reactions (infiltration of polymorphonuclear cells and lymphocytes) of the vessels in the subarachnoid space following SAH. Since inflammatory changes were seen within a few days after SAH and preceded the occurence of other structural changes such as intimal edema, thrombus formation, and myonecrosis, these authors surmised that the inflammatory reactions which follow the presence of subarachnoid clot initiate the sequence of events leading to the occurrence of chronic vasospasm.

Resorting to Russell Ross's theory on the pathogensis of atherosclerosis (Ross and Glomset 1976), Kassel et al. (1980) put forward a hypothesis that cerebral vasospasm is an acute proliferative vasculopathy resulting from a mitogenic substance released from the subarachnoid hematoma causing replication of smooth muscle cells. The accompanying paper (Peerless et al. 1980) reported that the intimal thickening found in those patients dying more than 10 days after the onset of spasm was caused by an asymmetrical accumulation of smooth muscle cells within and beneath the intimal layer. The proliferation of smooth muscle cells beneath the intima and between layers of elastica was also seen in an animal model of SAH using the monkey, although the change was far less pronounced than that seen in human SAH.

Similar pathological findings, i.e., subintimal proliferation of smooth muscle cells, were obtained in an animal model of SAH (Clower et al. 1981), and human autopsy cases as well (Smith et al. 1983). The latter authors also found a positive relationship between posthemorrhage survival time and the severity of angiopathy. The earliest

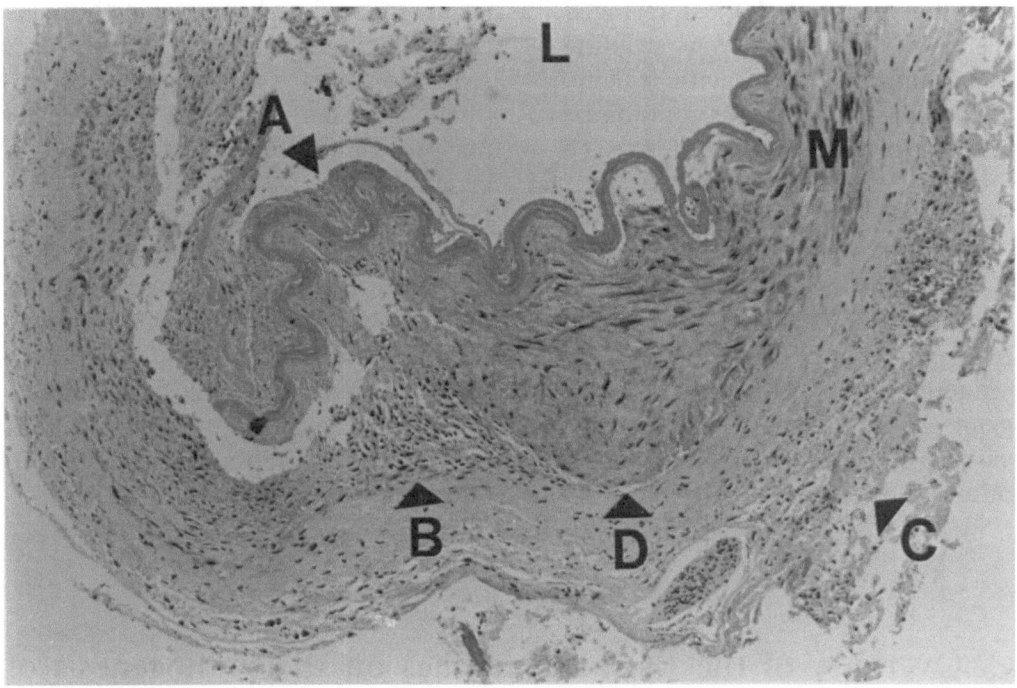

Fig. III-1. The organic changes associated with vasospasm (the internal carotid artery; 8 days after aneurysmal rupture). Severe vasospasm was shown in angiograms taken before death of the patient. *A* Accumulation of cells in the fold of the endothelium, *B* cellular proliferation in the medial layer, *C* cell infiltration in the adventitial layer, *D* edema in the medial layer, *L* arterial lumen, *M* smooth muscles. Note the absence of smooth muscle cells in areas where organic changes took place

changes consisted of swelling and thickening of the intimal layer with adhesion of platelets and leukocytes (1–9 days after SAH), which then proceeded to the stage of subintimal cellular proliferation (4–28 days after SAH). The most severe changes, being observed later on (16–40 days after SAH), consisted of severe subintimal cellular proliferation, rupture and splitting of the internal elastic membrane, and intramural hemorrhage with fibrosis and necrosis (myonecrosis) of the medial layer of the vessel (Smith *et al.* 1983). The angiographical constriction seen after the second week was almost always related to morphological thickening of the arterial wall and medial necrosis. These authors further suggested that the mere presence of blood surrounding an artery has a little influence on vessel alterations, while intramural vascular hemorrhage was a common pathological feature in vessels showing severe pathology.

Those pathological studies have consistently shown that significant morphological alterations occur in the vessel wall following SAH, and that they are presumably involved in the pathogenesis of vasospasm (**Fig. III-1**). The train of events may be summarized as follows: morphological changes in the vessel wall starts from swelling of the intima associated with inflammatory reactions, proceeds to subintimal cellular proliferation, and then culminate in the combination of intimal thickening and myonecrosis, which causes severe luminal narrowing. Conway and MacDonald (1972), and Hughes and Schianti (1978) observed relatively late changes of the above sequence and concluded that these organic changes are caused by preexisting vasospasm, *i.e.*, early spasm. Their surmise is no longer tenable because the occurrence of early spasm in humans has so far been negated. Therefore, it seems apposite to assume that the morphological changes in the arterial wall may not be the result but the cause of cerebral vasospasm.

In summary of this section, recent morphological findings have cast serious doubt on the preceding concept of cerebral vasospasm. From the outset, these has been an *a priori* acceptance that the angiographical spasm represents prolonged smooth muscle constriction. This notion is now open to question.

C. Overview of the Basic Researches on the Pathogenesis of Vasospasm

As described in the preceding chapter, there had been long-standing controversies about the concept of vasospasm and it has not been long since the clinical features and significance of vasospasm was established on firm grounds. Concerning the pathogenesis of vasospasm, numerous studies were carried out in the past. The majority of those studies stood on the presumption that early spasm heralds cerebral vasospasm and the discovery of vasoconstrictive substance(s) acting on the smooth muscle of cerebral vessels would lead to solution of the problem. This presumption is no longer valid because early spasm is now known not to occur in humans and organic changes of the vessel wall instead of smooth muscle constriction may be the principal cause of luminal narrowing.

In the past, a number of vasoconstrictive substances have been proposed as possible candidates of spasmogens. Also, numerous modes of treatments or prevention of vasospasm by the use of pharmacological agents antagonizing or inhibiting putative spasmogens have been advocated. There are excellent reviews on this subject (Wilkins 1980, White 1979, 1983, Kassell *et al.* 1985, Wellum *et al.* 1985), and regrettably, all of them hold the view that neither pathogenesis nor treatment of vasospasm has been established until now. In this section, therefore, description will be limited only to those known facts that the author thinks to be relevant to future investigation of vasospasm.

1. The Current Concept of Cerebral Vasospasm; Implications to Basic Research

In the first place, the events which are apparently related to the occurrence of vasospasm must be assembled in their chronological sequence. Hitherto con- strued hypothesis regarding the genesis of vasospasm was like the one shown below:

$$\text{subarachnoid clot} \xrightarrow{\text{(spasmogens)}} \text{early spasm} \begin{array}{l} \rightarrow \text{ organic changes} \\ \rightarrow \text{ delayed spasm} \end{array}$$

(Conway and McDonald 1972, Hughes and Schianti 1978).

Implicated by the hypothesis was the importance of early spasm, which was presumed to be due to the action of spasmogen(s) released from the subarachnoid clot. Such an assumption guided many researchers to the investigation of possible spasmogens. Nevertheless, this line of investigation has not reached any definite conclusion other than that multifarious factors and mechanisms would be responsible for the pathogenesis of vasospasm (White 1983).

Recent investigations indicate that organic changes in the vessel wall takes place much earlier than previously thought and progresses in a particular pattern *vis-a-vis* the occurrence of de- layed vasospasm. As described in the previous section, the train of events so far shown are as follows: aneurysmal rupture (vessel injury) → formation of subarachnoid clot → inflammatory reactions in the vicinity of tunica adventitia accompanied by endothelial damages such as formation of subintimal edema, disruption of tight junctions, aggregation of platelets, and infiltration of lymphocytes, leukocytes, or macrophages → subintimal proliferation of smooth muscle cells → myonecrosis and subendothelial fibrosis with concentric narrowing of the vascular lumen → healing. This progression of pathological events in the vessel wall is overlapped by the delayed occurrence of angiographical spasm.

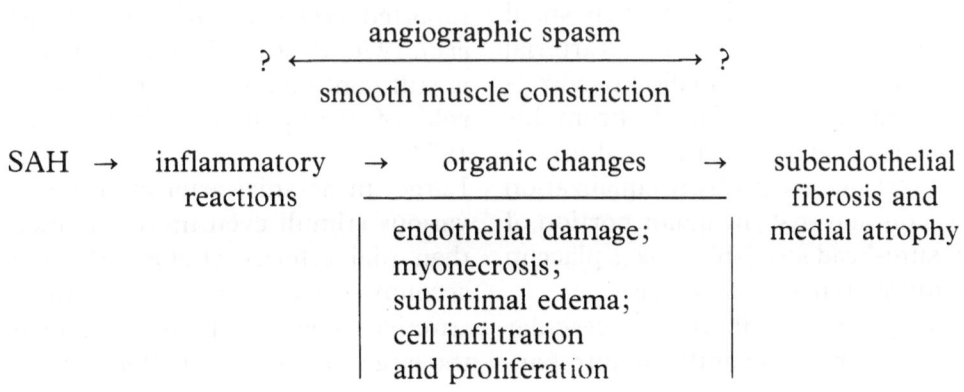

As shown above, the possibility that vasospasm represents a particular type of angiopathy which induces luminal narrowing appears more likely nowadays. Such a scheme will serve as a useful working hypothesis, but it certainly demands investigations of much broader spectrum than ever.

2. Smooth Muscle Constriction in Vasospasm

a) Role of Large Arteries in the Control of CBF

To understand the nature of cerebral vasospasm, it is necessary to know how large cerebral arteries function to regulate CBF in the normal brain. This can be estimated by measuring the pressure drop along the artery, which reflects the vascular resistance of that particular segment. From measurement of arterial casts of human cerebral vessels, Fukasawa (1969) concluded that the proportion of large arterial branches in the total arterial length is larger in cerebral arteries than in any other arterial system, causing a large pressure drop in the cerebral arteries. This prediction was confirmed by Shapiro et al. (1971), who used a servomicropipet system and directly measured the intraluminal pressure of pial vascular network. They found a substantial pressure drop along the arterial segment from the aorta to the larger arteries (200–455 μ od) of the pial vessel (39%) and a relatively small decrement (10%) across the pial arterial bed to the level of the smallest penetrating arterioles (about 25 μ od). From this result, they suggested that pial arterioles functions as a pressure equalization reservoir and that the major portion of pressure-head loss (46%) takes place in the intraparenchymal vessels.

Although the distribution of vascular resistanse predominantly in intrapar-

enchymal vessels is in harmony with the concept of metabolic control of CBF, controversy exists as to whether or not extraparenchymal vessels play a significant role in CBF regulation. It has been well established that extraparenchymal arteries respond to a variety of stimuli and their behavior can be accurately measured by direct observation (Purves 1972, Rosenblum and Kontos 1974). Further, extraparenchymal arteries have an abundant supply of adrenergic nerve fibers (Nelson and Rennels 1970, Nielsen and Owman 1967). Experimental results indicate that the cranial sympathetic nerves originating in the superior cervical ganglia play a significant role in the regulation of local cerebral blood flow, including its autoregulation, and in blood-brain barrier functions were reported (Purves 1972, Kushinsky and Wahl 1978, Edvinsson and MacKenzie 1977, Edvinsson et al. 1978). Nevertheless, discrepant results have also been reported (Harper et al. 1972, Heistad et al. 1976, 1977, 1978), and controversy is still continuing as to the functional role of sympathetic nerves (Purves 1978).

Large intracranial arteries respond to various stimuli even more remarkably than pial arteries (Echlin 1968). On account of its length and potent constrictive capacity, which correspond to the large pressure drop along this segment, large arteries was considered to be an important site of CBF regulation (Mchedlishvili et al. 1967, 1979, Fukasawa 1969). Harper et al. (1972) suggested that sympathetic nerves may constrict large cerebral arteries but autoregulatory dilatation of distal vessels might maintain a constant flow. Contrarily, Heistad et al. (1977, 1978) showed that large arteries do not constrict during sympathetic stimulation, but autoregulate during hemorrhagic hypotension. According to the method of Rapela and Martin (1975), catheters were inserted into both vertebral arteries in dogs and advanced until they wedged. With bilateral vertebral arteries thus occluded, the pressure gradient along the large arteries was determined by subtracting vertebral artery wedge pressure from carotid artery pressure. Resistance of large arteries, being calculated by dividing this pressure gradient by CBF, markedly increased on infusion of serotonin into both carotid arteries, but did not change on bilateral stimulation of both superior cervical ganglia (Heistad 1977). During graded hemorrhagic hypotension, autoregulation of CBF was maintained due to decrease in the total cerebral resistance. Resistance of large cerebral arteries decreased significantly during hypotension, which means that large cerebral arteries autoregulate during changes in arterial pressure.

Although discrepancy exists as to the function of sympathetic nerves, above studies indicate that large cerebral arteries play an important role in physiological regulation of cerebral resistance (Heistad et al. 1978). The fact that SAH causes an immediate loss of autoregulation (Jakubowski et al. 1982) may be related to loss of the regulating function of large cerebral arteries as indicated above. Thus, not only an increased vascular resistance due to luminal narrowing but also derangement of CBF regulation mechanism in large cerebral arteries need to be considered as factors causing ischemic brain damage.

b) Vasoconstrictive Substances

The number of substances which have been shown to constrict cerebral arteries in vitro or in vivo and suggested as spasmogens are numerous (White 1983). Among them are neurotransmitter amines such as epinephrine, norepinephrine and serotonin (Raynor et al. 1961, Nielsen and Owman 1971, Toda and Fujita 1973, Rosenblum and Giulianti 1973, Nagai et al. 1974, Alksne and Greenhoot 1974, Allen and Gross 1976, Zervas et al. 1975, Owman et al. 1979, Lobato et al. 1980), hemoglobin (Miyaoka et al. 1976, Osaka 1977, Sonobe and Suzuki 1978, Ozaki and Mullan 1979, Tanishima 1980, Wellum et al. 1982), thrombin (White et al. 1980), fibrin degradation products (Ito 1980), potassium (Wilkins and Levitt 1971, Shiguma 1982), classical prostaglandins (Yamamoto et al. 1972, Handa et al. 1974, Pickard et al. 1975, White et al. 1979, Maeda et al. 1982, White and Hagen 1982, Pickard and Walker 1984), thromboxane A_2 and prostacyclin (Ellis et al. 1977, Asano et al. 1978, Boullin et al. 1979, Sasaki et al. 1982 a, b, von Holst et al. 1982, Chan et al. 1984), hydroperoxides

(Asano *et al.* 1980, Sasaki *et al.* 1981, Koide *et al.* 1981, 1982, Asano *et al.* 1984), unidentified polypeptide (Kapp *et al.* 1968), and others. As none of these substances has yet been shown to be solely responsible for the occurrence of vasospasm, involvement of multiple substances has been surmised because they usually work in an additive fashion (White 1979 a, 1983).

Nevertheless, there have been a lot of discrepant results concerning each of the putative spasmogens. Further, the idea that hitherto known vasconstrictive substances either solely or in combination cause vasospasm is hindered by the following facts. First, there is a several day's delay in the occurrence of vasospasm after SAH, which precludes the involvement of any vasoconstrictive substances originally contained in the blood. Secondly, although potent vasoconstrictive action is usually found *in vitro* in the CSF obtained from SAH

patients (Boullin *et al.* 1981, Brandt *et al.* 1981 a, b, Sasaki *et al.* 1984), no close relationship between vasoconstrictor activity in postoperative CSF samples and the patient's clinical condition or angiographic vasospasm has been found (Boullin *et al.* 1981). Thirdly, in chamber studies using the rat stomach fundus (Boullin *et al.* 1981) or the canine basilar artery segment (Sasaki *et al.* 1984), the vasoconstrictive activity of the CSF obtained from SAH patients was not suppressed to any significant degree by cumulative addition of antagonists to serotonin, histamine, norepinephrine, epinephrine, acetylcholine, and angiotensin **(Fig. III-2)**. Only addition of dithiothreitol, dithioerythritol, or AVS [1,2-bis (nicotinamide)-propane] significantly suppressed the contraction, indicating the involvement of prostaglandins and/or lipid peroxides (Sasaki *et al.* 1982 c). Fourthly, therapeutic

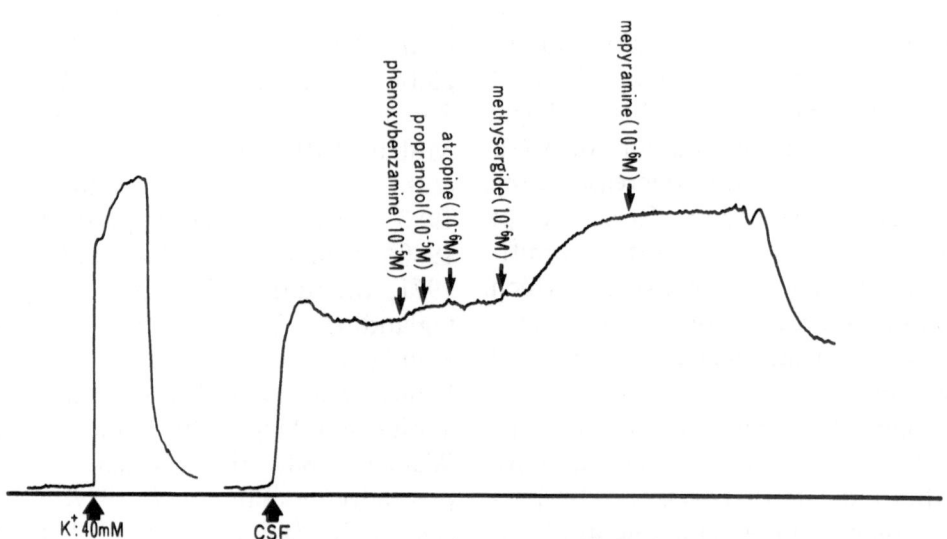

Fig. III-2. Vasoconstriction induced by bloody CSF. No vasodilation occurred on addition of each antagonist (from Sasaki *et al.* 1984)

trials using β-adrenergic stimulators, α-adrenergic blocking agents, catecholamine and/or serotonin depletors, phosphodiesterase inhibitors, nitrates, calcium antagonists, or many other drugs have invariably failed to prevent or reverse vasospasm (Wilkins 1980, Varsos et al. 1983, Grotenhuis et al. 1984, Espinosa et al. 1984 a, b).

c) Miscellaneous Mechanisms Possibly Involved in the Pathogenesis of Vasospasm

The past failure to identify particular spasmogenic substances does not necessarily imply that smooth muscle constriction does not take place in vasospasm at all. At acute aneurysm operation, in fact, we often observe the real constriction of large arteries, i.e., the decrease in their external diameters with whitish appearance of the vessel wall, in which the occurrence of vasospasm was angiographically confirmed (Fig. III-3). Therefore, it still holds true that some unknown mechanisms and/or substances are in operation to induce sustained smooth muscle constriction.

Regarding the role of neurotransmitter amines such as norepinephrine and serotonin, it has been proposed that SAH induces an increased sensitivity of cerebral arteries to these amines (Svendgaard et al. 1977, Lobato et al. 1980 b, Pickard and Perry 1984). Since the catecholamine fluorescence in the adventitia of large cerebral arteries is reduced following SAH (Fraser et al. 1970, Rosenblum and Giulianti 1973), and the norepinephrine content as well as the dopamine beta-hydroxylase ac-

tivity of cerebral arteries are decreased following SAH (Lobato et al. 1980 a), the increased sensitivity observed was interpreted as representing supersensitivity of a postdenervation type (Svendgaard et al. 1977, Lobato et al. 1980 b). In contrast, Toda et al. (1977) reported that the contractile response to vasoactive substances such as serotonin, norepinephrine, histamine, and K^+ of the middle cerebral artery showing vasospasm in situ decreased following SAH in dogs. The reason for this discrepancy of experimental results is not clear, but it may be pointed out that in the experiments of Lobato et al. (1980 b) and Svendgaard et al. (1977) using cats and rabbits respectively, the occurrence of vasospasm in the examined arteries was not angiographically confirmed. In addition, the concentration of norepinephrine in the cisternal CSF after SAH was in the order of 10^{-9} in molar concentration in humans (Shigeno 1982), which seems to be far too low to cause significant constriction of cerebral arteries even in the presence of supersensitivity.

On the other hand, a high urinary excretion of catecholamines (Neil-Dwyer et al. 1974) as well as elevated plasma level of norepinephrine (Peerless and Griffiths 1975) following SAH was reported. Also, it has been estimated that a high concentration of serotonin (1×10^{-6} M) is present in blood following SAH (Alksne et al. 1974, Allen et al. 1974). These and other vasoconstricting substances present in plasma may penetrate into the smooth muscle layer following SAH because of barrier disruption in the major cerebral arteries (Sasaki et al.

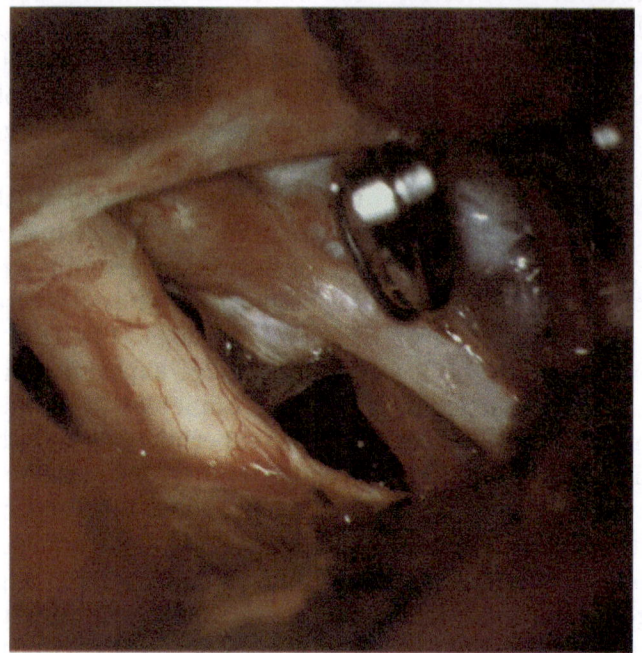

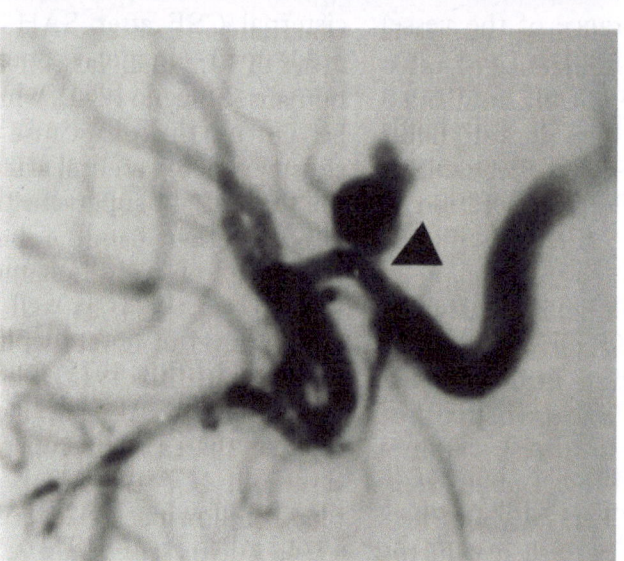

Fig. III-3. Pronounced vasospasm of the internal carotid artery. Left: The carotid angiogram taken before the day of operation. The arrow points moderate vasospasm of the internal carotid artery. Right: The intraoperative photograph of the same patient after aneurysmal clipping. The reduction of the external caliber of the internal carotid artery was present before operative manipulation

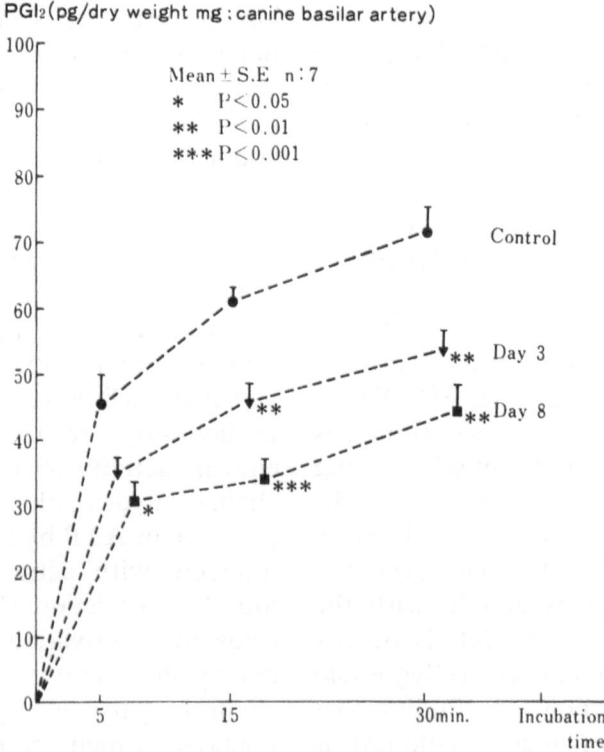

PGI₂(pg/dry weight mg : canine basilar artery)

Fig. III-4. PGI$_2$ synthetic capacity of the canine basilar artery following SAH

1985 a). Such an event may induce a pronounced arterial constriction since the smooth muscle cells closest to the lumen are the most sensitive to constrictor agonist (Garland and Keatinge 1982, Kassel et al. 1985). Also, penetration of plasma factors stimulating smooth muscle proliferation (Ross et al. 1974) may herald subsequent organic changes in the vessel wall (Peerless et al. 1980).

Related to the barrier disruption is the degenerative changes in endothelial cells (Fein et al. 1974, Alksne and Smith 1977, Tanabe et al. 1978, Joris and Majno 1981, Sasaki et al. 1985 a). It is known that endothelial cells produce PGI$_2$, which has a dilatory action on the

vascular smooth muscle and an anti-aggregatory action to platelets. Sasaki et al. (1982 b) has reported that the PGI$_2$ synthetic capacity of the canine basilar artery significantly decreases following SAH (Fig. III-4). Such a diminution in the endogenous PGI$_2$ synthesis in the endothelium may lead to uninhibited action of thromboxane A$_2$, culminating in smooth muscle constriction and thrombus formation (Boullin 1980, Sasaki et al. 1981 a, 1982 a).

During the past few years, attention has been directed toward the role of endothelial cells as mediators of vasodilation of arteries. Furchgott and Zawadzki (1980) found that relaxation of isolated preparations of rabbit aorta and other blood vessels by acetylcholine (ACh) requires the presence of endothelial cells, and that ACh, acting on muscarinic receptors of these cells, stimulates release of a substance(s) that causes relaxation of the vascular smooth muscle. The chemical nature of this endothelium-derived relaxing factor (EDRF) released by ACh and other agents has not yet been identified. It has been suggested that EDRF may be a labile free radical which stimulates guanylate cyclase in the muscle, leading to an increase in cyclic GMP that somehow activates relaxation (Furchgott et al. 1984). However, the above suggestion is at odd with the recent finding that EDRF is deactivated by activated oxygen (Glygrewski et al. 1986). Anyhow, the finding that hemoglobin (not methemoglobin) is an effective, rapidly acting inhibitor of ACh- and A 23187 (calcium ionophore)-induced relaxations in rabbit aorta (Furchgott 1984) is of particular interest because oxyhemoglobin is one of the putative spasmogens of cerebral arteries. A similar effect of oxyhemoglobin in the rabbit basilar artery has recently been shown (Fujiwara et al. 1986).

Attention has also been directed towards the possible influence of SAH on the molecular mechanisms of vascular smooth muscle constriction (Wellum et al. 1985). It has been shown that smooth muscle constriction and dilatation by Ca^{++} is controlled through phosphorylation (by myosin light chain kinase) and dephosphorylation (by myosin light chain phosphatase) of the myosin light chain. Myosin light chain kinase (MLCK) is composed of two polypeptides, one with molecular weight of 80,000–130,000 dalton and the other with molecular weight of about 20,000 dalton (Dabrowska et al. 1977, Yazawa and Yaki 1977, Uchiwa et al. 1982), the latter being the calcium-binding protein, i.e., calmodulin (Kakiuchi and Yamazaki 1970, Yazawa et al. 1977). The chain of events causing smooth muscle constriction is explained as follows; Ca^{++} first bind to calmodulin, which activates the enzyme activity of MLCK. The light chain of myosin, phosphorylated in the presence of ATP by activated MLCK, interacts with actin causing smooth muscle constriction. Depletion of Ca^{++} leads to inactivation of MLCK and dephosphorylation of myosin light chain by myosin light chain phosphatase, which results in smooth muscle dilatation (Sobieszek and Smoll 1977, Aksoy et al. 1976, Chacko et al. 1977, Ikebe et al. 1977). The activity of MLCK is further regulated by a cAMP-dependent protein kinase (Conti and Adelstein 1980), which is stimulated by cyclic AMP (cAMP). The stimulation of β-adrenergic receptors in the cellular membrane, which activates adenylate cyclase increasing the production of cAMP, therefore results in arterial relaxation. Wellum et al. (1985) have raised the possibility that vasospasm may be due to some derangement of the above mechanisms of smooth muscle constriction and dilatation. For instance, proteolytic digestion of the myosin light chain with papain has been shown to cause the binding of myosin to actin without the occurrence of normal calcium dependent activation or consumption of ATP, and consequent development of smooth muscle shortening. If a similar derangement in the contractile mechanism ever occurred in cerebral arteries following SAH, cerebrovascular smooth muscle would be able to maintain contraction without external calcium and with virtually no ATP utilization. In addition, SAH may induce derangement not only of contractile proteins but also of other systems regulating smooth muscle constriction, such as the distribution of membrane receptors and the activity of membrane enzymes. Pickard and Perry (1985) have recently shown that the activity of Na^+, K^+ ATPase in the cellular membrane of vascular smooth muscle may be increased following SAH. It is hard to interprete their finding in terms of spasmogenesis at present, but it may well implicate that SAH exerts a broad-spectrum influence on the mechanisms involved in the regulation of smooth muscle constriction. Those intriguing ideas as described above would certainly merit further investigation.

3. The Significance of Inflammatory Changes in the Vessel Wall

a) The Meningeal Inflammatory Reactions Due to Blood

The inflammatory changes in the subarachnoid space following SAH have been well documented (Bagley 1928, Hammes 1944, Jackson 1949, Crompton 1964 b). Hammes (1944) studied necropsy material from 114 cases of SAH due to rupture of a cerebral aneurysm, obtaining the following findings. Polymorphonuclear cells appeared around the pial blood vessels as early as a few hours after SAH. A greater diffusion of polymorphonuclear cells with appearance of lymphocytes was noted during the next 16–32 hours. With the appearance of mesothelial cells after the first 24 hours, the breakdown products of blood became to be recognized in the form of brown pigment and iron, free and within the cytoplasm of the multinuclear cells, and leukocytes in the subarachnoid space. Whereas all cells increased in number during the first few days, the percentage of multinuclear cells decreased and the activity of phagocytes and mesothelial cells became more evident. The polymorphonuclear reaction subsided within a week, but the pigment and iron were increased, The first appearance of fibrosis of the meninges occurred in half of the patients who died 10 days after SAH. Phagocytes and lymphocytes were present as long as any breakdown products of blood was present.

To identify the component of whole blood responsible for meningeal inflammatory reactions, Jackson examined the severity of meningeal responses, *i.e.*, the leukocyte (WBC) counts and protein content in the CSF as well as clinical signs of meningeal irritation, following intracisternal injection of various fractions of fresh or incubated blood in dogs (Jackson 1949). Injection of each 'fraction caused pleocytosis and elevation of protein content in the CSF of varying degrees, the most pronounced response being obtained with the blood incubated three days or longer. Hemolyzed red cells, which contain free hemoglobin, caused a greater cellular response than fresh whole blood but considerably less than degenerated blood or oxyhemoglobin. Based on the further comparative analysis, he concluded that the most toxic substance of the blood to the subarachnoid space was contained in the heme component of hemoglobin, probably being bilirubin or a like pigment.

The major portion (75%) of extravasated erythrocytes in the CSF becomes enmeshed and fixed in the arachnoid (Adams and Prawirohardjo 1959). The lysis of erythrocytes in the subarachnoid space proceeds gradually to reach the maximum at 5–10 days after SAH (Matthews and Frommeyer 1955). Most of the erythrocytes are cleared of CSF within about 30 days (Tourtellotte et al. 1964). Parallel to the lysis of erythrocytes, pigments in the CSF increase. Regarding the time course of CSF concentration of each pigment, Barrows et al. (1955) has shown the followings: oxyhemoglobin appears at the onset of SAH, progresses to a maximum the first few days and then gradually diminishes in amount; bilirubin, which can be produced only by the action of living cells, appears in two to three days and increases in amount as oxyhemoglobin decreases. Methemoglobin is absent or present only in very small amounts in the CSF of patients with SAH (Kajikawa et al. 1979).

Thus it seems well established that inflammatory responses start immediately after SAH and components of blood, i.e., probably breakdown products of hemoglobin, trigger the reaction. In regard to the pathogenetic mechanism of vasospasm, however, the toxic effects of these pigments causing meningeal inflammatory reaction have been rather neglected, and attention has first been directed towards their vasocontractile capacity. The blood incubated for 2 to 8 days develops vasoconstrictive activity (Wilkins and Levitt 1970), which has been attributed to release of oxyhemoglobin by erythrocyte lysis (Osaka 1977, Endo and Suzuki 1977, Miyaoka et al. 1976). Despite detail studies (Tanishima 1980, Wellum et al. 1980), the mechanism of its vasocontractile action remains unclear. The vasocontractile activity of oxyhemoglobin is relatively mild but more long-lasting than that of serotonin (Weir et al. 1980). Nevertheless, no definite correlation was found between the concentration of hemoglobin in the CSF and the incidence of vasospasm (Ohta et al. 1980), nor was there any correlation between the in vitro vasoconstrictive activity of CSF obtained from SAH patients and its oxyhemoglobin content (Usui and Asano, unpublished data). To date, there has not been firm evidence to show that oxyhemoglobin and related pigments cause vasospasm solely on account of their vasocontractile activities.

Mizukami et al. (1980), resorting to Hammes' observation (Hammes 1944) and his own observation of early contrast enhancement in the region of the circle of Willis following SAH, for the first time proposed that chronic spasm might be caused by the inflammatory reactions which follow the presence of blood clot in the subarachnoid space. He postulated that the vasa vasorum of large cerebral arteries play a significant role in the process, because hyperpermeability at the venular side causes edema and ischemia of the smooth muscle cells. Although the existence of vasa vasorum in cerebral arteries has been questioned (Hughes 1978, Clower et al. 1984), the importance of inflammatory changes appears to be more and more increasing in the light of recent knowledge on the chemical mechanism of inflammatory reactions.

b) Chemical Mediators and Vasospasm

Inflammation is characterized by clinical manifestations such as increased local blood flow (calor and rubor), increased vascular permeability and/or cellular infiltration (tumor), and release of substances at the site of inflammation that induce pain (dolor). These manifestations are produced by a variety of substances (chemical mediators) the sources of which may be plasma, infiltrating cells, or the injured tissue. The plasma contains three major mediator-producing systems (kinin, coagulation, complement), which interact in defined manners to generate phlogistic compounds. Other mediators are cell derived and, within the cells of origin, may be preformed and stored in granules (histamine in mast cells, cationic proteins in neutrophils) or may be newly synthesized

by the cells (interleukin 1, prostaglandins, leukotrienes, oxygen radicals, platelet-activating factor) (Larsen and Henson 1983). They show variable actions of different potency in vivo depending on the site of action, interacting each other in a very complex fashion. Growing evidence suggests that eicosanoids and free (oxygen) radicals play major roles in the chemical mechanism of inflammation (Kuehl and Egan 1980, Goetzl 1981, Fantone and Ward 1982, Samuelsson 1983).

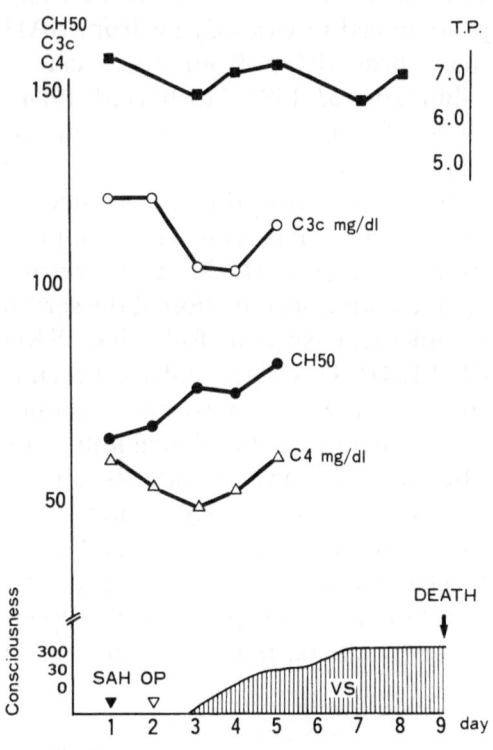

M.H. 27y M. IC-PC aneurysm

Fig. III-5. A representative time courses of the serum CH 50 value, C 3, C 4, and total protein following SAH. Note the sudden drop of serum complements and rise of CH 50 concomitant with the occurrence of vasospasm

Table III-1. Serum CH-50 titer in patients with ruptured aneurysms

Angiographic spasm		CH-50 (N: 35–45)
Diffuse, severe	(11)*	74.6 ± 6.7
Diffuse, mild or segmental	(8)	60.3 ± 7.1
No spasm (acute)	(6)	52.5 ± 4.8
No spasm (chronic)	(4)	43.0 ± 3.9

* Depression of C 3 c component to 90.8 ± 8.0 mg/100 ml in 8 of 9 patients (C 3 c, N: 130 mg/100 ml).
  N: normal level.

Regarding the pathomechanism of vasospasm, deposition of immunoglobulins and complement 3 (C3) in the cerebral artery was found to be more pronounced in cases dying from SAH than those dying from other causes (Shimizu *et al.* 1982, Hoshi *et al.* 1984). In SAH patients, the serum concentration of C3 significantly decreased concomitant with the occurrence of vasospasm (Matsutani *et al.* unpublished data) (**Fig. III-5**). These results are indicative of activation of the serum complement system following SAH (**Table III-1**). Of particular concern is the fact that C 5 a-activated granulocytes may induce endothelial injury, by liberation of oxygen radicals and/or lysosomal enzymes (Ryan and Ryan 1983, Fantone and Ward 1982). Such a mechanism might explain the occurrence of early disruption of the blood-vessel wall barrier following SAH

(Sasaki *et al.* 1985 a). Likewise, each of the pathological features of vasospasm such as edema, smooth muscle constriction, myonecrosis, and subintimal fibrosis might be causally related to successive production of particular chemical mediators in the process of inflammatory reactions. In this line of investigation, we envisaged to identify those substances which are produced in or around cerebral arteries following SAH and trigger the inflammatory responses. As already reported (Asano *et al.* 1980, Sasaki *et al.* 1981 a, b), we have obtained preliminary data showing that lipid peroxidation due to non-enzymatic, free radical reactions and/or the arachidonate cascade is involved in the pathogenesis of vasospasm. Our recent findings on this subject will be later presented together with our current hypothesis.

IV. RELEVANCE OF THE METABOLISM OF MEMBRANE LIPIDS TO CEREBRAL VASOSPASM AND ISCHEMIC BRAIN DAMAGE

The structural and functional integrity of biomembrane is essential for a cell to maintain its life and activity. An anoxic-ischemic insult induces profound alterations in the cerebral metabolism, which would in all likelihood adversely influence the environment of the biomembrane, at the same time interrupting the metabolic pathways necessary to maintain its structural integrity. Membrane damage thus incurred may constitute one of the major pathomechanisms underlying ischemic brain damage. In fact, all of the hitherto investigated topics regarding the biochemical mechanisms of ischemic brain damage, such as the release of neurotransmitters, ATP synthesis in the mitochondria, protein synthesis, and regulation of ionic compositions in the extra- and intracellular fluids, are related to differing aspects of membrane functions (Siesjo 1982, Raichle 1984). Thus, interest has been focused on the effect of anoxic-ischemic insult on the metabolism of membrane lipids.

The alterations in the metabolism of membrane lipids induced by ischemia and other insults would have dual implications to the pathomechanism of cell damage. One is the structural disruption of the membrane, *i.e.*, its compositional and/or conformational alterations, which would cause changes in the permeability to ions or other substances, the functions of surface receptors, the activities of membrane-bound enzymes and so on. The other is related to the components of membrane phospholipids, which are released from the membrane upon physiological or noxious stimuli and then transformed into a variety of substances, acquiring potent biological actions.

In the first part of this chapter, recent findings on the synthetic and catabolic pathways of membrane phospholipids are summarized. Particular emphasis is laid on the peroxidative processes of polyunsaturated fatty acids, either by the enzymatic reactions (the arachidonate cascade) or by the nonenzymatic reactions (free radical reactions). Based on these pieces of knowledge, the pathogenetic mechanisms underlying vasospasm and brain edema are separately discussed in the following sections. Our hypotheses concerning the pathogenetic mechanism of vasospasm and ischemic brain edema are presented.

A. Overview on the Metabolism of Membrane Lipids with Reference to Cerebral Ischemia

1. The Membrane Phospholipids

a) The Structure of Membrane Phospholipids

The phospholipids, which compose the structural matrix of cell membranes, have a polar "head" composed of a glycerophosphoryl ester and a "tail" composed of two fatty-acid hydrocarbon chains. As shown in **Fig. IV-1**, R 1 and R 2 represent long-chain fatty acids (tails), and the last alcohol group of the glycerol backbone is linked to a phosphate group (head). One of the hydroxyl groups of phosphate is substituted with a substituent, X, to form various *phosphatidyl* substances, such as phosphatidylethanolamine, phosphatidylcholine (lecithin), and phosphatidylserine. *Phosphatidal* compounds or plasmalogens are characterized by the presence of an α,β-unsaturated ether linkage, which replaces the acyl ester at C-1 of phosphatidyl compounds. In phosphatidyl and phosphatidal compounds, the fatty acids in the C-1 position are mainly saturated ones (C_{14}–C_{22}), whereas those in the C-2 position are predominantly unsaturated ones (C_{14}–C_{22}). Cardiolipin, *i.e.*, diphosphatidylglycerol in which phosphatidylglycerol is linked to the basic phosphatidyl unit, is present only in mitochondria.

Except for cardiolipin, these phospholipids are found throughout the nervous system, and the distribution of different kinds of phospholipids in neurones and glia seems to be very similar (Ansell and Spanner 1977). For example, in rat neurones and glia, the predominant phospholipid is phosphatidylcholine (50–60% of total phos-

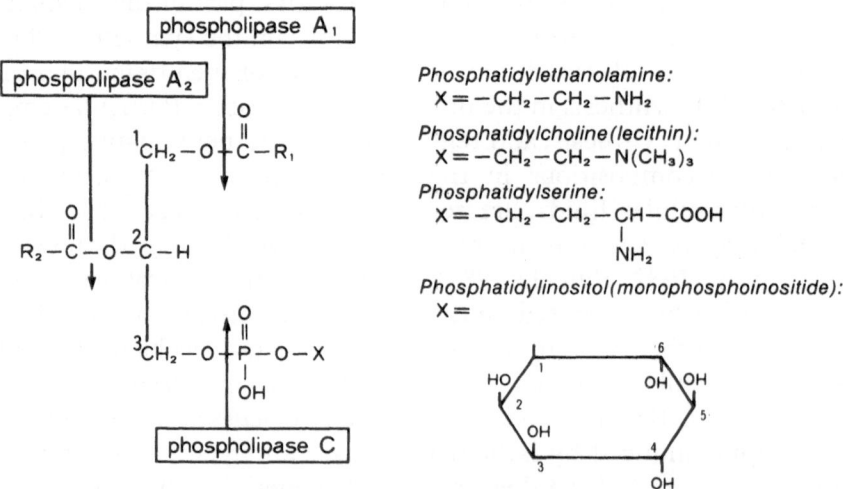

Fig. IV-1. The molecular structure of phospholipids and the site of action of phospholipases

pholipid), the remaining fraction of total phospholipid being occupied by phosphatidyl- and phosphatidalethanolamine (23–28%), phosphatidylinositol (5–7%), phosphatidylserine (5–7%), and sphingomyelin (4–5%). On the other hand, there are differences in the fatty acid content of the various phospholipid particularly between the synaptosomal phospholipids and those of the myelin. Both in the synaptosomal plasma membranes and in the synaptic vesicles, 33–48% of the fatty acids of phosphatidylserine, phosphatidylethanolamine, and phosphatidylinositol are polyunsaturated, mainly $C_{20:4}$ and $C_{22:6}$. In myelin, on the other hand, the unsaturated fatty acids in the phospholipids are largely $C_{18:1}$ (Ansell and Spanner 1977).

b) The Synthesis of Membrane Phospholipids

The biosynthesis in the brain of important components of membrane phospholipids, i.e., long-chain fatty acids such as palmitic ($C_{16:0}$), stearic ($C_{18:0}$), oleic ($C_{18:1}$), linoleic ($C_{18:2}$), linolenic ($C_{18:3}$), and arachidonic ($C_{20:4}$: eicosatetraenoic) acids are essentially identical with that in other organs. Palmitic, oleic, linoleic, and linolenic acids are rapidly taken up by the brain from the blood and rapidly incorporated into complex lipids. Significant amounts of palmitic acid are synthesized de novo from acetyl-CoA and malonyl-CoA, and further elongation takes place in mitochondria. The essential fatty acids, i.e., linoleic and linolenic acids, cannot be synthesized

in the brain, and they must be provided from dietary sources. Since the brain contains relatively large amounts of polyunsaturated fatty acids (PUFAs), the essential fatty acids are metabolically significant. Like fatty acids taken up from the blood, intracerebrally or intracisternally injected fatty acids can be rapidly assimilated into phospholipids (Sun and Horrocks 1969). Nevertheless, the relative roles of transported fatty acids and those synthesized in situ in the synthesis of phospholipids in the brain are unknown (Ansell and Spanner 1977).

The main pathways involving major brain phosphatidyl compounds are shown in **Fig. IV-2**. The pathway III is an important preliminary step which provides phosphatidic acid for the synthesis of mono- and polyphosphoinositides (pathway II) and diglyceride (diacylglycerol) for the synthesis of phosphatidyl-ethanolamine, phosphatidylcholine, and phosphatidylserine (pathway III). In the pathway I, choline or ethanolamine is first phosphorylated and is then converted to an active form, CDP choline or CDP ethanolamine, by a reaction with cytidine triphosphate (CTP). The activated form of choline or ethanolamine then forms each phosphatidyl compounds with diglycerids derived from phosphatidic acid (Suzuki 1981). In some organs, e.g., platelets, the direct conversion of phosphatidylethanolamine to lecithin by methylation with a methyl donor, S-adenosyl-L-methionine (SAM) was shown (Hirata and Axelrod 1980). However, this reaction apparently does not occur in the brain (Suzuki 1981).

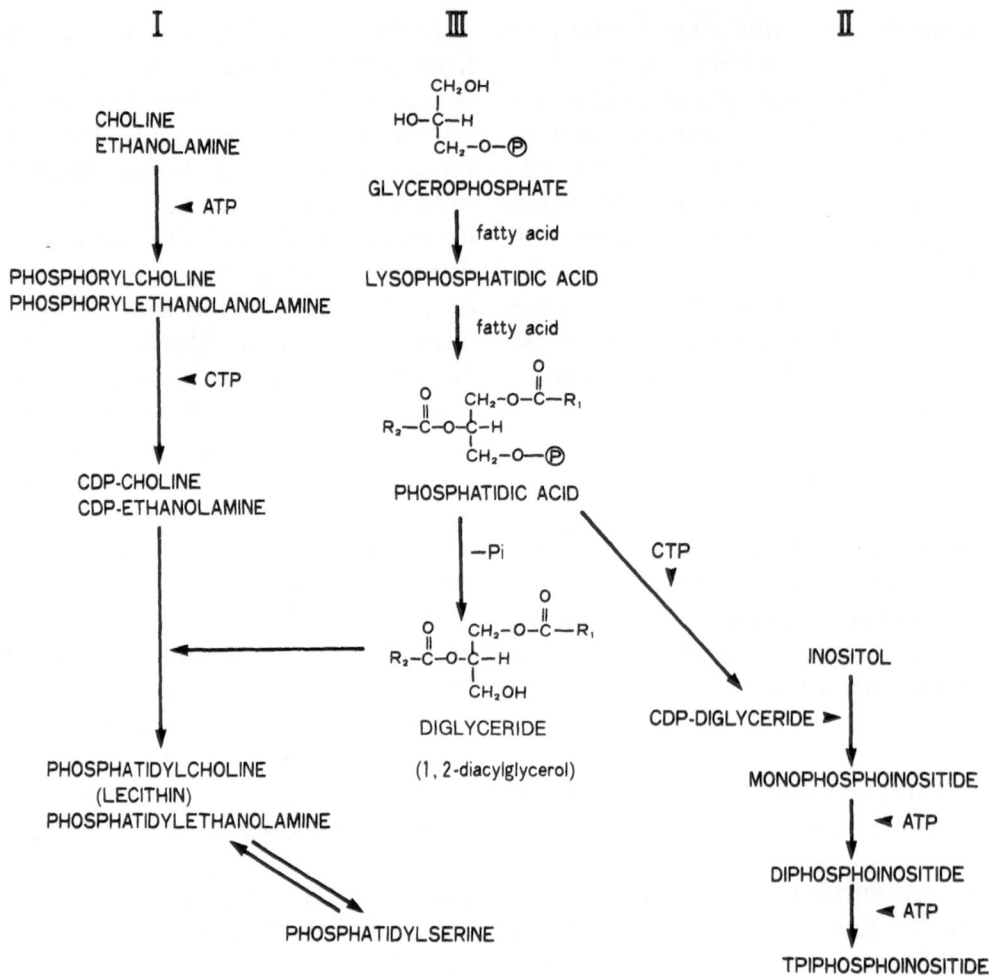

Fig. IV-2. The synthetic pathway of phospholipids

The enzymes responsible for the synthesis of major phospholipid are found in the endoplasmic reticulum, *i.e.*, the microsomal fraction. It appears that phospholipids generated in the endoplasmic reticulum are transferred to all membrane organelles by a phospholipid exchange protein. Therefore, modification of the phospholipid composition of membranes may well be a result of some alteration in the system which transfers phospholipids, as well as a change in the synthetic capacity of the endoplasmic reticulum (Ansell and Spanner 1977).

c) The Regulatory Mechanism of the Free Fatty Acid Level in the Brain

In the brain, there are phospholipases which split the phosphatidyl compounds at various sites (Fig. IV-1). Phospholipase A_1 and A_2, which are mainly present in the cell membrane, hydrolyze the ester linkages at C-1 and C-2 positions, respectively. Phospholipase C specifically hydrolyzes the ester linkage at C-3 position of phosphatidylinositides and releases 1,2-diacyl glycerol (diglyceride) and inositol-1-phosphate plus some inositol-1,2,-cyclic phosphate. Diglyceride is then phosphorylated to phos-

hydrolases(s), 2. the release of polyunsaturated fatty acids from membrane phosphoglycerides via phospholipase A_2 and subsequent reacylation of the lysophosphoglycerides by the acyltransferases (**Fig. IV-3**). Activity of the acyltransferase was found to be generally higher than that of the phospholipase A_2. However, since activities of the acylCoA ligase and hydrolase in

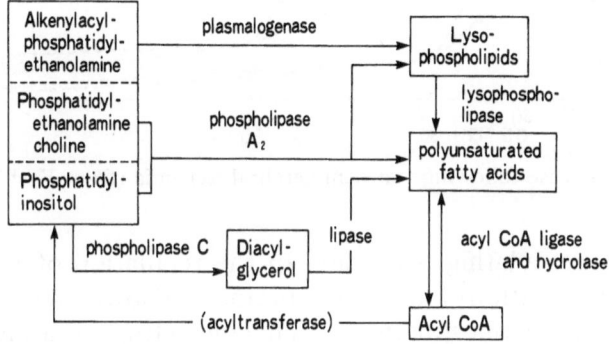

Fig. IV-3. The dynamic equilibrium between phospholipids and FFAs (from Sun *et al.* 1979; partially modified)

phatidic acid, which then acts as a substrate for the formation of CDP diglyceride, the precursor of phosphatidylinositol (Michell 1975). FFAs are liberated from diglyceride and phosphatidic acid by diglyceride lipase and phospholipase A_2, respectively. This sequential reaction is known as phosphatidyl inositol cycle and its biological significance will be discussed later.

Regarding the regulation of the brain FFA level, Sun *et al.* (1979), showed that two types of cyclic events are relevant: 1. the ATP-dependent activation of fatty acids to their acylCoA via the acylCoA ligase and subsequent hydrolysis of acylCoA by acylCoA

brain are higher than the phospholipase and acyltransferase, regulation by the first cyclic event was considered more pertinent in maintaining a constant level of FFAs in the brain.

d) The Release of FFAs Following Cerebral Ischemia

It has been shown that the free fatty acid (FFA) level in the normal brain is very low, but it rapidly increases during ischemia and during stimulation by electro-convulsive shock. Since PUFAs, such as arachidonate, were preferentially released under these conditions, it was surmised that not only an

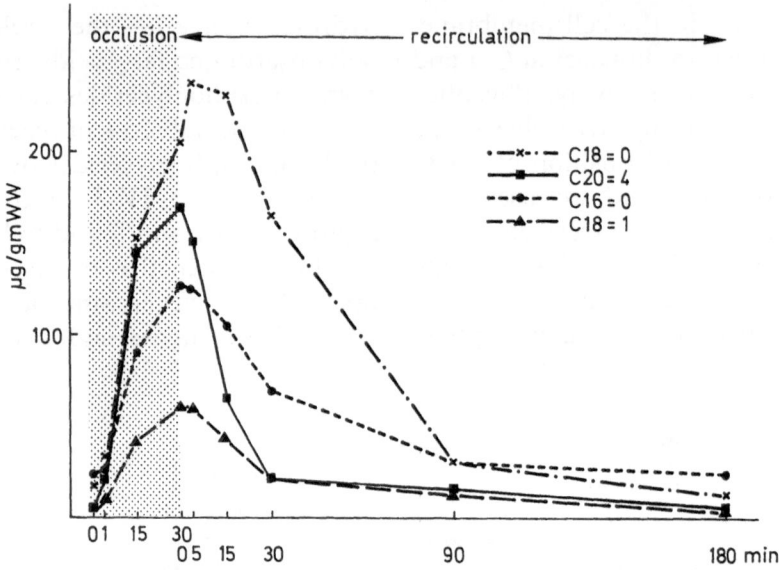

Fig. IV-4. FFA release following transient cerebral ischemia (from Yoshida *et al.* 1980)

inactivity of the reacylating enzyme system but also an activation of lipolytic enzyme systems such as phospholipase A₂ were involved in the process (Bazán 1970, Bazán *et al.* 1971). Bazán (1976) further suggested that the release of norepinephrine following ischemia which causes an increase in cyclic AMP and subsequent activation of protein kinase, might be responsible for the activation of phospholipase A₂. Essentially similar patterns of FFA release following cerebral ischemia were then reported by Yoshida *et al.* (1980) **(Fig. IV-4)** and Rehncrona *et al.* (1982). Regarding the changes in particular fractions of brain phospholipids, Yoshida *et al.* (1980) found a significant decrease in phosphatidyl choline, whereas Rehncrona *et al.* (1982) did not find any measurable changes, exept for a small decrease of inositol plus serine phospholycerides in the early post-ischemic period.

In the regulation of FFA level in the brain, anyhow, the activity of each enzyme acting on a particular pool of lipids as well as the availability of ATP are interrelated in a complex fashion as indicated by the paradigm of Sun *et al.* (1978). Activation of phospholipase A₁ and A₂ (Edgar *et al.* 1982), and also of phospholipase C (Hirashima *et al.* 1985) in the early postischemic period were reported. On the other hand, Yoshida *et al.* (1984) examined whether cerebral noradrenaline (NA) might play a central role in mediating the increased production of FFA during cerebral ischemia. The depletion of unilateral NA by prior destruction of locus coeruleus caused a diminished transient rise of cyclic AMP in response to ischemia, but did not affect cocommitant decline of energy state or the magnitude of FFA increase. An inverse correlation between FFA levels and total adenylate pool, and a preferential rise in the levels of stearic and arachidonic acids were also found. These results led to a conclusion that NA or cyclic AMP did not play a major role in increasing cortical FFAs during complete ischemia. The roles of activation of deacylating enzymes as well as of disturbance of reacylation due to impaired oxidative phosphorylation and energy depletion appeared to be more significant.

However, since the above studies employed models of severe or complete cerebral ischemia of relatively long duration (30 minutes), they did not specifically deal with events occurring in a very early ischemic period. In this respect, Rodriguez de Turco *et al.* (1983) showed a selective increase in arachidonic and stearic acids following bicuculline-induced seizures and therefore suggested that hydrolysis of phospholipid enriched in stearoyl-arachidonoyl groups, such as phosphatidylinositol of excitable membranes, might be stimulated during seizures. Similar results indicating early hydrolysis of phosphatidylinositol have also been obtained with models of hypoglycemia (Wieloch *et al.* 1984) and total brain ischemia by decapitation (Aveldano and Bazan 1975, Yoshida *et al.* 1984, Yasuda *et al.* 1985). Recently, Ikeda *et al.* (1986) carried out a detailed study on the quantitative relationship between phosphoinositides and FFAs. During the first 1–3 minutes following complete cerebral ischemia due to decapitation, there were rapid decreases in triphosphoinositide (PIP_2) and diphosphoinositide (PIP) together with preferential production of stearic and arach-

idonic acids in diglyceride (DG) and the FFA pool. After 10 minutes of ischemia, TPI, DPI, and DG approached plateau levels, but all FFAs continued to increase. These results were interpreted to indicate that, at the onset of ischemia, polyphosphoinositides (PIP_2 and PIP) are primarily responsible for the preferential increase in free stearic and arachidonic acids. Later than 10 minutes, in contrast, hydrolysis of other lipids, possibly biostructural phospholipids such as phosphatidylcholine, appeared to be primarily responsible for the increase in FFAs. These results indicating the fast degradation of inositol-containing phospholipids following ischemia are consonant to its high turnover in the brain (Sun and Su 1979).

Of particular interest is the fact that the turnover of inositol phospholipids is now considered to be a key event in signal transduction, *i.e.*, the process which pertains to activation of cellular functions in response to external stimuli (Michell 1975, Nishizuka 1984). Both PIP and PIP_2 are produced from PI *in situ* through sequential phosphorylation of its inositol moiety by the actions of PI and PIP kinases. A wide variety of extracellular signals, such as monoamine transmitters, acetylcholine, vasopressin, angiotensin II, bradykinin, thrombin, K^+ depolarization, and electrical stimulation, are known to induce inositol phospholipid turnover in their respective target tissues. It has recently been suggested that after stimulation of the receptor, PIP_2 rather than PI and PIP is degraded immediately to produce 1,2-diacylglycerol (DG) and *demyo*-inositol

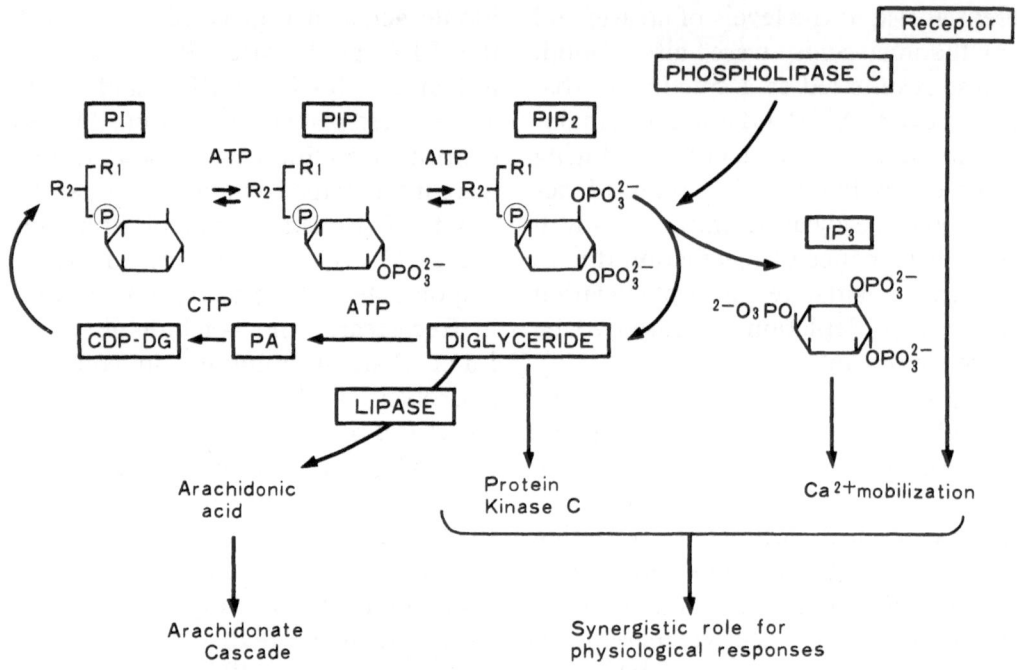

Fig. IV-5. The PI cycle and its role in signal transduction

1,4,5-triphosphate (IP_3) (Berridge *et al.* 1984). Thus produced DG initiated the activation of a specialized protein kinase C, whereas IP_3 may serve as a mediator for Ca^{2+} release from intracellular, most likely nonmitochondrial stores, increasing Ca^{2+} levels within the cells (**Fig. IV-5**). These two routes of information flow, *i.e.*, protein kinase C activation and Ca^{2+} mobilization, evokes subsequent cellular responses, such as release reaction in platelets. In nervous tissues, such a turnover of inositol phospholipid which is provoked by a single receptor stimulation as well as by depolarization may play a crucial role in neurotransmitter release. Insomuch as protein phosphorylation is known to play a central role in the regulatory mechanism in neurones

(Nestler *et al.* 1984), the role of protein kinase C might be extended to the modulation of membrane conductance, channels and active transport, signal-receptor interaction, axoplasmic flow, neurotransmitter biosynthesis, and other neuronal functions by phosphorylating the proteins involved (Nishizuka 1984).

Thus, the fact that the turnover of membrane phosphoinositide is stimulated in a very early period of ischemia raises the intriguing possibility that protein kinase C activation and Ca^{2+} mobilization may be at least partly responsible for the subsequent alterations in the function as well as the metabolism of the ischemic brain. As shown in the figure, however, generation of intracellular mediators, DG and IP$_3$, by cycling of inositol phospholipid turnover requires ATP. Yasuda *et al.* (1985) reported that the specific and rapid liberation of arachidonic and stearic acids following ischemia depended on the availability of ATP. Whether this specific liberation represents a self-limited, adaptive response to ischemia or it heralds the catastrophic train of events leading to cell death, remains to be elucidated.

2. Arachidonate Cascade

As described in the previous section, the turnover of membrane phospholipid is enhanced by external stimuli, trauma, and ischemic-anoxic insults, being followed by liberation of FFAs, particularly of arachidonic acid. Recent studies have shown that free arachidonic acid is incorporated into the arachidonate cascade, an enzyme system comprizing oxygenases, peroxidases, and converting enzymes, and is transformed to a variety of C$_{20}$ derivatives, *i.e.*, eicosanoids, which are immediately discharged out of the cell and exert their actions, combining to receptors present in the cell membrane. Although potent biological actions of eicosanoids have been shown in almost every organ in the body, each of them may show different actions depending on the concentration and the kind of target cells.

It is not the purpose of this section to review the already vast literature but merely to deal with some fundamental aspects of arachidonate metabolism which the author think pertinent to the pathomechanism(s) operating in the acute stage of SAH. For more extended knowledge, readers are referred to excellent reviews by Samuelsson *et al.* (1978), Pickard (1981), Wolfe (1982), Hammarstrom (1983), Walker and Pickard (1985), and many others.

a) Synthetic Pathways of Eicosanoids

The arachidonate cascade is largely devided into two pathways, *i.e.*, cyclooxygenase and lipoxygenase pathways (**Fig. IV-6**). In the former, the enzyme, cyclooxygenase, catalyzes incorporation of two oxygen molecules into arachidonic acid, forming an unstable endoperoxide, PGG$_2$. The same enzyme catalyzes reduction of the hydroperoxy group (OOH) at C 15 to a hydroxyl group (OH), generating another unstable endoperoxide, PGH$_2$. In the conversion of PGG$_2$ to PGH$_2$, oxygen-centered radicals are generated (Kuehl and Egan 1980). PGH$_2$ is then converted to each PG by corresponding PG synthase.

In the latter, lipoxygenases incorporate a molecule of oxygen into various positions (C 5, 8, 9, 11, 12, or 15) of arachidonic acid, forming hydroperoxides (HPETEs: hydroperoxy eicosatetraenoic acids), which are rapidly reduced to corresponding HETEs (hydroxy eicosatetraenoic acids). Particularly in leukocytes, the lipoxygenase product, 5-HPETE, is further transformed to various leukotrienes (LTs), which were shown to compose the slow-reacting substances of anaphylaxis (SRS-A) (Parker 1982,

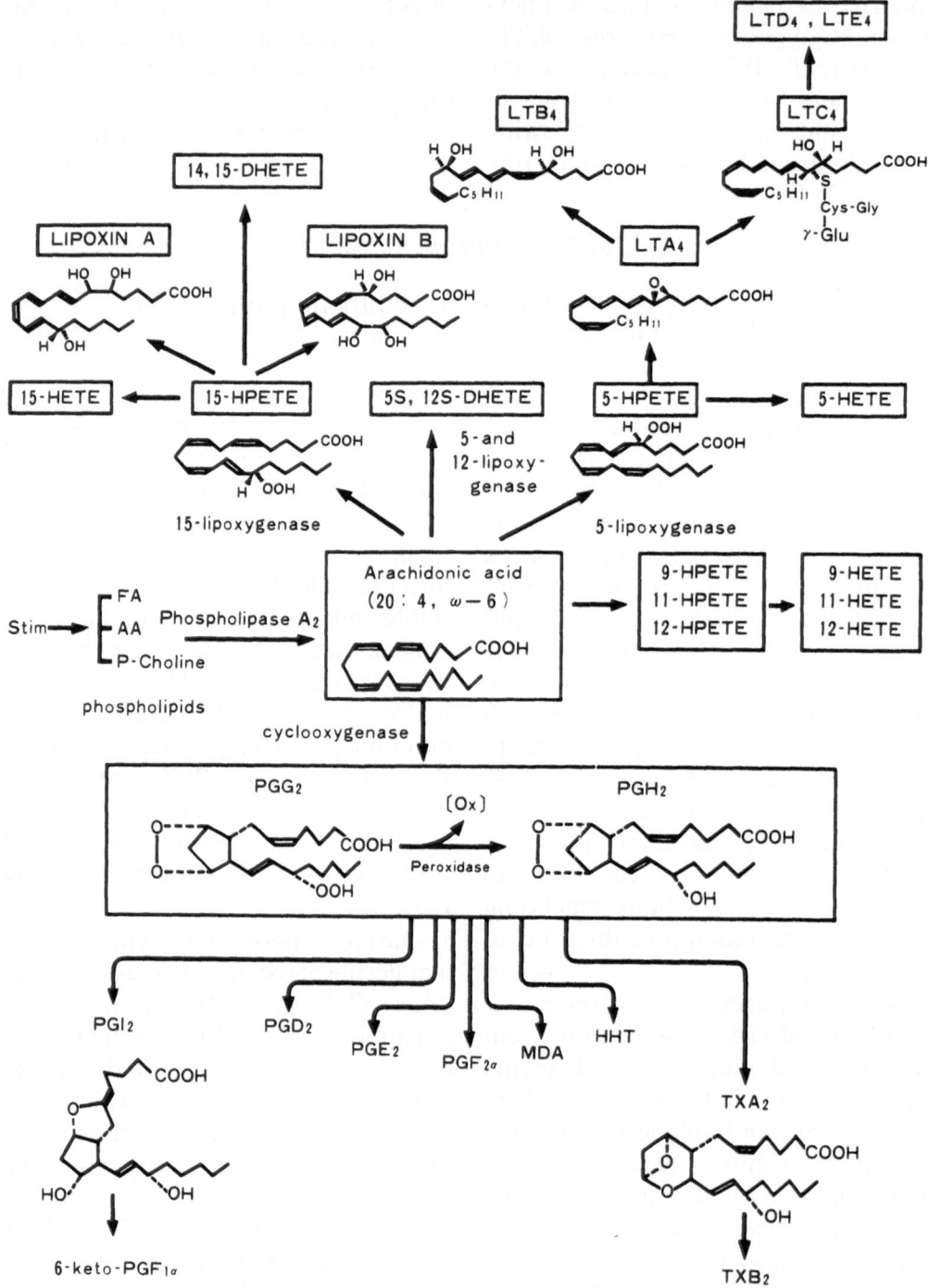

Fig. IV-6. The AA cascade. The upper and lower halves of the figure denote the lipoxygenase and the cyclooxygenase pathways, respectively

Samuelsson 1983). The next major metabolite are 5, 12-DHETE and 14,15-DHETE, which are generated by oxygenation of arachidonic acid at C-12 and C-15 (Samuelsson et al. 1980). Recently discovered lipoxins (lipoxin A and lipoxin B) formed by the action of 5- and 15-lipoxygenases and additional reactions, contain a conjugated tetraene structure and three alcohol groups (THETEs: trihydroxy eicosatetraenoic acids) (Serhan et al. 1984).

b) Eicosanoid Synthesis in the Normal Brain

The capacity of the brain tissue to synthesize prostaglandins has been evaluated by measuring each PG product in incubated brain homogenates or brain slices, with or without addition of exogenous arachidonic acid or PGH_2. It has thus been shown that the mammalian brain produce PGD_2, PGE_2, $PGF_{2\alpha}$, 6-keto $PGF_{1\alpha}$, and TXB_2 (Wolfe et al. 1976 a, b, Sun et al. 1977, Abdel-Halim et al. 1980 a, b, Bishai and Coceani 1981). In these studies, the predominant PGs in the rat and mouse brains were found to be PGD_2 and TXB_2 (Abdel-Halim et al. 1980 b), whereas in human nonvascular brain tissues, the major product was $PGF_{2\alpha}$ (Abdel-Halim et al. 1980 a). In cat and human brain tissues, the synthesis of PGD_2 was considered to be exceedingly low or absent (Wolfe 1982). Nevertheless, the PG profiles revealed by incubation techniques are obviously not those occurring normally in vivo.

Since the brain PG levels rapidly change following stimulation or damage, extreme caution is necessary to evaluate the PG levels in vivo. Gaudet et al. (1980) measured brain PG levels of mongolian gerbils killed by liquid nitrogen. The values for PGD_2, PGE_2, and $PGF_{2\alpha}$ in sham-operated animals were 1.4 ± 0.1, 0.5 ± 0.2, and 0.3 ± 0.1 ng/g tissue, respectively. Narumiya et al. (1982) measured the PG levels in the rat brain in which the activities to synthesize and catabolize PGD had been known to become zero after microwave irradiation. The "basal" levels of PGs thus measured (PGD_2, 1.72 ± 0.30; PGE_2, 0.66 ± 0.12; $PGF_{2\alpha}$, 0.49 ± 0.10 ng/g tissue) were in good agreement with those of Gaudet et al. (1980).

Not only the synthesis but also the metabolism are relevant to the maintenance of the basal levels of PGs in the brain. In this respect, it has been suggested that the brain has very low enzyme activity to metabolize PGs (Nakano et al. 1973). Bito et al. (1976) showed that PGs formed in the brain in vivo are cleared into the general circulation by transport across the blood-CSF and blood-brain barriers. However, recent studies indicate the presence of several enzymes which catabolize and inactivate the synthesized PGs in situ (Lee and Levine 1974, Wermuth 1981, Watanabe et al. 1980). The existence of a mechanism to utilize released PGs such as the uptake and release of monoamines transmitters in neurones, has not been known.

The profile, levels, and regional distribution of several PGs in postmortem human brain were examined by Ogorochi et al. (1984). It was revealed that PGD_2, PGE_2, and $PGF_{2\alpha}$ are present in essentially similar concentrations with variations in regional distribution similar to those in rat brain (Cseh et al. 1978, Narumiya et al. 1982). Thus, their results stand in clear opposition to the suggested species difference of brain PG profile and metabolism (Wolfe 1982). Although postmortem alteration in PG metabolism have been shown (Galli et al. 1980), the obtained values were considered to represent the "end point" concentration of PGs reached after declining from their peak values in postmortem brain and interpreted to imply that PG profile and metabolism are essentially similar among species.

Also important are the findings that human brain blood vessels can generate more PGs than the neuronal and glial tissue, and that the formation of PGs seemed to be oriented toward the prostacyclin pathway in the vascular tissue (Abdel-Halim et al. 1980 a, b). Large cerebral arteries such as the canine basilar and bovine cerebral arteries, were found to have the capacity of generating PGI_2 (Boullin et al. 1979, Sasaki et al. 1981, Maeda et al. 1981, Walker et al. 1983). Although PGI_2 (6-keto-$PGF_{1\alpha}$) was the most abundant PG, synthesis of PGE_2, $PGF_{2\alpha}$, and TXB_2 was also observed (Hagen et al. 1979, Maeda et al. 1981, Walker et al. 1983). Preponderant synthesis of 6-keto-$PGF_{1\alpha}$ was also shown with the choroid plexus (Goehlert et al. 1981, Abdel-Halim et al. 1981 b, Ogorochi et al. 1984), arachnoid, and dura mater (Ogorochi et al. 1984).

There has been some discrepancy of experimental results regarding the PG profile of brain microvessels. Gerritsen et al. (1980) examined PG formation of bovine brain microvessels with PGH_2 added in the incubation medium. Although 6-keto-$PGF_{1\alpha}$ was also formed, formation of PGE_2 was favored. Using bovine brain microvessels, on the other hand, the report of Maurer et al. (1980) showed that among the endogenously formed PGs (6-keto-$PGF_{1\alpha}$, $PGF_{2\alpha}$, TXB_2, PGE_2), the formation of PGI_2 was by far the greatest. Similar preponderance of PGI_2 formation was shown also with cat (Birkle et al. 1981) and rabbit (Rodrigues and Gerritsen 1984) brain microvessels. Studies using rat brain microvessels yielded more discrepant results. In the study of Goehlert et al. (1980), the major PG formed was PGI_2 regardless of whether the PG synthesis was endogenous or from exogenous endoperoxides. In contrast, other studies (Gerritsen and Printz 1981, Gecse et al. 1982, Asano et al. 1983) yielded results suggesting that not PGI_2 but PGD_2 is the major product of rat brain microvessels. Although this discrepancy may be partly due to the use of different methods of microvessel preparation and the use of different precursors (endoperoxides, arachidonic acid), further confirmation is obviously required.

Regarding the products of lipoxygenase pathway, available information is scanty yet. Sautebin et al. (1978) and Spagnuolo et al. (1979) reported that the gerbil brain has lipoxygenase activity to produce 12-HETE. Recently, Wolfe and Pappius (1984) showed that the incubated rat cerebral cortex releases HHT (12-hydroxyheptadecatrienoic acid), 5-, 11-, 12-, and 15-HETEs upon stimulation by the calcium ionophore A 23187. Lindgren *et al.* (1984) also showed that leukotrienes (LTs) C_4, D_4, and E_4 were isolated after incubation of rat brain tissue *in vitro* with the ionophore A 23187 and arachidonic acid. The biosynthesis of LTC_4 was found in most brain regions, being highest in the hypothalamus and the median eminence. *In vitro* synthesis of 5- and 12-HETE in the bovine retina was also reported (Birkle and Bazan 1984). Generation of leukotrienes by vascular tissue has recently been indicated (Piper *et al.* 1983, Wölbling *et al.* 1983). Koide *et al.* (1984) suggested that the rat brain microvessel has lipoxygenase activity which is enhanced following regional ischemia and stimulation by a lipid hydroperoxide, 15-HPETE. The brain levels of LTC_4 and LTD_4 were shown to be significantly increased on recirculation following temporary global ischemia (Moskowitz *et al.* 1984). The alteration of the eicosanoid synthesis in the brain and cerebral vasculature following cerebral ischemia or SAH, and its relevance to the subsequent brain damage will be dealt with in the following sections.

c) Regulation of the Activity of Arachidonate Cascade

The brain is composed of various tissues and cells, such as neuronal cells, glia, extra- and intraparenchymal vessels, and the choroid plexus. As described in the former section, different activities and preferential pathways of AA cascade have been found in each tissue. This is apparently based on the particular pattern of distribution of cyclooxygenase, lipoxygenase, and other converting enzymes, which seems to be consonant to the biological role of each kind of cells.

It has been shown that the rate-limiting step of eicosanoid synthesis resides in the availability of free arachidonic acid liberated from membrane phospholipid (Flower and Blackwell 1976, Lands 1979, Wolfe 1982). As already described, liberation of AA from membrane phospholipid may be due to activation of calcium-dependent (Siesjö 1982, Moskowitz *et al.* 1983) or calcium-independent (Edgar *et al.* 1982) phospholipase A_2, and of phosphatidylinositol-specific phospholipase C (Aveldano and Bazan 1975, Lapetina *et al.* 1981, Rodriguez de Turco *et al.* 1983, Yoshida *et al.* 1984, Ikeda *et al.* 1986). Trauma or anoxia-ischemia leads to accumulation of considerable amount of free arachidonic acid in the brain. However, that only a small portion of liberated arachidonic acid, possibly representing a specific pool(s) is utilized to form eicosanoids is indicated by the following facts: 1. under various conditions, the concentration of arachidonic acid is 100–1,000 times greater than that of simultaneously existing eicosanoids (Bosisio *et al.* 1976, Marion and Wolfe 1978, Yoshida *et al.* 1980, Gaudet *et al.* 1980); 2. because of their hydrophobic nature, most FFAs are bound to membrane or hydrophobic sites on cellular proteins or extracellular serum albumin, thus being inaccessible to cyclooxygenase in microsomes (Lunt and Lowe 1968, Spector 1975, Lands 1979); recently, it was shown that the major portion of arachidonic acid liberated following cerebral ischemia is incorporated into triglycerids (Yoshida *et al.* 1986). In any event, recent studies indicate that the mechanism of arachidonic acid release from membrane phospholipids in the very early period of ischemia is different from that in later periods (Yasuda *et al.* 1985, Ikeda *et al.* 1986). How these mechanisms are coupled to eicosanoid synthesis remains as a problem of fundamental importance.

Since molecular oxygen is required for oxygenation of arachidonic acid, its concentration should affect the rate of PG synthesis. Lands (1979) showed that the level of oxygen needed to give appreciable oxygenation activity of cyclooxygenase is around $5\,\mu M$. At tissue O_2 levels greater than $20\,\mu M$, synthesis seems limited by substrate acid availability. Thus the arachidonate accumulating in hypoxic-ischemic conditions can elevate the rate of PG synthesis as long as O_2 concentration are still sufficient ($5\text{--}10\,\mu M$) to maintain near-optimal cyclooxygenase rates. Nevertheless, as recently suggested by Rodrigues and Gerritsen (1984), oxygen tension may differentially affect PG synthesizing activities leading to PGI_2 versus PGE_2 production. Such a result may implicate the existence of two distinct forms of cyclooxygenase in brain tissue (Lysz and Needleman 1982), each of them being coupled to synthesis of each PG.

Further, there is ample evidence showing that the activity of cyclooxygenase as well as that of lipoxygenase are regulated by the level of lipid peroxides (Gale and Egan 1984, Lands *et al.* 1984). As described in the next section, generation of oxygen free radicals as well as the activity of lipoxygenase lead to formation of hydroperoxides within cells. Although the hydroperoxides thus formed may be small in amount, they act as activators of cyclooxygenase. In high concentrations, they conversely inactivate cyclooxygenase. Hydroperoxides are effectively catabolized by cytosolic enzymes such as glutathione peroxidase and glutathione S transferase. Thus, Lands *et al.* (1984) suggested that the balance between the generation and removal of hydroperoxides in a tissue determines the ambient level of peroxide activator and the proportion of cyclooxygenase which is catalytically active.

As shown above, the activity of eicosanoid cascade seems to be regulated by multiple factors. The next section deals with another aspects of PUFA metabolism, *i.e.*, lipid peroxidation due to nonenzymatic reactions.

3. *Free Radicals and Lipid Peroxidation*

The reduction of oxygen to water tends to proceed by a series of single electron transfer, generating highly reactive intermediates. The deleterious effects of these "activated oxygen" such as superoxide radical, hydrogen peroxide, and hydroxyl radical are diverse, involving all the interactions with proteins, lipids, polysaccharides, nucleic acids, and other biochemicals. Among those toxic effects, interaction of activated oxygen with membrane lipids causing lipid peroxidation and membrane disruption has been thought to be of prime importance, so far as the pathomechanism of anoxic-ischemic brain damage is concerned. A brief account of mechanisms involved in this "free radical damage" will be given in the present section. The whole spectrum of oxygen toxicity and free radicals have been reviewed (Fridovich 1976, Pryor (ed) 1976–1984, Chance et al. 1979, Lewis and del Maestro (eds) 1980, Halliwell and Gutteridge 1984, Siesjö and Wieloch 1985).

a) Oxygen Free Radicals

A "free radical" is defined as any atom or molecule that has one or more un-

paired electrons occupying an outer orbital. The lone electron has a strong tendency to interact with other electrons to form an electron pair, i.e., a chemical bond. As O_2 has two unpaired electrons located in a different π orbitals, it is a "biradical". A series of single electron transfer to the ground state O_2 molecule generates extremely reactive intermediates such as superoxide radical (O_2^-), peroxide ion (O_2^{2-}) which is immediately protonated to hydrogen peroxide (H_2O_2), and hydroxyl radical ($\cdot OH$). O_2^- and $\cdot OH$ have an unpaired electron and therefore are "oxygen free radicals".

$$O_2 \xrightarrow{e^-} O_2^- \xrightarrow[]{e^- + 2H^+} H_2O_2 \xrightarrow[-H_2O]{e^- + H^+} \cdot OH \xrightarrow{e^- + H^+} H_2O$$

In addition to those intermediates, there are other species of activated oxygen. The outer orbitals of O_2 have two unpaired electrons and is a triplet in its ground state ($^3\Delta g+$) because the electron spins are parallel or unpaired. This arrangement of electron can be excited to two singlet states. In one ($^1\Delta g$), both electrons occupy the same orbital with paired electron spins. The other form of singlet oxygen ($^1\Delta g+$) usually decays to the $^1\Delta g$ state before it has time to react with anything. Although O_2 ($^1\Delta g$) is not a free radical, it may be considered a related species since it is very reactive and some radical reactions result in its generation (del Maestro 1980, Halliwell and Gutteridge 1984).

Generation of O_2^- leads to subsequent production of H_2O_2 and $\cdot OH$ through reactions as shown below. First, in aqueous solution, O_2^- undergoes so-

called dismutation reaction to form H_2O_2 and O_2. This reaction is catalyzed by superoxide dismutase (SOD) (Fridovich 1976).

$$2 O_2^- + 2 H^+ \rightarrow H_2O_2 + O_2$$

H_2O_2 is reduced by catalase to H_2O.

$$2 H_2O_2 \rightarrow O_2 + 2 H_2O$$

In the presence of an iron(II) salt, $\cdot OH$ radical is generated from H_2O_2 (Fenton-type reaction).

$$Fe^{2+} + H_2O_2 \rightarrow Fe^{3+} + \cdot OH + OH^-$$

Hydroxyl radical is also formed by interaction between O_2^- and H_2O_2 (Haber-Weiss reaction).

$$H_2O_2 + O_2^- \rightarrow O_2 + OH^- + \cdot OH$$

The rate constant for the above reaction in aqueous solution is virtually zero. However, this reaction is catalysed by traces of transition metal ions (an iron-catalyzed Haber-Weiss reaction) and accounts for $\cdot OH$ formation *in vivo* (Halliwell and Gutteridge 1984).

$$Fe^{3+} + O_2^{\cdot-} \rightarrow Fe^{2+} + O_2$$
$$Fe^{2+} + H_2O_2 \rightarrow Fe^{3+} + \cdot OH + OH^-$$
(Fenton reaction)

Most of the O_2 consumed by aerobic cells is reduced to H_2O without generation of $O_2^{\cdot-}$ or H_2O_2. This is because of the divalent and tetravalent reduction of O_2 carried out by cytochrome c oxidase (Fridovich 1976). However, isolated mitochondria release H_2O_2 at rates that depend on the metabolic state (Boveris 1977). This has been ascribed to generation of $O_2^{\cdot-}$ through the non-enzymatic reduction of O_2 by ubisemi-quinone. $O_2^{\cdot-}$ formation by this route is proportional to oxygen tension (Boveris 1977).

It has been shown that in addition to the nonenzymatic reactions which is catalyzed by a transition metal ion and a reducing agent such as ascorbic acid, the enzymatic reaction is also involved in lipid peroxidation (Mead 1976). The enzymatic reaction may be the usual drug-hydroxylating system comprizing cytochrome P-450, Cyt P-450 reductase, a reducing cofactor (as NADPH or ascorbic acid), and molecular oxygen. Hochstein and Ernster (1963) suggested that the nucleotide-chelated $Fe^{2+}O_2$ plays a central role as the initiating agent. In addition to this, a variety of enzymes, such as xanthine oxidase, aldehyde oxidase, dihydro-orotic dehydrogenases, flavin dehydrogenases, and peroxidases release $O_2^{\cdot-}$. The autooxidation of a large group of compounds including catecholamines and ferredoxin also involve the release

of $O_2^{\cdot-}$ (Fridovich 1976, Chance *et al.* 1979, del Maestro 1980). Further, prostaglandin cyclooxygenase-peroxidase release oxygen-centered radical (Kuehl and Egan 1980). Inflammatory cells, such as polymorphonuclear leukocytes, macrophages and monocytes release a great amount of $O_2^{\cdot-}$ and H_2O_2 utilizing NADPH oxidase (Fantone and Ward 1982). Since generation of $O_2^{\cdot-}$ is followed by an even more reactive species, *i.e.*, $\cdot OH$, survival of an organism would be impossible if it were not for an array of defense mechanisms which effectively quenches these reactive intermediates.

b) Protective Mechanisms—Free Radical Scavengers

SOD catalyzes the dismutation reaction of $O_2^{\cdot-}$ to form H_2O_2. It has been shown that oxygen-metabolizing organisms contain SOD, whereas those that do not use oxygen do not. Thus, SOD is regarded as the primary defense against $O_2^{\cdot-}$ (Fridovich 1976). H_2O_2 is scavenged by two classes of related enzymes—the catalases and the peroxidases. Since H_2O_2 is relatively stable, not all the respiring organisms possess catalase. In mammals, erythrocytes contain high levels of catalase whereas some tissues are devoid of the enzyme. This is because the circulating blood can remove and decompose the H_2O_2 produced in such tissues. Human brain has very little catalase, although H_2O_2 is generated by mitochondrial monoamine oxidase (Sinet *et al.* 1980). Parenthetically, in specific tissues which lack catalase, H_2O_2 serves as a useful intermediate, as known with thyroid and phagocytic leukocytes (Fridovich 1976). In the former, H_2O_2 is used for organification of I^- to thyroxin. In the latter, myeloperoxidase-H_2O_2-halide system serves to enhance the bactericidal activity (Fantone and Ward 1982).

Peroxidases, especially glutathione peroxidase can act on lipid hydroperoxides as well as on H_2O_2, utilizing a variety of intracellular electron donors and reduced glutathione (GSH). Oxidized glutathione (GSSG) is reduced by the NADPH-dependent glutathione reductase (Chance et al. 1979).

$$RH_2 + H_2O_2 \rightarrow R + 2H_2O.$$
$$2GSH + H_2O_2 \rightarrow GSSG + 2H_2O.$$
$$2GSH + ROOH \rightarrow ROH + GSSG + H_2O.$$
$$GSSG + 2NADPH \rightarrow 2GSH + 2NADP^+.$$

The release of $NADP^+$ subsequent to reduction of H_2O_2 was shown to lead to stimulation of the pentose phosphate pathway in rat brain synaptosomes (Hothersall et al. 1982). It has been shown that the brain has a very large latent capacity of the pentose phosphate pathway, and correspondingly high activity of the oxidative enzymes of the pathway and of glutathione peroxidase. Vitamin E (α-tocopherol), intercalated in cellular membranes, scavenges free radicals by providing hydrogen atoms, and prevents chain progating reactions (Pryor 1976). Free radicals are removed via formation of tocopherol quinone or dimer (Witting 1980).

In addition to those described above, ascorbic acid, cysteine, ceruloplasmin, and transferrin probably have scavenging capacity in hydrophilic environ-

ment (del Maestro 1980). Cholesterol intercalated in membranes was found to protect against the free radical damage to membrane fatty acids (Demopoulos et al. 1976).

c) Lipid Peroxidation and Resultant Membrane Damage

As shown above, generation of O_2^- is linked to production of other species of activated oxygen, i.e., H_2O_2 and $\cdot OH$. Among these, the reactivity of hydroxyl radical is so great that they react immediately with whatever biological molecule is in their vicinity, producing secondary radicals (Halliwell and Gutteridge 1984). In respect to membrane phospholipids, hydrogen abstraction from the polyunsaturated fatty acid (PUFA) chains becomes the initiating step of lipid peroxidation. Protonated O_2^- (HO_2^-) as well as singlet O_2 ($^1\Delta g$) possibly can attack fatty acids directly. Hydrogen abstraction from fatty acids (R-OH) leaves behind an unpaired electron on the carbon atom, i.e., carbon-centered or in this case, alkyl radical ($R \cdot$).

Initiation:

$$R–H + \cdot OH \rightarrow R \cdot + H_2O.$$

The double bonds in PUFAs tends to be stabilized by a molecular rearrangement to produce a conjugated diene, which rapidly reacts with O_2 to produce a hydroperoxy radical ($ROO \cdot$). $ROO \cdot$ abstracts hydrogen atoms from other lipids, generating another radical ($R \cdot$), and this chain reaction continues. $ROO \cdot$ combines with the hydrogen atom to give a lipid hydroperoxide

(ROOH). Propagation of this chain reaction is evidently facilitated by more or less parallel and contiguous arrangement of fatty acid chains in the membrane, being maximal when the unsaturated centers of the chains are in close ordered proximity (Mead 1976).

Propagation:

$$R \cdot + O_2 \rightarrow ROO \cdot.$$
$$ROO \cdot + R\text{--}H \rightarrow ROOH + R \cdot.$$

As the above chain reaction propagates within the membrane, lipid hydroperoxides (ROOH) accumulate. These lipid hydroperoxides are then decomposed by many metal complexes present *in vivo*, which include simple complexes of iron salts, heme, hemoproteins such as hemoglobin, methaemoglobin, peroxidase, cytochrome P-450, other cytochromes and non-heme iron proteins. Insomuch as iron (II)- and iron (III) compounds lead to formation of alkoxy (RO·) and peroxy (ROO·) radicals, respectively, the former stimulate peroxidation more than does the latter. This is probably because alkoxy radicals formed by Fe^{2+}-complexes is more reactive than peroxy radicals in initiating peroxidation (Halliwell and Gutteridge 1984).

Decomposition of lipid hydroperoxide:

$$ROOH + Fe^{2+}\text{-complex} \rightarrow Fe^{3+}\text{-complex} + OH^- + RO \cdot.$$
$$ROOH + Fe^{3+}\text{-complex} \rightarrow Fe^{2+}\text{-complex} + H^+ + ROO \cdot.$$

O_2^- and perhaps $HO_2 \cdot$ decompose lipid hydroperoxides in a similar fashion (Thomas *et al.* 1982). Further, the hydroperoxide of arachidonic acid (HPETE) may be decomposed by prostaglandin hydroperoxidase to yield an oxidant and a hydroxy acid (HETE) (Kuehl and Egan 1980).

$$ROOH + O_2^- \rightarrow RO \cdot + OH^- + O_2.$$
$$ROOH + HO_2 \cdot \rightarrow RO \cdot + H_2O + O_2.$$
$$HPETE \xrightarrow{\text{Peroxidase}} HETE + [Ox].$$

Termination of lipid peroxidation involves a variety of processes. Examples of homo- and cross-terminations in PUFAs are shown below.

Termination:

$$2\,R \cdot \rightarrow R\text{--}R;\ R \cdot + ROO \cdot \rightarrow ROOR;$$
$$2\,ROO \cdot \rightarrow ROOR + O_2.$$

Formation of these new bonds in addition to peroxidation of PUFAs will produce an alteration of membrane structure. Of further importance is the fact that there are a number of membrane enzymes the activities of which are dependent on the integrity of associated phospholipid (Fourcans 1974, Farias *et al.* 1979). The activities of those enzymes, for instance Na^+, K^+ATPase, might be inhibited by hydroperoxides. Also, hydroperoxides and its breakdown product, malonaldehyde, can cause cross-linking in a variety of enzymes, proteins, and nucleic acids, producing fluorescent pigments (Tappel 1980). Thus, unless effectively inhibited, lipid peroxidation is destructive in biological systems. In a variety of pathological conditions as well as aging, distortion of the balance between generation of free radicals and the capacity of scavenging systems was suggested as an underlying pathomechanism (Harman 1962, Barber and Bernheim 1967, Slater 1972, del Maestro

1980, Halliwell and Gutteridge 1984). In this line of thinking, Demopoulos *et al.* (1977, 1980) and Flamm *et al.* (1978) put forward an attractive hypothesis that in cerebral ischemia or trauma, cell damage is incurred by generation of free radicals with subsequent peroxidative degradation of cell membranes. On the basis of hitherto accumulated experimental findings, the occurrence of lipid peroxidation in brain tissue and its possible role in ischemic brain damage will be discussed next.

d) Lipid Peroxidation in Brain Tissue *in vitro*

Theoretically, there are a number of methods to detect the occurrence of lipid peroxidation in tissues. They consist of detection or measurement of the following parameters: 1. the activated oxygen species, lipid hydroperoxides, and other free radicals by the use of electron spin resonance, chemiluminescence, UV absorbance, and etc; 2. the membrane damage which is known to be characteristic to free radical injury, such as alterations in phospholipid content, fatty acid composition, activities of membrane-bound enzymes, and cholesterol content; 3. degradation products of hydroperoxides such as malonaldehyde, ethane, and penthane; 4. formation of lipid-soluble fluorescent products due to malonaldehyde addition to amino acids (Schiff base), amine groups of phospholipids, enzymes, proteins, and nucleic acids; 5. the increase in the ratio of oxidized form or consumption of tissue antioxidants, such as α-tocopherol, ascorbic

acid, glutathione, and cysteine (Barber and Bernheim 1967, Tappel 1975, 1980, Pryor (ed) 1976–1984, Demopoulos *et al.* 1979).

Among the above-listed methods, malonaldehyde has been most widely used as an indicator of lipid peroxidation (Willis 1966, Barber and Bernheim 1967). Lipid hydroperoxides, when heated in acidic condition in the presence of thiobarbituric acid (TBA), yield products which have a pink color (TBA test) (Dahle *et al.* 1962). Pryor *et al.* (1976) showed that in this condition, autooxidation of either the free acid or esterified PUFA leads to formation of prostaglandin-like endoperoxides. Thermal decomposition of those endoperoxides yields malondialdehyde (MDA), which then forms adduct (chromogen) with TBA. This test has been used to examine the occurrence of lipid peroxidation in tissues including the brain. However, since MDA is derived from endoperoxides, TBA test detects not only lipid peroxides generated via free radical reactions but also PG endoperoxides generated by AA cascade. Shimizu *et al.* (1981) showed that the efficiency of reaction with TBA was much greater in PGG_2 and PGH_2 than in a lipid hydroperoxide, 15-HPAA, and that the serum level of TBAR was decreased to a significant extent (25–60%) by aspirin administration causing inhibition of cyclooxygenase.

There is a general agreement that it is impossible to get a positive TBA test in undamaged freshly isolated cells and tissues under any of the experimental conditions thus far tested. If the tissue is homogenized and aerobically incubated, however, the TBA test becomes positive and the amount of reactive material increases with continued incubations (Barber and Bernheim 1967). Past experiments using rats indicate that brain homogenates undergo considerable lipid peroxidation in comparison to other tissues presumably due to its high content of lipids and unsaturated fatty acids, and that peroxidation is markedly enhanced in the presence of ascorbic acid and Fe^{2+} as well as in low pH (Willis 1966, Barber and Bernheim 1967, Boehme *et al.* 1977). In this regard, however, it seems worth while mentioning that the rate at which a tissue homogenate undergoes autooxidation, as determined by the TBA test, may be regarded as a reflection of the intrinsic susceptibility of that tissue homogenate to reaction with active oxygen species. The human brain homogenate is extraordinarily resistant to autooxidation and that of shorter lived species, *e.g.*, rats, are more susceptible to autooxidation (Cutler 1984).

Rehncrona *et al.* (1980) studied the relationship between the occurrence of lipid peroxidation and changes in brain cortical fatty acids and phospholipids using rat brain homogenates incubated with or without oxygen. As parameters of lipid peroxidation, which was fortified with addition of Fe^{2+} and ascorbic acid, multiple methods such as determination of TBA-reactive substances (TBAR), conjugated diene (UV absorbance), and fluorescence intensity of the lipid extracts were employed. These parameters consistently showed changes indicating the occurrence of lipid peroxidation under aerobic con-

dition but not under anaerobic condition. Lipid peroxidation was followed by selective losses of arachidonic acid $(C_{20:4})$ and docosahexaenoic acid $(C_{22:6})$ with a decrease in ethanolamine phosphoglyceride. These changes in fatty acid and phospholipid composition of brain tissue was considered to characterize the occurrence of lipid peroxidation not only *in vitro* but also *in vivo*.

Kovachich and Mishra (1980) reported that lipid peroxidation in incubated rat brain slices (formation of TBAR) was expressed as a function of [atm oxygen × min], and was accompanied with partial inactivation of tissue Na^+, K^+ ATPase (Kovachich and Mishra 1981). Addition of PUFAs ($C_{18:2}$, $C_{18:3}$, $C_{20:4}$, $C_{22:6}$) in the incubation medium of rat brain homogenates or slices significantly increased the formation of MDA and superoxide (Chan and Fishman 1980). Further, the possibility that the increases in intracellular Ca^{2+} may magnify lipid peroxidation was suggested (Braughler *et al.* 1985, Siesjö and Wieloch 1985 a).

Recently, Siesjö *et al.* (1985 a) examined the effect of acidosis on lipid peroxidation of brain tissues *in vitro*. The pH of the medium for incubation of rat brain homogenates, being fortified with Fe^{2+} and ascorbic acid, was altered from 7.0 to 5.0 and subsequent measurements of TBAR, water- and lipid-soluble antioxidants (glutathione, ascorbate, and α-tocopherol), and phospholipid-bound fatty acids (FAs) were carried out. It was shown that a lowering of pH from 7 to 6 enhanced the formation of TBAR and the degradation of phospholipid-bound polyenoic

FAs, but that it had no influence on the oxidation of the water-soluble free radical scavengers (glutathione and ascorbate). In contrast, changes in α-tocopherol content mirrored those in TBAR material. These results led to following speculations that: 1. at pH 6.0, further reduction of ascorbate was not possible and reduction of hydroperoxy FAs by glutathione system operate at maximal rates already at pH of 7.0; 2. the hydrogenated form of O_2^- ($HO_2 \cdot$), being a stronger oxidant than O_2^- and more soluble in the lipid phase, become more preponderant in low pH, 3. low pH releases protein-bound iron.

In this regard, we measured formation of HETEs in incubated rat brain homogenates, not fortified with Fe^{2+} or ascorbic acid. At pH of 7.0, the level of each HETE progressively increased, whereas it progressively decreased at pH of 5.5 (Noguchi et al., unpublished data). This result was contrary to our expectation that low pH would lead to enhancement of lipid peroxidation, hence formation of more HETEs. This result may indicate that hydroperoxides generated in those conditions cannot be effectively reduced to hydroxyacids, possibly due to suppression of hydroperoxide phospholipase present in the membrane (Tappel 1980), or inability of cytosolic glutathione peroxidase to reduce hydroperoxides present in membranes (McCay et al. 1976). Or it may be that general peroxidation reactions yielding TBAR has little relevance to specific peroxidation reaction in the arachidonate cascade (Morisaki et al. 1984). In any event, accumulated data show that brain tissue undergoes peroxida-

tion on incubation, which is fortified by Fe^{2+}, ascorbic acid, O_2, low pH, or NADPH (Kogure et al. 1985). However, whether or not thus exaggerated lipid peroxidation in incubated brain tissues has any relevance to in vivo situations is not clear at present.

e) Lipid Peroxidation in vivo

It has been well shown that all respiring cells generate oxygen radicals at certain rate and that they have arrays of protective systems to prevent the subsequent occurrence of lipid peroxidation. More than merely to defend themselves, they have developed systems to utilize activated oxygen species as exemplified in the production of eicosanoids via the arachidonate cascade and H_2O_2 generating capacity of leukocytes. It has been shown that the activity of arachidonate cascade is dependent on the concentrations of free arachidonic acid as well as of free radicals, i.e., active oxygen species and hydroperoxides which stimulate or deactivate cyclooxygenase and lipoxygenases (Lands et al. 1984). The level of antioxidants is thus crucial to maintain a proper level of free radicals, hence a proper level of activity of arachidonate cascade. Vice versa, the level of intracellular free radicals may exert an influence on the level of antioxidants. Cutler (1984) demonstrated that there is a good correlation between the longevity and the antioxidative capacity of each mammalian species. He suggested that each species has a unique set point to maintain a given mean level of active oxygen species within their cells about which the compensational regulation of antioxidant levels would operate. This regulation of tissue levels of antioxidants was ascribed to the function of a system comprizing arachidonic acid, prostaglandin cyclooxygenase, and guanylate cyclase (GAC model). In brief, the activity of arachidonate cascade is stimulated by arachidonic acid and the tissue oxidant level. PGs produced in the cyclooxygenase pathway work to elevate the tissue level of cAMP, whereas PGG_2 and other hydroperoxides stimulate guanylate cyclase, elevating the level of cGMP. He proposed that the ratio of intracellular levels of cGMP to those of cAMP determines the rate of antioxidant synthesis, which in turn regulates the activity of arachidonate cascade and the level of active oxygen species.

Probably because of the fact that the mammalian species are endowed with so well developed protective systems against active oxygen species and that free radicals are reactive transients which do not have a sufficiently high steady state concentration, demonstration of the occurrence of free radical reactions or subsequent lipid peroxidation *in vivo* has been regarded as a most difficult and controversial task (Tappel 1980). In spite of such a difficulty, however, the hypothesis that lipid peroxidation may trigger irreversible tissue damage by any type of injury is still attractive and appears to deserve verification.

Regarding ischemic or traumatic brain damage, Demopoulos *et al.* (1977, 1980) proposed that in hypoxic conditions, liberation of O_2^- from CoQ present in the mitochondrial electron transport chain rather increases because the auto-

oxidation of CoQ proceeds more easily than the multivalent electron transfer by enzymes. As evidence for the occurrence of lipid peroxidation *in vivo*, he showed significant decreases in the contents of antioxidants such as cholesterol and ascorbic acid as well as a selective decrease in PUFAs of membrane phospholipids in the cortex of the cat's brain exposed to five hours of middle cerebral artery occlusion. Nevertheless, as already pointed out by others (Rehncrona 1984, Kogure *et al.* 1985), the total contents of these antioxidants, not the ratios of their oxidized to reduced forms, would not exclusively reflect their comsumption due to free radical reactions.

Regarding the changes in the fatty acid composition of membrane phospholipids, we examined the level of FFAs in the cat cerebral cortex exposed to 4 hour's regional ischemia due to MCA occlusion. The brain was frozen *in situ* while respiration and systemic arterial pressure were maintained. After sacrifice, the cortical samples were chiselled out from the MCA territory in a cold room. The result is shown in **Table IV-1**. It is noteworthy that only PUFAs ($C_{18:2}$, $C_{20:4}$, and $C_{22:6}$) significantly increased following ischemia. Saturated fatty acids showed no significant changes. The prominent increase of docosahexaenoic acid ($C_{22:6}$) seemed to characterize the mode of FFA liberation in this model. Thus our result is consistent with the selective loss of PUFA from the membrane phospholipids as previously shown in a similar cat MCA occlusion model (Demopoulos *et al.* 1980) and in the study using rat brain homogenates

Table IV-1. Alteration of FFA contents during 4-hour MCA occlusion in cats

Experimental conditions	Normal control (n = 6)	MCA + saline (n = 7)
$C_{16:0}$	0.235 ± 0.025	0.267 ± 0.029
$C_{18:0}$	0.225 ± 0.018	0.253 ± 0.027
$C_{18:1}$	0.090 ± 0.027	0.094 ± 0.010
$C_{18:2}$	0.065 ± 0.011	0.164 ± 0.014**
$C_{20:4}$	0.108 ± 0.010	0.237 ± 0.024**
$C_{22:6}$	0.074 ± 0.009	0.347 ± 0.051**
Total	0.796 ± 0.062	1.362 ± 0.111**

** $p < 0.01$ (μmol. g^{-1}).

(Rehncrona et al. 1980). It is also apparent that the pattern of released FFAs following regional ischemia is markedly different from that following global ischemia (Yoshida et al. 1980, Rehncrona et al. 1982). This might implicate that the mechanism of phospholipase activation in regional ischemia is considerably different from that in global ischemia. The metabolic fate of docosahexaenoic acid which showed the maximal change is also of interest, although our knowledge on this subject is scanty. Anyhow, the selective change in PUFAs of membrane phospholipids is not in itself a direct proof of the occurrence of lipid peroxidation.

In opposition to the hypothesis of Demopoulos et al., Kogure et al. (1979) proposed that a lipid peroxidation takes place in ischemic brain due to restoration of oxygen to the ischemic tissue by collateral circulation, or recirculation. This proposition was based on the experimental results that not in the ischemic hypoxic conditions but in the postischemic hyperoxic conditions, there was an elevation in the level of MDA and GSSG/GSH ratio, and a decrease in vitamin E (Kogure et al. 1979). Further in support of the view, the study using NADH tissue fluorescent photography demonstrated that in some brain regions, the electron transport system became overoxidized, i.e., devoid of electrons, following reperfusion. Although this result is a qualitative one, such a condition of the electron transport system is considered to facilitate the incomplete reduction of O_2, leading to increased generation of active oxygen species (Kogure et al. 1985).

In accordance with Kogure's theory, a significant elevation of TBAR on recirculation following transient ischemia due to bilateral carotid ligation in gerbils (Yoshida et al. 1980), and plug compression in rats (Yoshida et al. 1982) has been reported. Using the permanent middle cerebral artery occlusion model in rats, we examined the cerebral concentrations of glutathione following ischemia (Gotoh et al. 1983). As shown in **Table IV-2**, the cerebral concentration of GSH was significantly reduced and the ratio of GSSG to the total glutathione was significantly increased 48 hours after the onset of ischemia in the affected hemisphere. Recently, Watson et al.

either ischemia or recirculation in rat models of incomplete and complete global ischemia. Cooper et al. (1980) examined the ratio of oxidized form to total pool of glutathione and ascorbate during four vessel occlusion in rats and following recirculation. Since neither of these parameters showed significant

Table IV-2. Cerebral concentrations of glutathione in rats after forty-eight hours of ischemia

Conditions (n)	GSH (μmole/g)	GSSG (nmole/g)	% GSSG
Normal control (9)	2.15 ± 0.15	25.1 ± 1.4	1.21 ± 0.08
Sham-operated (3)	2.15 ± 0.10	27.4 ± 1.0	1.28 ± 0.09
Ischemic hemisphere (9)	1.71 ± 0.12[a]	28.8 ± 1.4	1.72 ± 0.10[b]
Nonischemic hemisphere (9)	2.04 ± 0.10	25.9 ± 1.9	1.28 ± 0.09

Each value represents the mean ± SEM. Values which are significantly different from normal control ones at [a]$p < 0.05$ and [b]$p < 0.01$ (Student's t test), respectively.

(1984), using the four vessel occlusion model of Pulsinelli and Brierley (1979), showed a significant increase in conjugated diene formation following reperfusion. Kogure et al. (1985) also showed that the chemiluminescence from the brain surface increased immediately after injecting minute emboli into the internal carotid artery in rats. This increase in chemiluminescence was ascribed to generation of some radical species in the periphery of ischemic areas.

Results contradicting the occurrence of lipid peroxidation during either ischemia or recirculation have also been reported. Rehncrona et al. (1980) did not find any increase in GSSG or the ratio of GSSG/GSSG + GSH during

changes, their study failed to provide evidence for increased O_2^- production during and following cerebral ischemia. The activity of Na^+, K^+ ATPase, which is expected to be suppressed by lipid peroxidation (Demopoulos et al. 1979, Kovachich and Mishra 1981, Chan et al. 1983), was shown not to be suppressed but rather increased during and following global ischemia in rats (MacMillan 1982) and gerbils (Enseleit et al. 1984). Only after six hours of unilateral carotid occlusion and on recirculation following 60 minutes bilateral carotid occlusion in gerbils, a significant decrease in the Na^+, K^+ ATPase was observed (Palmer et al. 1985). In addition, the level of water soluble antioxidant or TBAR showed no significant change during and following ischemia (MacMillan 1982).

Thus, the current status of investigation on lipid peroxidation in cerebral ischemia, has been summarized that "Neither production of active oxygen species, nor the process of lipid peroxidation by active oxygen has been experimentally proved" (Kogure et al. 1985, Siesjö and Wieloch 1985). Similarly, Halliwell and Gutteridge (1984) admitted that no human disease caused by increased radical formation has been discovered as yet. But, the failure to detect free radicals and subsequent lipid peroxidation might be due to the limitations of available techniques. Obivously, it seems too early to dismiss the possibility that active oxygen species exert some role, whatever it might be, in physiological and pathological conditions. Of equal importance is the activity of arachidonate cascade which has been shown to be intimately related to the ambient level of free radicals. Fortunately, recent remarkable advances in this research field have enabled to detect very minute amounts of each eicosanoid with great accuracy.

B. Vasospasm

It is now apparent that cerebral vasospasm is not merely a prolonged contraction of the arterial smooth muscle, but represents a particular type of arteriopathy in which smooth muscle constriction is a consequence of antecedant metabolic and organic changes in and around the artery. The occurrence of smooth muscle constriction may in turn provoke further organic changes of the vessel wall due to distortion of the vessel wall structure and subsequent derangements in the nutritional conditions. Without exception, past efforts to prevent or ameliorate vasospasm by the use of a variety of vasodilating drugs have failed. Therefore, it would be apposite to direct attention to the not yet well-defined conditions which antecede the occurrence of prolonged smooth muscle constriction and subsequent organic changes.

In all likelihood, the whole events which precedes the occurrence of either angiographic or symptomatic vasospasm are related to the existence of blood clot around the artery. Available morphological and biochemical evidence suggests that those events are inflammatory in nature. Assuming that vasospasm is a manifestation of inflammatory processes provoked by blood components, it then becomes necessary to identify relevant chemical mediators and their sources. Although all the possible mediator-producing systems need to be considered, we have confined our attention to the roles of eicosanoids and free radicals. In the following sections, pertinent findings of our own and others together with our current working hypothesis will be presented.

1. Vascular Actions of Eicosanoids

a) Classical PGs

The vascular actions of classical PGs (PGA ~ F) have been studied using various species such as cats, dogs, monkeys, and humans. The experimental methods employed are also variable, including the chamber study using arterial strips (Yashon et al. 1977, Toda and Miyazaki 1981) or segments (Allen et al. 1974, Miller 1980, Asano et al. 1982), topical application to the exposed basilar artery (Kapp et al. 1976, Handa et al. 1974) or pial arteries (Yamamoto et al. 1972, Rosenblum 1975, Ellis et al. 1979), angiographic (Pennink et al. 1972, White et al. 1975) or CBF studies (Nakano et al. 1973, Pickard et al. 1977) following cisternal, intraarterial or intravenous injection, and others. The vasoconstrictive properties of PGD_2, PGE_2, and $PGF_{2\alpha}$ which are known to be produced in the mammalian brain and blood vessels, have been most extensively studied.

Irrespective of the species and methods employed, $PGF_{2\alpha}$ was shown to have the most potent vasoconstrictive action among other classical PGs and known agonists (White 1983, Miller 1980, Asano et al. 1982), although the sensitivity of the cerebral artery to each agonist considerably differs between species. The vasoconstrictive effects of PGs ($PGF_{2\alpha}$ and PGB_2) and norepinephrine or serotonin were found to be additive (Rosenblum 1975 a, b). PGE_2 and PGD_2 also constricted the canine (Handa et al. 1974, Toda and Miyazaki 1978, Asano et al. 1982) and human (Yashon et al. 1977) basilar arteries in vitro. In contrast to above studies, Ellis

et al. (1979) reported dilation of the cat cerebral arterioles by PGE_2 and PGD_2 in an experiment using the cranial window technique. This discrepancy may be due to the difference in species and the regional difference of the artery used. In addition, the difference of experimental methods must be considered since in the cranial window technique, the changes in the arteriolar caliber on application of the test drug would reflect not only the direct action on the smooth muscle but also the sum of actions on the more proximal arterial segments, vascular nerves, and the metabolic as well as functional alterations in the exposed cortical area.

b) Endoperoxides, PGI_2, and TxA_2

The unstable endoperoxides, PGG_2 and PGH_2 were shown to exert a biphasic effect on extracranial and intracranial arteries, i.e., the initial small contraction followed by a longer lasting relaxation (Hamberg et al. 1975, Bunting et al. 1976, Boullin et al. 1979, Toda 1980). The relaxation phase was attributed to the conversion of endoperoxides to PGI_2, since inhibition of PGI_2 synthase by prior application of 15-HPAA reversed the relaxation to contraction (Moncada et al. 1976, Bunting et al. 1976, Toda 1980).

PGI_2 produces a biphasic response in canine basilar artery in vitro (Chapleau and White 1979, Asano et al. 1982). In low concentrations (10^{-8}–10^{-6} M), it causes slight dilation whereas in high concentrations ($> 10^{-5}$ M), it causes contraction. The most remarkable

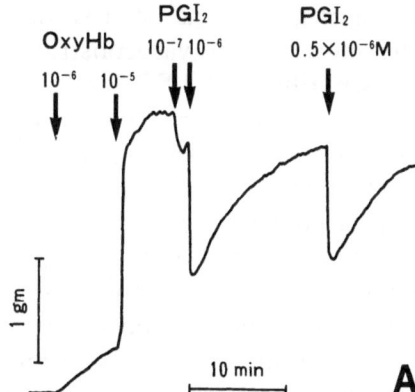

Fig. IV-7. A) Inhibition of oxyHb-induced vasoconstriction by PGI_2, B) inhibitory effect of PGI_2
on vasoconstriction induced by a variety of agonists

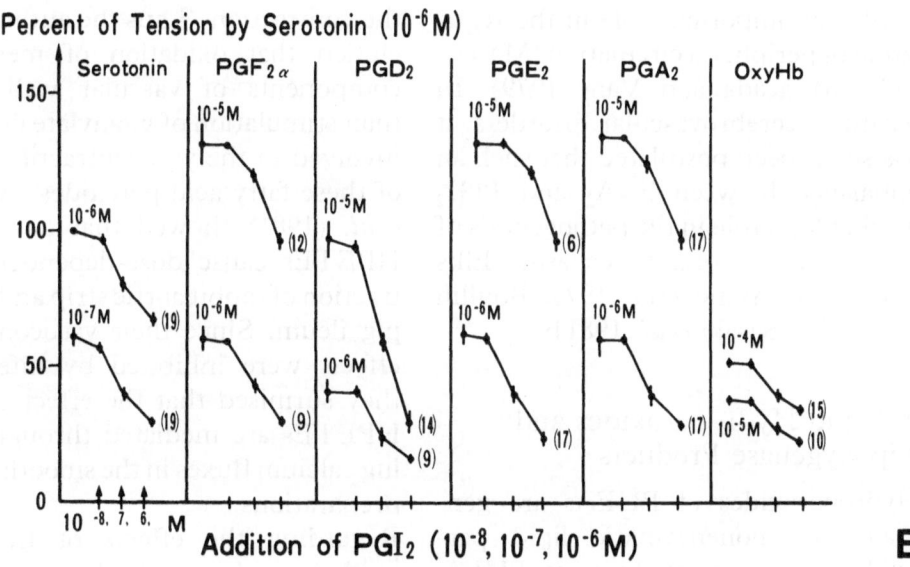

action of PGI_2 is its antagonizing action to contractions induced by other agents, such as the vasospastic CSF obtained from SAH patients (Boullin et al. 1979), serotonin, $PGF_{2\alpha}$, PGD_2, PGE_2, PGA_2, oxyhemoglobin (Asano et al. 1982: Fig. IV-7-A, B), angiotensin II, TxA_2, U-46619 (a selective TxA_2 mimetic), and norepinephrine (Paul et al. 1982). Uski et al. (1983) reported that the antagonizing action of PGI_2 to vasoconstriction is more remarkable in the human pial artery than in the cat basilar artery, and that there is also a regional difference between the cat basilar and middle cerebral arteries. These species and regional differences in the action of PGI_2 is of importance in the interpretation of experimental results.

On the other hand, TxA_2 which is preponderantly released from platelets on aggregation has been shown to be the most potent vasoconstrictor ever known (Ellis *et al.* 1977, Paul *et al.* 1982, Fujiwara and Kuriyama 1984). TxA_2 also induces platelet aggregation. Both of the platelet aggregating and smooth muscle contracting actions of TxA_2 are antagonized by PGI_2 which is predominantly produced in the vessel wall. Therefore, the chemical balance between productions of TxA_2 and PGI_2 has been thought to represent the blood-vessel wall interrelationship, and to play an important role in the regulation of peripheral circulation (Marcus 1978, Moncada and Vane 1979). In regard to cerebrovascular disorders, it has so far been postulated that such an imbalance between TxA_2 and PGI_2 would play a role in the pathogenesis of TIAs and cerebral vasospasm (Ellis *et al.* 1977, Asano *et al.* 1978, Boullin *et al.* 1979, Sasaki *et al.* 1981 b).

c) Lipid Hydroperoxides and Lipoxygenase Products

Hydroperoxides of PUFAs are generated by nonenzymatic lipid peroxidation. The most abundant PUFA, *i.e.*, arachidonic acid (AA), however, is peroxidized also by lipoxygenase producing a variety of isomers of HPETEs. As later described, those hydroperoxides are considered to be involved in the pathogenesis of vasospasm, hence a brief comment on their actions on vascular smooth muscles will be given in this section.

Hydroperoxides of PUFAs such as linoleic, linolenic, eicosadienoic and arachidonic acids were shown to contract rabbit aortic strip or guinea pig

ileum (Asano and Hidaka 1979). The vasocontractile action of these hydroperoxides were not inhibited by phentolamine, diphenhydramine, atropin, or aspirin. Although it is known that both fatty acids and unsaturated fatty acid peroxides stimulate guanylate cyclase, the latter but not the former produced contraction of aortic strips. Sulfhydril compounds such as 2-mercaptoethanol and dithioerythritol or an SH blocking agent such as N-ethylmaleimide significantly inhibited the contraction. Thus the authors concluded that oxidation of membrane components of vascular wall rather than stimulation of guanylate cyclase is involved in the vasocontractile actions of these fatty acid peroxides. Aharony *et al.* (1981) showed that 5- and 12-HPETEs cause dose-dependent contraction of rabbit aortic strip and guinea pig ileum. Since their vasocontractile effects were inhibited by nifedipine, they surmised that the effect of these HPETEs are mediated through altering calcium fluxes in the smooth muscle preparations.

Regarding the effects of hydroperoxides on the cerebral artery, Tanishima *et al.* (1979) reported a dose-dependent contraction of the canine basilar artery by hodrogen peroxide, 15-hydroperoxyarachidonic acid (15-HPAA), and linoleic acid hydroperoxide. Similar contractile effects of hydroperoxides of linoleic acid and arachidonic acid (13-HPLA and 15-HPAA) on the canine basilar artery was reported by Koide *et al.* (1981). It was also shown that the maximal contractile force produced by 15-HPAA *in vitro* was equivalent to that of $PGF_{2\alpha}$, being

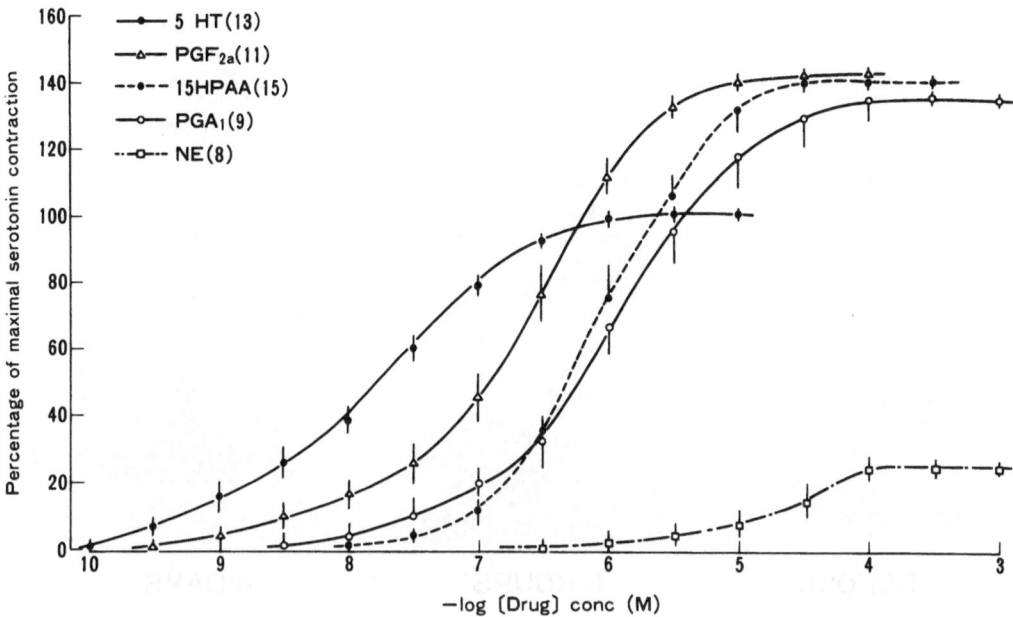

Fig. IV-8. Vasocontractile capacity of 15-HPAA compared to those of other agents

1.5 times that of serotonin (Koide *et al.* 1982. **Fig. IV-8**). Like rabbit aortic strips, the contraction of canine basilar artery by hydroperoxides was not suppressed by phenoxybenzamine, methysergide, diphenhydramine, or atropine, suggesting that α-adrenergic, serotonergic, histaminergic or cholinergic mechanisms are not involved in the contraction mechanism.

Sasaki *et al.* (1981 a) injected 15-HPAA into the cisterna magna of the dog and observed the alteration of the caliber of the basilar artery by repeated angiography. The injection of 0.2 mg of 15-HPAA caused a mild constriction of the basilar artery which lasted about 7 hours. The injection of 2 mg of 15-HPAA caused a biphasic constriction, *i.e.*, the initial moderate narrowing lasting about 10 hours followed by a more

pronounced constriction lasting until sacrifice on the 7th day after injection. This prolonged constriction was accompanied by marked degenerative changes in the endothelium (**Fig. IV-9-A, B**).

Thus far, it has been shown that hydroperoxides of PUFAs, especially of AA, possess pronounced contractile actions on the cerebral artery both *in vitro* and *in vivo*.

Parenthetically, Kontos *et al.* (1981) reported that topical application of PGG_2 which has the structures of hydro- and endoperoxides dilates cat cortical arterioles. This dilation was accompanied by endothelial damage. They suggested that oxygen free radicals generated in the conversion of PGG_2 to PGH_2 are responsible for the endothelial damage and ˏvasodilation

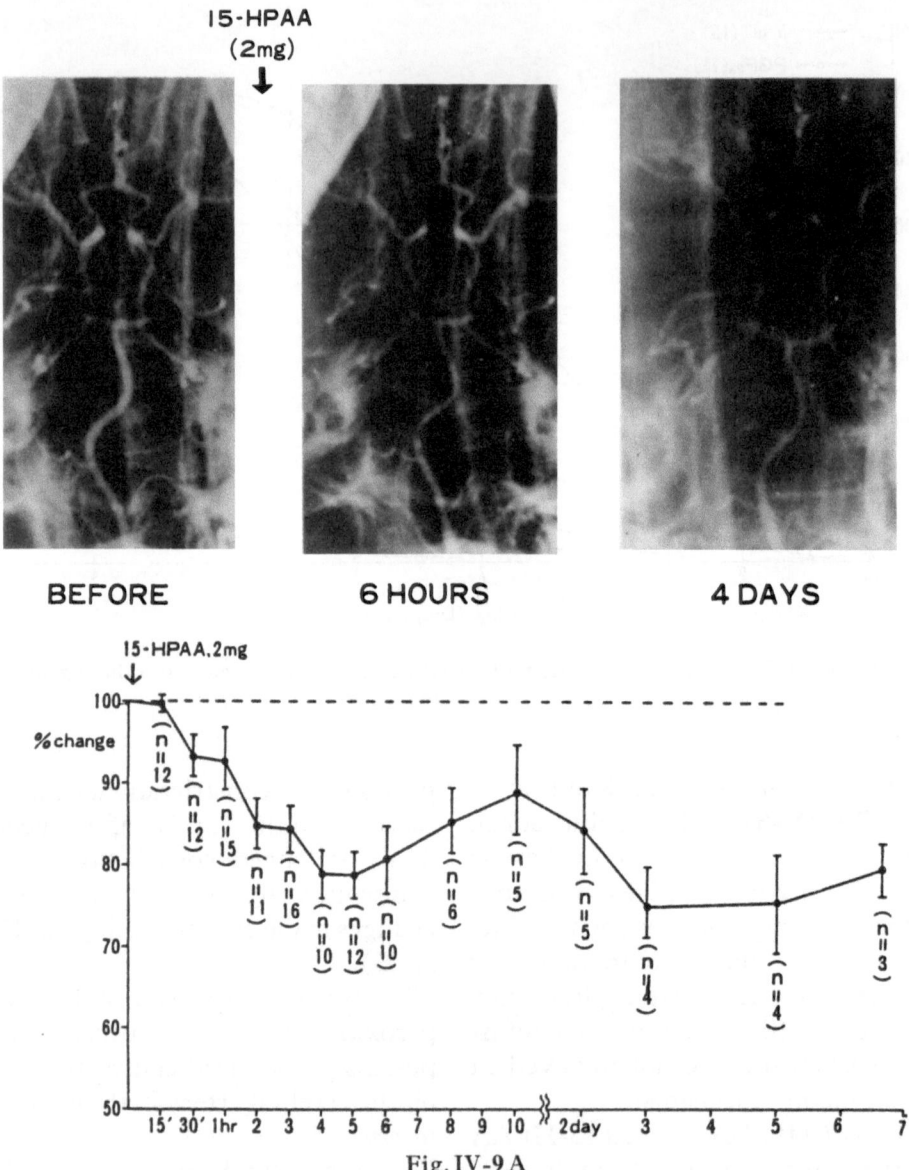

Fig. IV-9 A

Fig. IV-9. A) The time course of basilar artery spasm following intrathecal injection of 15-HPAA in dogs. B) Electron microscopical appearance of the basilar artery 5 days after the intrathecal injection of 15-HPAA. The arrow indicates the separation of the endothelial junction

(Kontos *et al.* 1980), and may be relevant to the arteriolar damage as seen in concussive brain injury (Ellis *et al.* 1981). It is of interest that both 15-HPAA and putative oxygen free radicals caused endothelial damage, and that the former contracted the canine basilar artery whereas the latter dilated the feline cortical arteriole. Therefore, it seems likely that the endothelial

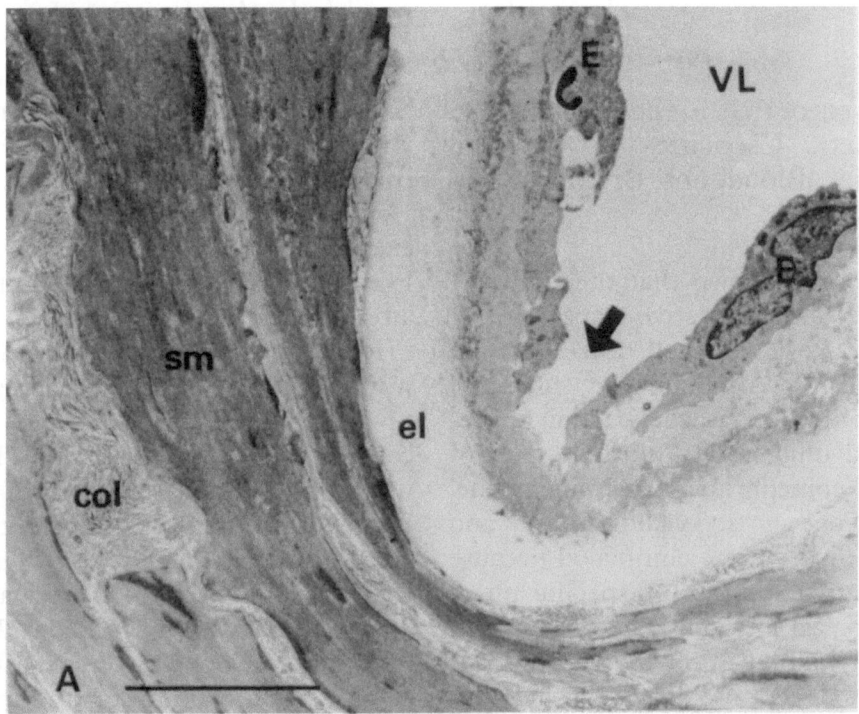

Fig. IV-9 B

damage is incurred by free radicals released from either PGG_2 or 15-HPAA. The discrepant response of the arterial smooth muscle to these two hydroperoxide may be due to species and regional difference. Or it may be surmised that they are bound to different metabolic pathways or compartments, thus exerting different influences on the mechanism of smooth muscle contraction. Recently, Gryglewski et al. (1986) reported that superoxide anion is involved in the breakdown of endothelium-derived vascular relaxing factor (EDRF), although the implication of the finding to cerebral vasospasm remains to be elucidated.

The lipoxygenase products of leukocytes, i.e., leukotrienes (LTs) C_4 and D_4, were shown to cause pain and vasodilation when locally injected in human skin (Bisgaard et al. 1982). LTC_4 or LTD_4 neither contracted nor relaxed isolated human cerebral arteries (von Holst et al. 1982), or rabbit pial arteries (Kamitani et al. 1985). In contrast with above results, Tagari et al. (1983) reported that intracarotid injection of LTD_4 constricted the internal carotid and other large arteries in rats and that LTD_4 (10^{-9}–10^{-7} M) caused sustained contraction of isolated human intracranial arteries. We observed that LTC_4 only in a high concentration (10^{-5} M) moderately contracted the canine basilar artery segment, whereas LTD_4 (10^{-5} M) did not (Asano 1983).

2. The Synthesis of Eicosanoids Following SAH and Its Pharmacological Modulation

a) Levels of PGs in the CSF and PG Synthetic Capacities of Cerebral Arteries, Blood, and Brain Following SAH

It has been known that the CSF develops vasoconstrictive capacity following SAH (Buckell 1964, Boullin et al. 1976). Although the substance responsible for vasoconstriction has remained unidentified until now, serotonin, epinephrine, norephineprine, histamine, acetylcholine, and angiotensin $_{II}$ were eliminated because cumulative addition of specific pharmacological antagonists to each of them did not inhibit vasoconstriction by CSF (**Fig. III-2**). Potassium was also eliminated because its concentration in the obtained CSF samples was not high enough to cause significant vasoconstriction by itself (Boullin et al. 1981). Recently, Kaye et al. (1984) showed that there is a relationship between smooth-muscle constrictor substance in the CSF after SAH and both the degree of angiographic vasospasm and the outcome. As to the nature of the vasoconstrictive substance in CSF, Sasaki et al. (1984) further showed that the vasoconstrictive effect was significantly suppressed by application of disulfide bond-reducing agents such as dithiothreitol and dithioerythritol. From this result, possible involvement of prostaglandins, hemoglobin or lipid hydroperoxides in the CSF-induced vasoconstriction was suggested.

The CSF level of $PGF_{2\alpha}$ have been shown to be significantly elevated after SAH, although it is poorly correlated with the severity of vasospasm or neurological deficits (la Torre et al. 1974, Hagen et al. 1977). Walker et al. (1983) examined the cisternal CSF levels of $PGF_{2\alpha}$, PGE_2, 6 oxo $PGF_{1\alpha}$, and TxB_2 after induction of SAH in dogs. Three to four days after SAH, all the PG levels were increased. The increase was most pronounced in PGE_2 (25.5 times of the control level) while in the other PGs it was 2–4 times of the control level. However, the combined concentrations of the two vasoconstrictive PGs, i.e., PGE_2 and $PGF_{2\alpha}$, which were around 2 nmol/l at maximum were considered a little low for potent vasoconstrictor action, since in both dog and man, in vitro contractions of cerebral arteries are produced with concentrations above 100 nmol/l (Allen et al. 1974, Toda and Miyazaki 1978, Asano et al. 1982). The CSF level of TxB_2 was much lower than these PGs. The reported CSF levels of $PGF_{2\alpha}$ in SAH patients (la Torre et al. 1974, Hagen et al. 1977) fell in the similar ranges as those in dogs. Thus, the vasocontractile capacity of CSF obtained from SAH patients cannot be solely ascribed to vasoconstrictive PGs or TxA_2. Clearly, more study is needed.

Anyhow, the levels of eicosanoids in CSF samples would not accurately reflect those in the vicinity of cerebral arteries. Further, those eicosanoids relevant to vasospasm may not be conveyed via the CSF but may be generated within the vessel wall or at the blood-vessel wall interface. It was reported that the synthetic capacity of PGE$_2$ but not of PGF$_{2\alpha}$ was significantly increased in the canine basilar artery exposed to SAH (Maeda et al. 1981). Walker et al. (1983) also showed that SAH did not alter the production of PGs by dog whole cortex or choroid plexuses in vitro, but production by pooled dissected cerebral arteries of PGE$_2$ was increased. It remains to be seen, however, whether or not PGE$_2$ generated in the vessel wall attains to the concentration enough to cause smooth muscle constriction.

Of equal importance to production of vasoconstrictive PGs is the possible alteration in the synthesis of PGI$_2$ following SAH. It has been shown that the PGI$_2$ synthetic capacity of the canine basilar artery progressively decreases after induction of SAH (Sasaki et al. 1981 b, Maeda et al. 1981, Walker et al. 1983). Again, it is not certain whether or not the shown diminution in PGI$_2$ synthetic capacity results in a real PGI$_2$ deficiency. But, it seems to be a very likely event, considering the pathological findings which showed the occurrence of degenerative changes in the endothelial cells after SAH, and the finding that the intrathecal injection of 15-HPAA, an inhibitor of PGI$_2$ synthase, caused a prolonged vasoconstriction in dogs (Sasaki et al.).

There is a paucity of literature regarding the eicosanoid synthesis in the circulating blood after SAH. Only Denton et al. (1971) reported a pronounced decrease in the number of platelets present in the jugular blood following experimental SAH. Since this result is suggestive of platelet aggregation, liberation of TxA$_2$ may also be surmised.

So far, however, there has not been any reliable data showing an enhanced liberation of TxA$_2$ in the cerebral vascular beds following SAH.

b) Pharmacological Modulation of Eicosanoid Synthesis and Its Effects on in vitro Vasoconstriction and Vasospasm

Pharmacological agents, such as indomethacin, aspirin, and other nonsteroidal antiinflammatory drugs are known to inhibit cyclooxygenase. In vitro, aspirin potentiated the contractile response of the dog cerebral artery to PGF$_{2\alpha}$ and PGE$_2$ but did not alter the relaxation induced by PGE$_1$ (Toda and Miyazaki 1978) or PGI$_2$ (Chapleau et al. 1980). A prostaglandin synthetase inhibitors, meclofenamate and indomethacin, enhanced the relaxant action of PGI$_2$ and markedly suppressed contractions induced by arachidonic acid or PGF$_{2\alpha}$ (Chapleau et al. 1980). However, in small human cerebral arteries preincubated with indomethacin, contractions induced by CSF obtained from SAH patients and NE were markedly increased, probably due to inhibition of PGI$_2$ synthesis (Brandt et al. 1981 b). Those reported data indicate that different prostaglandin synthesis inhibitors have different effects on cerebral vasomotion in vitro.

In vivo, indomethacin (20 mg/kg i.v.) failed to prevent the occurrence of vasospasm in the canine SAH model (White et al. 1975). Since indomethacin has a very short half-life in dogs, the effects of prostaglandin synthetase inhibitors with longer half-lives on cerebral vasospasm were studied

(White *et al.* 1979). Sudoxicam was found to significantly reduce the incidence and the magnitude of the vasospasm, and to prevent the behavioral changes caused by SAH. In this study, however, the observation was confined within one day after induction of SAH. Considering that indomethacin and other cyclooxygenase inhibitors suppress the generation of both vasoconstrictive PGs and vasodilatatory PGI_2, and that the biosynthetic capacity of PGI_2 in cerebral arteries are probably decreased after SAH, it seems more advantageous from the pharmacological standpoint, to selectively inhibit generation of vasoconstrictive PGs in order to prevent the occurrence of vasospasm. For this purpose, selective inhibitors of TxA_2 synthase such as OKY 1581, OKY 046, and other imidazol derivatives have been available. OKY 1581 was shown to effectively prevent the occurrence of chronic vasospasm in the canine (Sasaki *et al.* 1982 a) and the rabbit (Chan *et al.* 1984) SAH models. OKY 046 showed a similar beneficial effect on vasospasm (unpusblished data). Since open trials with OKY 046 indicated a beneficial effect as well as safety, a double-blind

test was carried out. The study revealed a significant amelioration of neurological deficits, as well as reduction in the severity of cerebral damage judged by CT scans (Sano *et al.* 1986). However, since the incidence of symptomatic vasospasm *per se* was not significantly reduced, the beneficial effect of OKY 046 might be more related to the improvement of cerebral microcirculation than the prevention of sustained smooth muscle constriction in major cerebral arteries.

The antiinflammatory effect of steroids has recently been attributed to induction of macrocortin and lipomodulin, endogenous inhibitors of phospholipase A_2 (Blackwell and Flower 1983). Chyatte *et al.* (1983), using the canine "two hemorrhage" SAH model, reported that a nonsteroidal antiinflammatory drug, ibuprofen and high-dose methylprednisolone similarly reduced vasospasm, meningismus and accelerated the rate of neurological recovery. We also examined the effect of high-dose methylprednisolone using the same experimental protocol as theirs, but could not confirm its beneficial effect (unpublished data). The reason for this discrepancy is unclear.

3. Relevance of Oxygen Free Radicals, Lipid Peroxidation, and Lipoxygenase Products to the Pathogenesis of Vasospasm

a) Summary of Our Experimental Findings

So far, it has been shown that SAH is followed by alterations in the synthesis of cyclooxygenase products, which may be linked to the pathogenesis of vasospasm. To date, however, informations

regarding the other important members of the eicosanoid familiy, *i.e.*, lipoxygenase products, are totally lacking. Since lipoxygenase products have been shown to act as important chemical mediators in inflammatory responses, their roles in the occurrence of cerebral vasospasm need to be elucidated.

As described in the preceding chapter, the activity of arachidonate cascade including both the cyclooxygenase and lipoxygenase pathways, is thought to be regulated by the concentration of ambient free radicals. In this respect, SAH appears to create an unusual condition in which generation of free oxygen radicals and subsequent lipid peroxidation are particularly liable to occur because: 1. the autooxidation of hemoglobin, *i.e.*, the conversion of oxyhemoglobin to methemoglobin, generates O_2^- (Misra and Fridovich 1972, Winterbourn *et al.* 1976, Fujita *et al.* 1980); 2. erythrocyte lysis of subarachnoid clot may represent the occurrence of lipid peroxidation (Kellog and Fridovich 1977); 3. catabolites of hemoglobin such as methemoglobin, hematin, heme, or free iron catalyze free radical reactions (Tappel 1953, Willis 1966, Barber and Bernheim 1967, Halliwell and Gutteridge 1984); 4. CSF is thought to be relatively devoid of intrinsic free radical scavengers due to low concentrations of proteins or lipids; 5. infiltrating cells, *e.g.*, leukocytes generate oxygen radicals; 6. hydroperoxides produced by either enzymatic or nonenzymatic reactions generates oxygen radicals (Kuehl and Egan 1980).

Preliminary data showing the occurrence of lipid peroxidation in the CSF following SAH and its correlation with symptomatic vasospasm in SAH patients has been reported (Asano *et al.* 1980). Together with them, brief accounts on our recent findings pertinent

to this line of investigation will be given below.

(i) TBAR and vasocontractile capacity of incubated blood

The whole blood develops vasocontractile capacity on incubation. A significant correlation was found between the vasocontractile capacity and the content of TBAR (Asano *et al.* 1980).

(ii) The vasocontractile actions of hydroperoxides

As described previously, hydroperoxides of linoleic and arachidonic acids potently constrict the cerebral arterial smooth muscle (Asano *et al.* 1980, Koide *et al.* 1981, 1982).

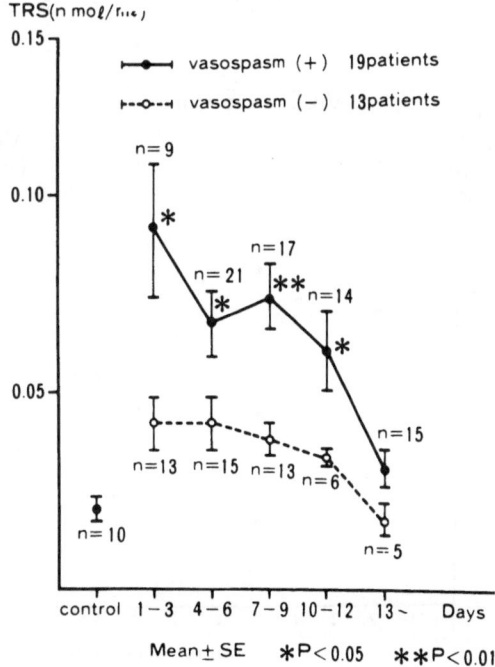

Fig. IV-10. The time course of TBA-reactive substance (TRS) in the CSF of SAH patients

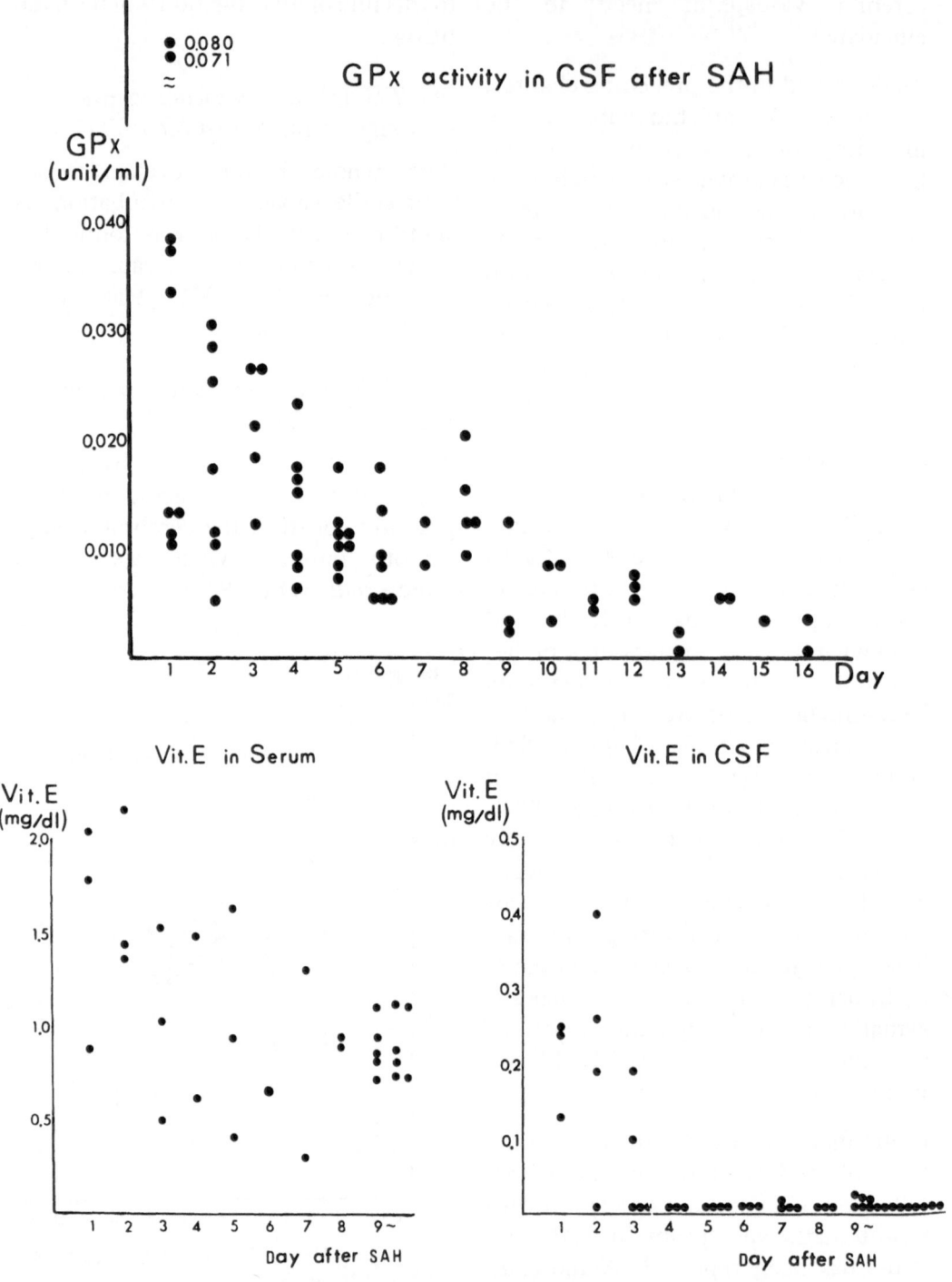

Fig. IV-11. The glutathione peroxidase (GPx) activity of the CSF (upper), the vitamin E contents in the serum and CSF (lower) after SAH

(iii) TBAR in the CSF of SAH patients

The level of TBAR in the normal CSF is extremely low. Following SAH, there was a pronounced increase of TBAR in the CSF. Examinations of the CSF samples serially obtained from each SAH patient revealed that CSF-TBAR levels were significantly higher in patients who developed symptomatic vasospasm than in those who did not (**Fig. IV-10**: Asano *et al.* 1980). There was no significant correlation between the CSF-TBAR level and the clinical grade of each patient.

(iv) Effect of intrathecally administered hydroperoxide

To see whether or not hydroperoxides cause vasospasm, 15-HPAA was injected into the cisterna magna in dogs. Serial angiography revealed a dose-dependent sustained contraction of the basilar artery, in which pronounced degenerative changes in the endothelium and the tunica media were observed by electron microscopy (Sasaki *et al.* 1981a, cf. **Fig. IV-9 A, B**).

(v) The levels of free radical scavengers in the CSF

In the normal CSF, the activity of glutathione peroxidase (Gpx) and the content of vitamin E are extremely low. In the CSF of SAH patients, they were increased in the first few days and then rapidly decreased (**Fig. IV-11**: Asano *et al.* 1981). Kuwabara *et al.* (1982) showed that in SAH patients (17 cases operated in acute stage), the occurrence of angiographic and symptomatic spasm as well correlated with the decrease in the activity of SOD in the cisternal and ventricular CSF (**Fig. IV-12**).

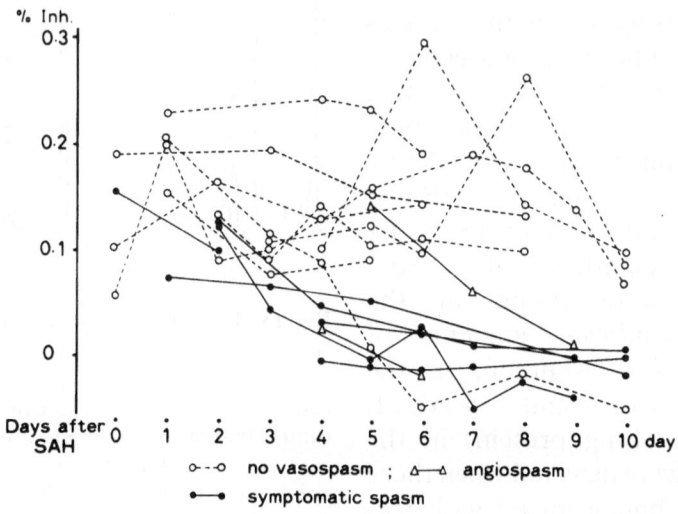

Fig. IV-12. The superoxide dismutase (SOD) activity in the CSF following SAH (from Kuwabara *et al.* 1982; with permission)

(vi) Detection of HETEs in the CSF of SAH patients

To detect and identify lipid peroxides in the CSF following SAH, CSF samples were obtained sequentially from 10 patients who developed vasospasm and were analyzed by high performance liquid chromatography (HPLC) and gas chromatography mass spectrometry (GC/MS). One of the peaks appearing on the 7th day after SAH was identified as 5-HETE, which showed a possible correlation with the occurrence of vasospasm (Suzuki et al. 1983).

(vii) Detection of HETEs in the canine SAH model

Using the "two-hemorrhage" canine SAH model as reported by Varsos et al. (1983) and Liszczak et al. (1983), we examined the HETEs present in the CSF, perivascular clot, and the basilar artery. Each sample was homogenized with methanol or ice-cold saline and the lipid extract was applied to the reverse-phase HPLC. The structural elucidation of these metabolites was performed by gas chromatography mass spectrometry after purification with HPLC. In normal control dogs, no HETEs were detected in the CSF or the basilar artery. On the eighth day after SAH, no HETEs were found in the CSF, but significant amounts of 12-HETE (5.74 ± 2.47 nmol/g) in the perivascular clot and 5-HETE (0.71 ± 0.47 nmol/mg protein) in the basilar artery were detected when those samples were homogenized with ice-cold saline. When homogenized with methanol, only 12-HETE was found in the perivascular clot, and no HETE was found in the basilar artery. The above results were interpreted as to indicate the accumulation of 12-HETE in the perivascular clot probably due to the action of 12-lipoxygenase of platelets, and an enhancement of 5-lipoxygenase activity in the wall of the basilar artery (Watanabe et al. 1986).

(viii) The activation of 5-lipoxygenase pathway in the canine basilar artery following SAH

Next, the capacity of leukotriene (LT) formation as well as enzyme activities involved in the synthesis was examined using the isolated canine basilar artery (Shimizu et al. 1986). The basilar artery segments were obtained from normal

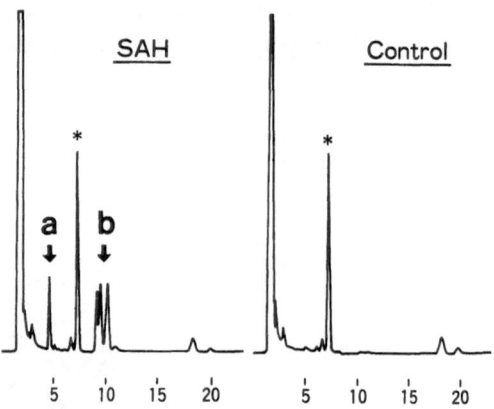

Fig. IV-13. Ca^{2+} ionophore (A 23187)-stimulated production of leukotrienes in the canine basilar artery. No synthesis in the normal canine basilar artery segment (right). In the canine basilar artery exposed to SAH, significant amounts of LTC_4 (a), 6-trans-LTB_4, 12-epi-6-trans-LTB_4, and LTB_4 (b) were detected (left). The peak designated by asterisk is the internal standard, LTB_2 (from Shimizu et al. 1986)

dogs and dogs exposed to SAH by the "two hemorrhage" method. When stimulated by Ca^{2+} ionophore (A23187), the artery exposed to SAH produced significant amounts of LTB$_4$ cerebral artery like that in other cells (Gale and Egan 1984, Lands et al. 1984) is stimulated by lipid hydroperoxides. Our previous finding that 15-HPETE increased the formation of hydroxy-

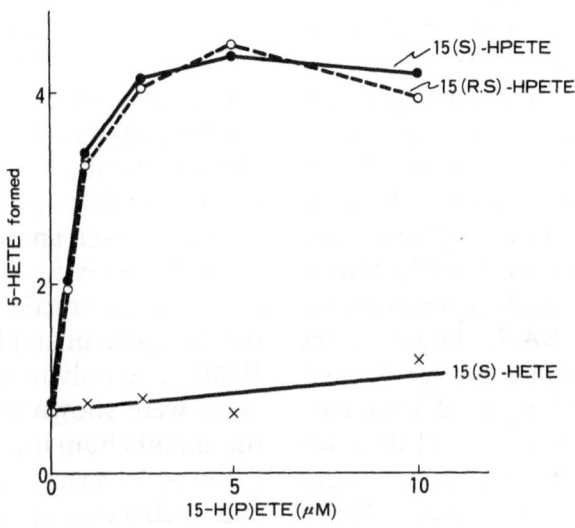

Fig. IV-14. The activation of 5-lipoxygenase in the canine basilar artery exposed to SAH by 15(s)-, and 15(R,S)-HPETEs. No activation by 15(s)-HETE (from Shimizu et al. 1986)

and LTC$_4$ in addition to 5-hydroxy-6,8,11,14-eicosatetraenoic acid (5-HETE), whereas the artery from normal dogs did not (**Fig. IV-13**). The enzymologic survey revealed that 5-lipoxygenase activity which was not detectable in the normal artery was prominently activated in the artery exposed to SAH. This enzyme activity was further activated by 6-fold in the presence of 5 µM 15-hydroperoxy-5,8,11,13-eicosatetraenoic acid (15-HPETE). No significant difference was observed in the activation between 15(S)-HPETE and 15(S,R)-HPETE (**Fig. IV-14**). This result indicates that the activity of 5-lipoxygenase in the

acids in the isolated canine basilar artery (Koide et al. 1982) and in brain microvessels (Koide et al. 1985) was thus confirmed. Since such a stimulatory action of 15-HPETE on lipoxygenase activity is considered to be due to generation of free radicals such as activated oxygen (Koide et al. 1982), a beneficial effect of antioxidants on vasospasm was anticipated.

(ix) Effect of administration of antioxidants on vasospasm and 5-HETE content of the canine basilar artery

AVS [1,2-bis(nicotinamide)-propane, Chugai Pharmaceu. Co.] has been shown to scavenge hydroxyl radicals *in vitro* (Koide *et al.* 1983) and to ameliorate vasospasm in the canine SAH model (Asano *et al.* 1984) as well as ischemic brain edema in rat (Gotoh *et al.* 1984) and cat (Asano *et al.* 1984) middle cerebral artery occlusion models. Using the canine "two hemorrhage" SAH model, the effects of AVS and another free radical scavenger, glutathione (GSH), on choronic vasospasm and the HETE content of the basilar artery were examined (Watanabe *et al.* 1985). In the saline-treated group, the mean diameter of the basilar artery was 41.4% of the control on the eighth day after SAH. Those in the glutathione-(200 mg/kg/day for 7 days) or AVS (800 mg/kg/day)-treated groups were 65.8% and 51.0%, respectively, each showing a significant amelioration of vasospasm. When samples were homogenated with ice-cold saline, the mean values of 5-HETE contents in the basilar artery in the saline-, glutathione-, and AVS-treated groups were 0.71, 0.04, and 0.042 nmol/mg protein, respectively. Thus, the beneficial effects of these free radical scavengers on vasospasm seemed to correlate with their effects to decrease the activity of 5-lipoxygenase in the basilar artery.

b) Suggested Pathogenetic Mechanism Underlying Cerebral Vasospasm

Based on the hitherto obtained experimental and clinical findings, we constructed a hypothesis regarding the involvement of free radicals, hydroperoxides and eicosanoids in the pathogenetic mechanism underlying cerebral vasospasm **(Fig. IV-15)**. It needs to be emphasized that the paradigm shown here would represent only a part of the whole mechanisms involved in the occurrence of cerebral vasospasm. Obviously, further work is needed to clarify the complex interrelations between these and other chemical mediators of inflammatory responses, such as those produced in the plasma (kinin, coagulation, complement), and those produced by a variety of cells (Larsen and Henson 1983). For example, both 5- and 12-HETEs are potent chemotactic factors for neutrophils (Kuehl and Egan 1980, Goetzl 1980). The polymorphonuclear leukocytes were shown to migrate through the endothelium transcellularly (Faustmann and Dermietzel 1985) and cause endothelial damage following intravascular complement activation and O_2^- generation (Movat 1979, Fantone and Ward 1982). 5-HETE also enhances the expression of $C3b$ receptors on these cell (Goetzl 1980). That 5-HETE was detected indicates simultaneous generation of di-HETEs, leukotrienes, and the platelet-activating factor (PAF). In fact, we showed that the canine basilar artery exposed to SAH has the capacity of leukotriene formation, although 5-lipoxygenase might be derived from leukocytes migrating into the vessel wall. Thus, the identification of lipoxygenase products would lead us to the already vast and still rapidly growing research field of inflammatory reactions. Presently, the factor(s) responsible for arterial constriction remains obscure, but it may at least be stated that an new approach to the understanding and therapy of cerebral vasospasm has been opened up.

Inferred from our paradigm is the possible usefulness of a variety of pharmacological approaches in the prevention TxA_2 synthetase inhibitor, OKY 046 on vasospasm both in experimental and clinical conditions. In addition, antioxidants such as AVS and glutathione significantly ameliorated vasospasm in

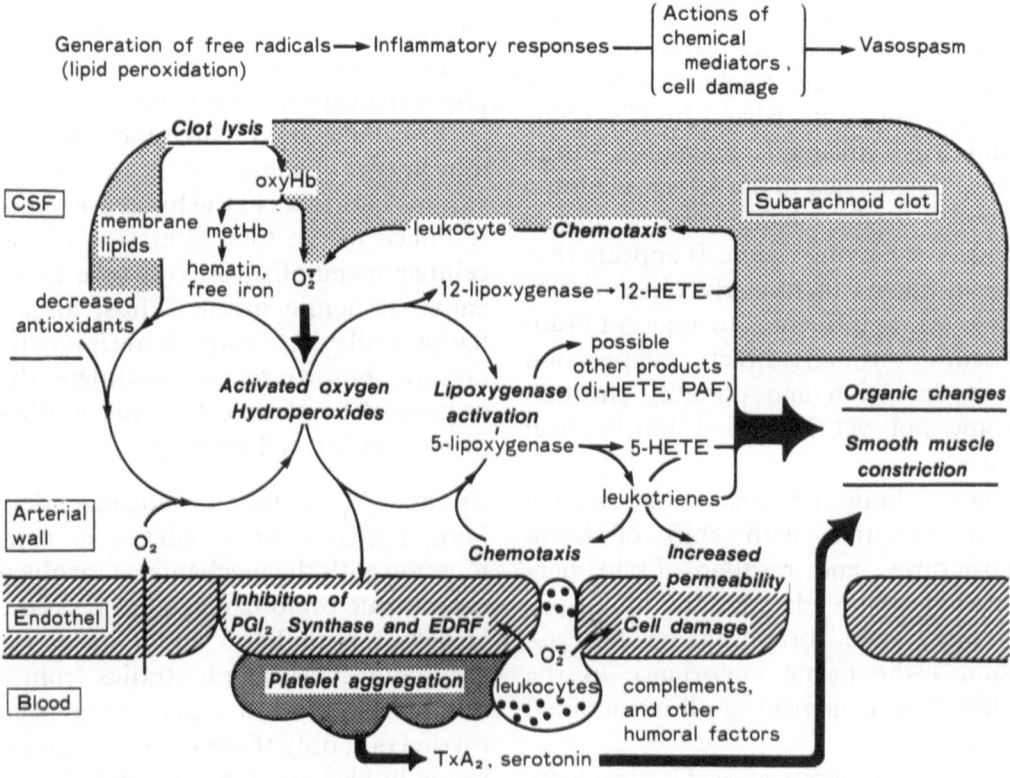

Fig. IV-15. Suggested pathogenetic mechanism of cerebral vasospasm

and treatment of vasospasm, such as inhibitors of phospholipases, lipoxygenase, or TxA_2 synthetase, PGI_2-like substances, antagonists of TxA_2 and leukotrienes, and antioxidants. To date, only a few of these pharmacological approaches have been subjected to experimental or clinical assessment. We showed the beneficial effect of a the canine "two hemorrhage" model. Insomuch as those preliminary results are supportive of the clinical feasibility of our approach, further study is obviously needed to determine which one or what combination of the above listed drugs would be most beneficial in the prevention and treatment of cerebral vasospasm.

C. Ischemic Brain Edema

Introduction

Cerebral edema which is defined as an increase in brain tissue volume resulting from an increase in its fluid content (Katzman and Pappius 1973), is an inevitable and the most dangerous complication of any kind of brain damage. As the whole sequence of events can be visualized by CT scans and MRI (magnetic resonance imaging), the increase in brain tissue volume due to brain edema behaves like a space-occupying lesion. It appears that brain edema in its early phase exerts deleterious influences to adjacent brain tissues through direct compression and distortion and possibly through some not yet identified biochemical effects.

Massive brain edema causes an extreme ICP elevation with shift of brain structures, and resultant brain herniation often endangers the life of patients. Thus, brain edema has assumed the prime importance in the clinical management of all intracranial lesions.

The clinical significance of brain edema is all the more apparent in the management of acute SAH because without exception, occurrence of massive brain edema means a poor outcome.

In spite of past enormous research efforts, the pathomechanism underlying ischemic brain edema has remained elusive, and we have not had any effective measures to control brain edema, except for inducing a temporary alleviation by the use of hypertonic solutions, CSF drainage, or hyperventilation. This is probably based on the very nature of the lesion, that in cerebral infarction, brain edema appears to be entirely an accompaniment of necrosis and not an independent or prenecrotic process (Plum and Posner 1963). This notion seems to have been more firmly substantiated by recent physiological and biochemical studies at the cellular level which indicate that movements of water and electrolytes at various interfaces in the brain are tightly linked to the total alteration of the cellular metabolism in response to an anoxic-ischemic insult. Thus, an attempt to alleviate brain edema might be nothing but to try to modulate the sequential biochemical events leading to irreversible cell damage.

Anyhow, the pathomechanism underlying ischemic brain edema in conjunction with the mechanisms involved in ischemic brain damage has been the target of both basic and clinical investigations. Although studies from a number of different aspects have been carried out, only those related to membrane lipids, free radicals, and eicosanoids on which we have concentrated our research effort for the past decade will be described below. For more extensive knowledge, reviews and monographs such as of Katzman *et al.* (1977), MacKnight and Leaf (1977), Bradbury (1979), Rapoport (1979), Hertz (1981), Goldstein and Betz (1983), and Siesjo (1985) as well as the Proceedings of International Symposium on Brain Edema (1965–1985) would be recommended.

1. Mechanisms Underlying the Occurrence of Ischemic Brain Edema

The classification of brain edema introduced by Klatzo (1967) which includes vasogenic and cytotoxic edema, has been widely accepted. Vasogenic edema is characterized by vascular changes that permit leakage of proteins and extravasation of intravenously injected BBB indicators into the extracellular spaces, particularly in the white matter. In contrast, cytotoxic edema affects either white or gray matter (or both) and is characterized by the accumulation of fluid, usually intracellular, without leakage of proteins or extravasation of BBB indicators (Katzman et al. 1977). Although it is true that ischemic brain edema begins as a cytotoxic type and in later periods converts to a vasogenic type, the nature of ischemic brain edema has not been wholly explained by a combination of or a transition between these edema types. Hence ischemic brain edema is currently thought to constitute a separate entity of brain edema, characterized by the specific pathophysiology and pattern of development.

In any event, glial swelling accompanied by increases in the brain contents of sodium and water marks the onset of brain edema due to anoxic-ischemic insults. For the sake of succeeding discussions, an overview of relevant mechanisms will be given in this section.

a) Mechanisms Relevant to Glial Swelling Following Ischemia

It has been established that the "double-Donnan system" (Leaf 1959) plays a predominant role in the regulation of cellular volume. In this system, sodium ions, maintained extracellularly by active transport out of cells, serve as the "impermeant" external ion to counterbalance the intracellular colloid osmotic pressure. Sufficient concentration of sodium, and with it chloride, will be kept extracellular in excess of its expected Donnan distribution to provide the balancing osmotic activity. Since the concentration gradient of sodium ion between the intra- and extracellular fluids is maintained by its pumping in the face of its leak in the opposite direction, the system is also described as the "pump-leak hypothesis". It has been shown that the rate of activity of sodium pump is adjusted by the intracellular concentration of sodium and potassium to balance sodium entry and thereby maintain cellular volume in a steady state.

Since the energy for this system comes from the metabolism of the cell, its depletion due to any causes should result in accumulation of sodium within cells. The resulting increase of sodium, with chloride to maintain electroneutrality, would draw in water as well, and the cells would swell. Alternatively, increased permeability to sodium, whether produced by physical or chemical factors may result in cellular swelling. In this regard, calcium plays an important role because increased cellular calcium has been shown to increase monovalent cation permeability (MacKnight and Leaf 1977).

That the "double-Donnan system" is also relevant to volume regulation of neuronal and glial cells has been supported by numerous reports showing that ischemia is followed by alterations of extra- and intracellular ion concentrations, especially of potassium and sodium, consistent to the theory (Hertz 1981, Siesjö 1985) and findings that intracerebral injection or topical application of an Na^+, K^+ ATPase inhibitor, ouabain caused astroglial or dendritic swelling (Cornog et al. 1967, Meier-Ruge et al. 1974, Lowe et al. 1975, Tanaka et al. 1977). Nevertheless, the well known fact that ischemia leads to swelling of glial cells but not of neurons (Kalimo et al. 1982) suggests the existence of other ion exchange mechanisms differently distributed in glial and neuronal cells. Elevated potassium concentration in the extracellular fluid per se was shown to increase glial swelling, and the involvement of exchanges or cotransport of ions such as Na^+, Cl^-, K^+, H^+, and HCO_3^- in this process has been suggested (Hertz 1981, Kimelberg and Bourke 1984, Siesjö 1985). Recently, Siesjö (1985) proposed an intriguing hypothesis that acidosis disrupts volume regulation in glial cells by leading to grossly enhanced Na^+/H^+ exchange, accompanied by Cl^-/HCO_3^- exchange. This can cause marked swelling only if the Na^+ gradient is maintained by continued ATPase activity. It was emphasized that such swelling requires that a delivery line is established from Na^+ and Cl^- in the direction blood $\rightarrow ECF \rightarrow ICF$. In this regard, Kimelberg and Bourke (1984), postulated that the presumptive Na^+/H^+ antiporter is located at the perivascular or ependymal attachment of astrocytes. By this setting, Na^+/H^+ antiport was assumed to take place between the blood or CSF and the astrocyte. So far, however, the location of Na^+/H^+ antiporter in astrocytic foot processes has not been supported by firm evidence. The existence of Na^+/H^+ antiporter in the abluminal membrane of endothelial cells has been suggested (Betz 1983 a, b), but in this case the direction of Na^+ transport is opposite to that of astrocytes. In addition, it seems clear that this system cannot function in case of severe energy depletion, since its energy source is the Na^+ gradient created by the Na^+, K^+ ATPase (Siesjö 1985).

b) Movements of Sodium and Water in Ischemic Brain Edema

It has been well shown that a severe reduction in rCBF exceeding certain threshold value is followed by immediate increases of K^+ and decreases of Na^+ and Ca^{2+} in the extracellular fluid, and by astrocytic swelling causing shrinkage of the extracellular space (van Harreveld and Ochs 1956, Branston et al. 1977, Hossmann et al. 1977, Matsuoka and Hossmann 1982). These early changes can be explained on the basis of a disruption of the "double-Donnan system" due to energy depletion, and are accompanied by a slight net gain of water and sodium in the ischemic brain tissue. However, it has previously been questioned whether or not the occurrence of acute astrocytic swelling can be equated with edema (Katzman et al. 1977). In this respect, Matsuoka and Hossmann (1982) examined the correlation between the changes in cerebral impedance (an indicator of the extracellular space) and in the tissue water and sodium contents using the cat MCA occlusion model. The result indicated that the narrowing of the extracellular compartment and ischemic brain edema are relatively independent consequences of cerebral ischemia.

Clearly, ion exchanges confined between extra- and intracellular fluids, no matter what kind they may be, cannot cause edema without concomittant ionic movements across the BBB. Although there are numerous reports showing the involvement of sodium in the occurrence of ischemic brain edema (Siegel *et al.* 1973, Brunson *et al.* 1973, Shibata *et al.* 1974, O'Brien *et al.* 1970, Hossmann 1985, Kogure *et al.* 1981, Hilal *et al.* 1983), it was not until the study of Gotoh *et al.* (1985) that an unequivocal quantitative relationship between the sodium influx across the BBB and edema development has been demonstrated.

In this study, the time courses of the brain water, sodium and potassium contents and BBB permeability to ^{125}I-albumin were examined using the middle cerebral artery (MCA) occlusion model in rats (Tamura *et al.* 1981 a, b). As shown in **Fig. IV-16**, the increase as well as the decrease in the water content of the ischemic cerebral hemisphere was closely paralleled by the changes in sodium and potassium contents. There was a clear discrepancy between the increases in water content and the BBB permeability to injected ^{125}I-RISA. Throughout the develop-

ment and resolution of edema, a highly significant linear correlation between the hemispheric contents of water and sodium was revealed (**Fig. IV-17**). Using these data, whether or not edema fluid may be regarded as plasma ultrafiltrate was examined as follows.

In normal conditions, the sodium influx across the BBB is extremely low (Oldendorf 1970). The massive increase in the brain content of sodium accompanying edema formation indicates a significant increase in permeability of the BBB to sodium. If it is assumed that the BBB acts like a semipermeable membrane permitting fluxes of water and electrolytes but not of protein molecules, the sodium concentration in the edema fluid must be a little lower than that in plasma according to Donnan equilibrium. In this condition, one to one exchange of sodium with potassium at the BBB does not accompany water movement. Thus, between the increments of hemispheric contents of water (ΔH_2O) and sodium (ΔNa^+) and the decrements of hemispheric content of potassium (ΔK^+), the relationship as shown in **Fig. IV-18-A** exists. In each specimen, the increment or decrement of water and electrolytes were obtained as the differences between the obtained values and the mean values in normal rat hemispheres. The value of theoretical ΔNa^+ (ΔNa^+_T) was calculated from the values of ΔH_2O and ΔK^+ using the formula, and it was plotted against the actual value of ΔNa^+ (**Fig. IV-18-B**). As the figure shows, a very close linear correlation between the actual and theoretical values of ΔNa^+ was revealed. Thus, this study clearly proved that the increased BBB

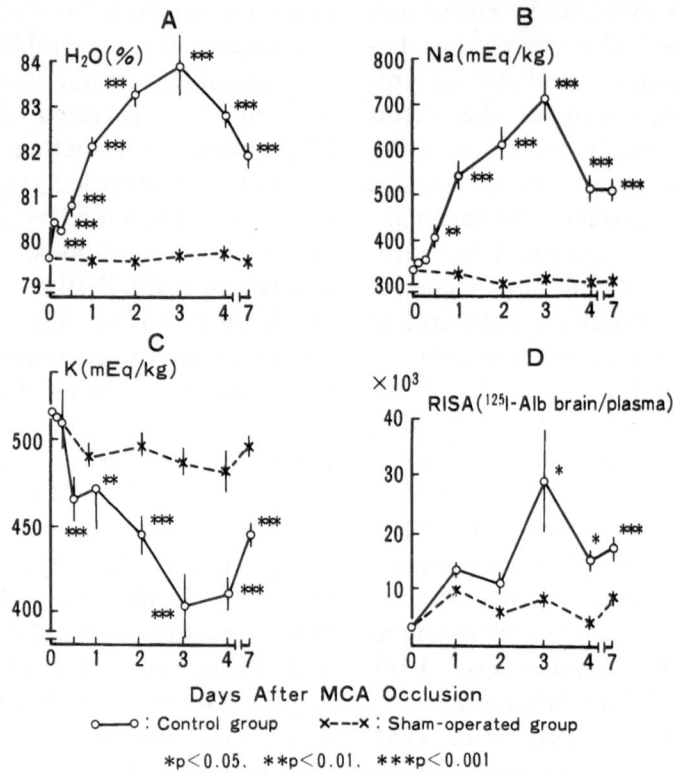

Fig. IV-16. The time-courses of the hemispheric contents of water, sodium, and potassium, and the BBB permeability of ^{125}I-albumin following permanent MCA occlusion in rats (from Gotoh *et al.* 1985)

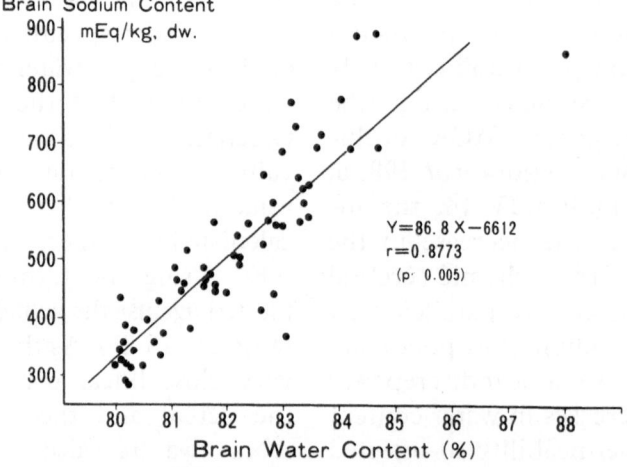

Fig. IV-17. The correlation between the water and sodium contents in the affected hemispheres

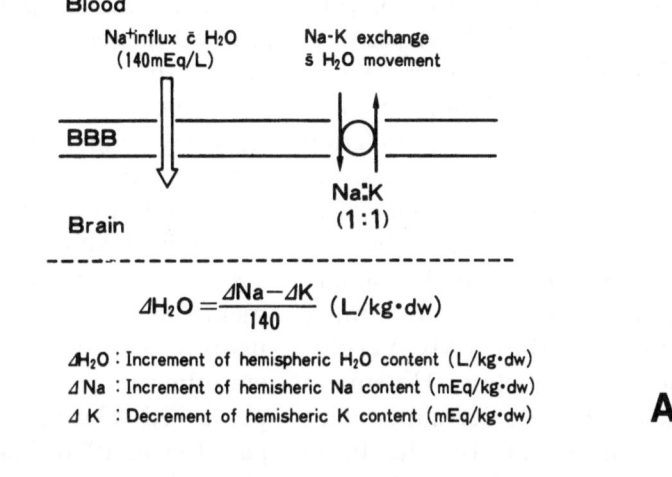

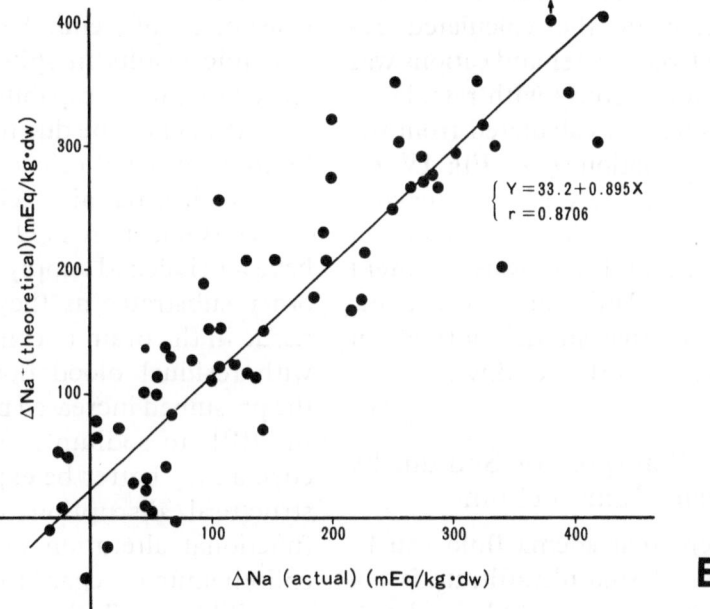

Fig. IV-18. A) The movements of sodium, potassium, and water across the BBB according to the Donnan equilibrium. In this condition, the sodium concentration in the edema fluid is lower than that in the serum and it was arbitrarily set as 140 mEq/L. B) The correlation between the actual (the abscissa) and the theoretical (the ordinate) values of ΔNa^+

permeability to sodium is the predominant cause of ischemic brain edema and that the composition of edema fluid can be equated with plasma ultrafiltrate. Considering that the same linear correlation between increments of water and sodium is maintained throughout the evolution of edema, hitherto suggested factors such as idiogenic osmols (Hossmann and Takagi 1976), hydrostatic pressure (Kogure *et al.* 1981), or protein leakage (Ito *et al.* 1976) seem to play only limited roles in edema development at least in models of permanent regional ischemia. A similar correlation between the cortical contents of water and sodium was found 4 hours after occlusion of the MCA in cats (Shigeno *et al.* 1986). Schuier and Hossmann (1980) reported an essentially similar result to ours using the cat MCA occlusion model. Nevertheless, Hossmann (1985) discussed that a certain amount of water is taken up by the brain without any net cation changes, because the intercept of the calculated regression between water and cations was negative. In our study with rats, however, the intercept calculated from the regression equation (*cf.* **Fig. IV-17**: $Y = 86.8\,X - 6612$; X: brain water content; Y: brain sodium content) at normal value of brain water content (80%) was 332 mEq/kg d.w., being very close to that in the normal rat hemisphere (334 mEq/kg d.w.).

c) Active Transport of Sodium by the Capillary Endothelium

The concept that edema fluid can be regarded as plasma ultrafiltrate is not new since it has originally been developed in relation to edema of vasogenic type, such as induced by cold injury. Insomuch as ischemic brain edema resembles vasogenic edema in this respect, they considerably differ from each other in regard to the nature of BBB injury. In vasogenic edema, the BBB is both functionally and morphologically destroyed from the outset, permitting free passage of plasma constituents including proteins. In contrast, evidence showing frank destruction of the BBB has not been obtained in ischemic brain edema, especially in its early stage conventionally designated as cytotoxic edema (Ito *et al.* 1976, Gotoh *et al.* 1985).

In cerebral ischemia, necrotic changes are confined to various cellular elements of brain parenchyma such as neurons and glia whereas the capillary endothelium remains strikingly well preserved (Garcia *et al.* 1971, Westergaard *et al.* 1976, Katzman *et al.* 1977, Dietrich *et al.* 1984). This pronounced resistance of endothelial cells to ischemic insults in spite of its apparently large work capability (Oldendorf *et al.* 1977) may be due to the facts that brain endothelial cells can utilize fatty acids such as palmitic acid as an energy source (Goldstein 1979) and that they have a privileged supply of oxygen and other substrates as they are the first tissue in the brain to come into contact with residual blood flow. Therefore, the presumed increased permeability of the BBB to sodium in early ischemic edema may better be explained not by structural disruption but by some functional alteration of the capillary endothelium in regard to sodium transport. This possibility appears plausible in consideration of pertinent findings as described below.

Recent studies using brain microvessels showed that sodium is actively transported across the BBB by Na^+, K^+ ATPase localized in cerebral capillaries (Eisenberg and Suddith 1979). Betz *et al.* (1980), demonstrating that Na^+, K^+ ATPase is localized only in the antiluminal membranes of brain capillaries, suggested that this polarity should permit active solute transport across the BBB. The time-honoured notion that the brain extracellular fluid, like the CSF, is a secretion dependent upon an active transport of Na^+ by brain capillaries (Davson 1956, Wright 1972, Bradbury 1979) has thus been substantiated. Further, it has been shown that in addition to Na^+, K^+ ATPase, Na^+/H^+ antiporter is localized in the antiluminal membrane (Betz 1983 a) and that two distinct transport systems, *i.e.*, furosemide-sensitive Na^+-Cl^+ cotransport system and amiloride-sensitive Na^+ transport system, exist in the luminal membrane (Betz 1983 b).

As previously described, alterations in cellular metabolism as well as ionic distribution in the intra- and extracellular fluids in fact lead to enhanced transport of sodium together with its anionic counterpart Cl^- and water across the BBB. This may well be a simple diffusional process of sodium due to an increased concentration gradient across the BBB accompanied by an increased permeability of the BBB (Schuier and Hossmann 1980). However, in the face of well documented viability of endothelial cells, it is tempting to speculate that the process is mediated by a preserved or even enhanced activity of Na^+, K^+ ATPase in endothelial cells. It must be emphasized that this notion does not contradict the concept that the BBB behaves as if it were a genuine semipermeable membrane following ischemia, because from the thermodynamic standpoint, Donnan equilibrium across the BBB does not preclude the participation of facilitated or active transport mechanism in endothelial cells, so far as the unipolar movement of sodium from blood to extracellular fluid is concerned.

d) Test of the Hypothesis

To test our novel hypothesis, we examined whether or not pharmacological suppression of brain capillary Na^+, K^+ ATPase by ouabain would lead to amelioration of edema due to regional ischemia *in vivo* (Shigeno et al. 1985, Asano et al. 1985). In cats, the MCA was occluded by the transorbital approach and a fine catheter was inserted into the MCA distal to the occlusion site so that the ischemic brain area was directly perfused. In order to preserve the natural circulatory condition, perfusion was maintained as intermittent pulse injections with a duration of 15 seconds, given every 5 or 2 minutes (**Fig. IV-19**). The cortical specific gravity at 6 locations over the ischemic area was measured 4 hours after MCA occlusion. The topographically corresponding local CBF (l CBF) before and during MCA occlusion was measured using the hydrogen clearance technique. The cortical contents of water, sodium, and potassium were also measured. In the control group, aerated Krebs-Ringer solution alone was intermittently perfused. Ouabain was dissolved in the solution in a concentration of 10^{-5} M. Obtained results show that ouabain perfused every 2 minutes significantly ameliorated edema formation, particularly in the moderately ischemia area (**Fig. IV-20**). In a separate series of animals, the Na^+ flux

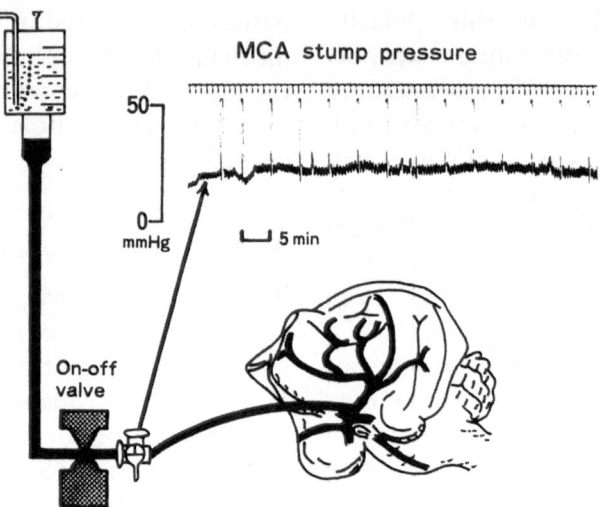

Fig. IV-19. The experimental design of intermittent perfusion of the MCA territory following MCA occlusion in cats. The on-off valve was controlled by the use of a computer

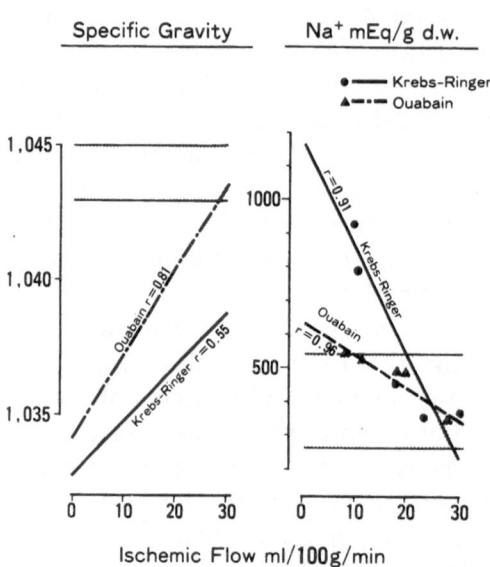

Fig. IV-20. Effect of ouabain administration on the cortical specific gravity and sodium content. The decrease in the cortical specific gravity and the increase in the sodium content were significantly inhibited by administration of ouabain (from Asano *et al.* 1985 a)

across the BBB was studied by injecting labeled ^{22}Na together with an intravascular reference (^{57}Co microsphere, 15 μ in diameter) into the ischemic area via the catheter in the MCA. The brain uptake index (BUI) of ^{22}Na thus obtained by a modified Oldendorf's method (Oldendorf 1970) showed a marked increase in the cortical area where the ischemic flow was below 30 ml/100 g/minute in control animals. The rise of BUI of ^{22}Na in the ischemic foci was remarkably supressed by ouabain (Asano *et al.* 1985 a, Shigeno *et al.* 1986).

Above results show in a most direct fashion that capillary Na$^+$, K$^+$ ATPase activity is involved in edema formation following regional ischemia. Apparently, this conclusion is in opposition to the hitherto known effect of ouabain to cause glial or dendritic swelling (Cornog *et al.* 1967, Meier-Ruge *et al.*

1974, Tanaka *et al*. 1977). But, glial swelling *per se* does not necessarily mean edema (Katzman *et al*. 1977, Matsuoka and Hossmann 1982). In preceding studies, convincing evidence which indicates the occurrence of edema, *i.e.*, net increases in water and sodium contents, was not presented. Since systemically administered ouabain crossed the BBB in sufficient amounts to inhibit Na^+, K^+ ATPase of intraparenchymal cells causing an immediate fall in $CMRO_2$ and a net potassium efflux (Astrup *et al*. 1981), Na^+, K^+ ATPase in capillary endothelium should have been similarly inhibited in our experiment. Gazendam *et al*. (1979) observed that the sodium concentration of edema fluid obtained from an edema focus induced by a freezing lesion was significantly decreased by ouabain injection. They mentioned to the possibility that the decrease of Na^+ in the edema fluid may be attributed to inhibition by ouabain of a Na^+, K^+ ATPase at the BBB. Thus, it appears rational to interpret our experimental results as indicating a significant role of capillary Na^+, K^+ ATPase in the pathomechanism of ischemic brain edema. This conclusion leads to the next important question: what biochemical mechanisms are relevant to the activation of capillary Na^+, K^+ ATPase following ischemia? This problem will be discussed in the following sections in relation to the recently demonstrated actions of eicosanoids.

2. Alterations of Eicosanoid Synthesis Following Cerebral Ischemia and Effects of Pharmacological Modulation

Ischemia causes immediate stimulation of phospholipase activities and release of FFAs. Chemically active PUFAs, especially arachidonic acid (AA), are for the most part conjugated to albumin and other protein molecules, or incorporated into triacylglycerol (Yoshida *et al*. 1986). Thus a very small fraction of free AA (one hundredth or thounsandth of the total liberated AA) remains for conversion to eicosanoids. However, since each eicosanoid albeit small in amount has a significant potential of biological activities, its role needs to be sought in relation to pathomechanisms operating during and after cerebral ischemia. In this section, we will see first what eicosanoids are generated in the brain exposed to ischemia, and then what effects pharmacological modulation of the arachidonate cascade exert on subsequent cerebral hemodynamics and edema formation.

a) Cyclooxygenase Products

It has been shown that brain levels of eicosanoid does not change during complete ischemia, moderately increases during incomplete, regional ischemia, and show a pronounced but transient increase following recirculation. Gaudet and Levine (1979, 1980) were the first to report on the brain levels of cyclooxygenase products during and after transient occlusion of bilateral carotid arteries in gerbils. They showed that no significant changes in each prostaglandin takes place so long as ischemia continued. Increased levels following reperfusion were highest at 5 minutes and then declined. Pretreatment with nonsteroidal antiinflammatory drugs such as aspirin, indomethacin, piroxicam, and flufenamic acid effectively inhibited increases of all the products following recirculation. The same authors also reported on the effect of unilateral common carotid artery occlusion in gerbils on levels of cyclooxygenase products in cerebral hemispheres (Gaudet and Levine 1980 b). Interestingly, in animals with positive neurological signs, PG levels increased in both hemispheres. The levels of PGD_2 and $PGF_{2\alpha}$ were elevated at 15 minutes and 2 hours after occlusion. The levels of 6-keto-$PGF_{1\alpha}$ increased only slightly, becoming significantly greater only at 15 minutes of occlusion in the left hemispheres. It is of interest that in this study, the levels of PGD_2 and $PGF_{2\alpha}$ in the non occluded (left) hemispheres were apparently higher than those in the occluded (right) hemispheres. This may be due to the well known fact that in gerbils, unilateral carotid artery occlusion often produces lesions in bilateral hemispheres. The authors however suggested the occurrence of diaschisis in this model as reported by Kempinski (1958).

Ellis et al. (1981) measured cyclooxygenase product in cat cerebral cortex after experimental concussive brain injury. During the 60-minute period after injury, only PGE_2 significantly elevated up to 181% of the control level. $PGF_{2\alpha}$ and 6-keto-$PGF_{1\alpha}$ did not show significant changes.

In the above studies, measurements were done with hemispheres or brain tissues containing blood. It has been shown that brain perfusion prior to sampling with saline containing indomethacin results in considerable decreases in levels of $PGF_{2\alpha}$ (26% of control) and PGE_2 (47% of control) (Bhakoo et al. 1984). Determination of PG levels using this "biochemical freezing" technique revealed that $PGF_{2\alpha}$ not PGE_2 significantly increased during bilateral carotid ligation in gerbils (Crockard et al. 1982). Following recirculation, levels of both PGs pronouncedly increased (Bhakoo et al. 1984). Although these authors attempted to correlate PG levels to regional changes in CBF, no significant regional differences in PG levels were found, probably because the rCBF drop in this model is very pronounced and diffuse, being less than 10 ml/100 g/minutes in all areas examined (Bhakoo et al. 1984). We examined the correlation between the regional changes in brain PG levels and CBF changes following regional ischemia and subsequent recirculation using the cat MCA occlusion model (Hanamura et al. 1986). Each animal was exposed either to 2 hours MCA occlusion, 2 hours MCA occlusion followed by 30 minutes recirculation, or 2 hours MCA occlusion followed by 2 hours recirculation. At the end of experiment, the brain was briefly perfused with ice-cold saline and frozen in situ using liquid nitrogen. Small brain samples in areas topographically corresponding to each site of local CBF (1CBF) monitoring by the hydrogen clearance technique were chiselled out. PGs were determined using the radio-

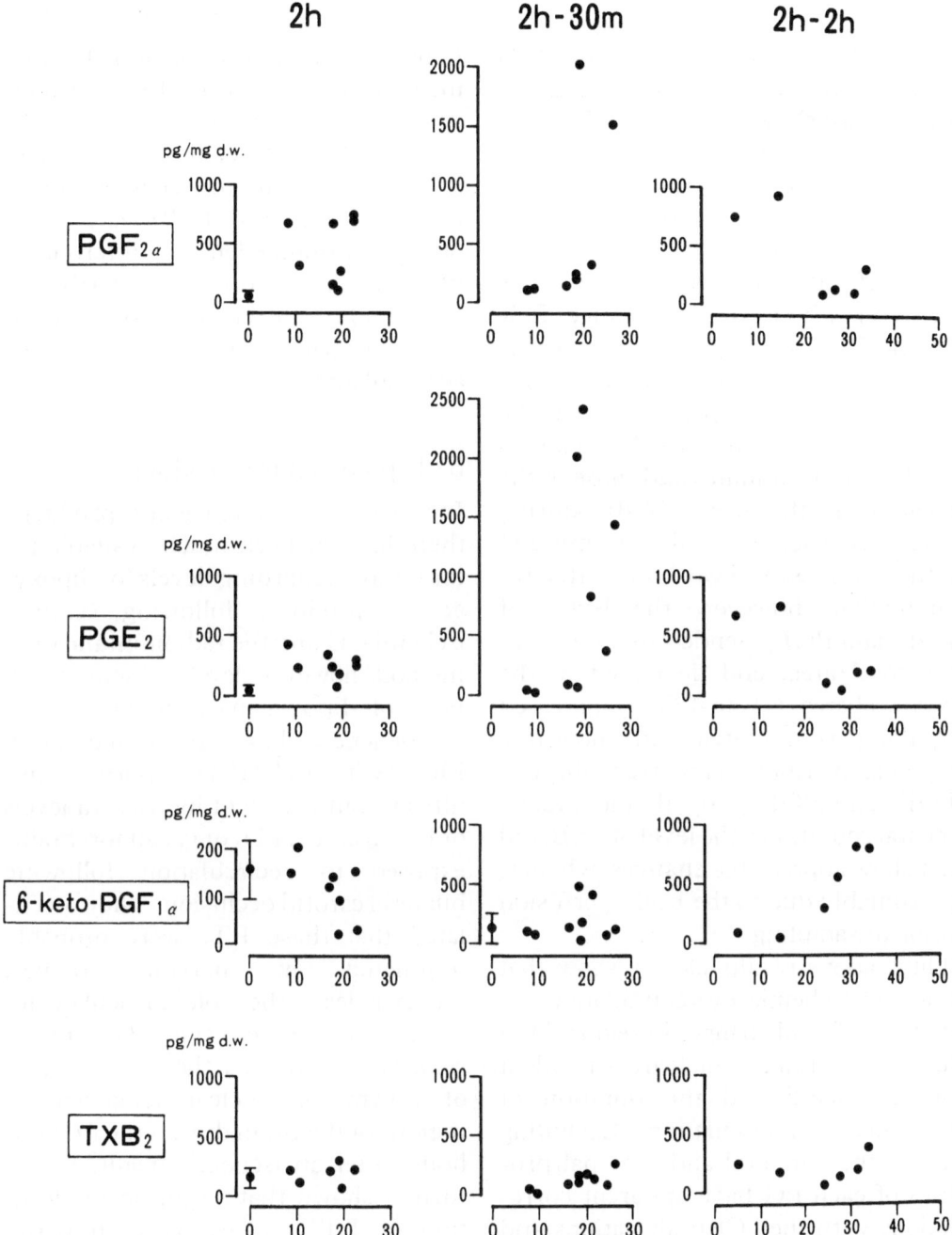

Fig. IV-21. The regional brain content of prostaglandins (ordinate) topographically correlated to the clip flow (the mean lCBF during MCA occlusion: abscissa). The data at 2 hours after MCA occlusion (2 hours), 2 hours MCA occlusion plus 30 minutes recirculation (2 hours—30 minutes), and 2 hours MCA occlusion plus 2 hours recirculation (2 hours—2 hours) are separately shown. The control value of $PGF_{2\alpha}$, PGE_2, TxB_2, and 6-keto-$PGF_{1\alpha}$ were 39.9 ± 31.5, 48.7 ± 37.2, 127.0 ± 120.7, 159.0 ± 136.2 pg/mg dry weight (mean $\pm$ SD), respectively (each shown as the bar in the left of each figure

immunoassay method after separation with thin layer chromatography. Results are shown in **Fig. IV-21**. Following 2 hours MCA occlusion, brain levels of $PGF_{2\alpha}$, PGE_2, and 6-keto-$PGF_{1\alpha}$ were significantly elevated in areas where the l CBF was depressed below 20–30 ml/100 g/minute, whereas the level of TxB_2 did not change. Following 30 minutes recirculation, levels of $PGF_{2\alpha}$ and PGE_2 remarkably increased only in the perifocal area (the mean l CBF during MCA occlusion > 20 ml/100 g/minute) and those in the focal area (the mean l CBF during MCA occlusion < 20 ml/100 g/minute) rather decreased. Two hours after recirculation, however, the levels of PGE_2 and $PGF_{2\alpha}$ tended to increase in the focal area, and decreased in the perifocal area. 6-keto-$PGF_{1\alpha}$ increased in the perifocal area and showed a regional distribution reverse to those of PGE_2 and $PGF_{2\alpha}$. In all the experimental conditions, the level of TxB_2 did not show appreciable changes, which is presumably due to the brain perfusion prior to sampling.

Thus, above results clearly show that regional ischemia or recirculation induces profound changes in regional PG levels of the brain, which are dependent on the depth and the duration of ischemia or recirculation. Excluding TxB_2, the temporal and regional profiles of each PG have apparent correlations with the l CBF alterations and edema development which have been disclosed in this model (**cf. Fig. IV-23 and IV-24**). In regard to the effect of recirculation, particularly noteworthy is the existence of a common CBF threshold by which subsequent altera-

tions in brain PG levels and development of edema in the focal and perifocal areas were clearly divided. It may be surmised, therefore, that the primary brain damage due to ischemia, alterations in PG levels or l CBF, and development of brain edema are interrelated phenomena. In order to clarify the cause and result relationship, however, more experimental data need to be accumulated.

b) Lipoxygenase Products

In contrast to cyclooxygenase products, there has not been much available information regarding levels of lipoxygenase products following cerebral ischemia. Using the radioimmunoassay method, however, levels of some products of lipoxygenase pathway, *i.e.*, leukotrienes, have been measured. Moskowitz *et al.* (1984) reported a significant but transient increase in levels of LTC_4 and LTD_4 of gerbil forebrains exposed to recirculation following bilateral carotid occlusion. They speculated that these LTs were probably originated in nervous tissues. Nevertheless, the role of leukocytes cannot be negated since Hallenbeck *et al.* (1986) showed the accumulation of polymorphonuclear leukocyte in regions of the brain during the first few hours after an ischemic insult. It was further shown that a significant elevation of LTC_4 takes place in gerbil models of ischemic insult, subarachnoid hemorrhage, and concussive injury (Kiwak *et al.* 1985). Also using the bilateral carotid occlusion model in gerbils, Dempsy *et al.* (1986) measured brain levels of cyclooxygenase products

Table IV-3. HETE contents in MCA occluded rat brain

Duration of Ischemia(hr)	5-HETE	9-HETE	8-and 12-HETE	11-HETE	15-HETE
0 (n=4) (control)	0.0±0.0	0.0±0.0	0.0±0.0	0.0±0.0	2.1±2.1
24(n=4)	0.0±0.0	0.0±0.0	0.0±0.0	1.6±1.6	9.1±3.8
72(n=4)	18.3±2.8**	16.6±3.0*	23.7±14.9	14.0±3.4*	19.8±5.5*

(ng/gr wet weight)

MCA : middle cerebral artery

Each value represents the mean ±S.E. for the number of independent determinations indicated in parenthesis

significantly different from the control at $P < 0.05$: * $P < 0.01$: ** (Student t-test)

and LTB$_4$ during ischemia and recirculation. Like preceding studies, levels of 6-keto-PGF$_{1\alpha}$, PGE$_2$, TXB$_2$, and LTB$_4$ were found to be increased on recirculation. Pretreatment with indomethacin inhibited the increase in these cyclooxygenase products, whereas formation of LTB$_4$ was conversely enhanced.

In the above studies, only leukotrienes were assayed using the radioimmunoassay technique. However, recent studies have shown the existence of a wide variety of lipoxygenase products, each of them having specific biological actions. Hence, studies to survey the whole spectrum of lipoxygenase products are indicated. Using the HPLC (revease and straigth phase) and GC/MS methods, we made an attempt to detect HETEs generated in the cerebral hemisphere exposed to focal

ischemia due to permanent MCA occlusion (Usui et al. 1985, 1987). The whole ischemic hemispheres frozen in situ without prior perfusion were used for analysis. Only a small amount of 15-HETE was found in control hemispheres. As shown in the Table IV-3, 11- and 15-HETE showed a slight increase 24 hours after MCA occlusion, whereas 72 hours thereafter, a variety of HETEs significantly increased, The detection of 5-HETE indicates simultaneous production of leukotrienes in a smaller amount since the convertion ratio of these substances from the common precursor 5-HPETE was shown to be 10 : 1 (Shimizu et al. 1986). The temporal profile of edema development in this model has previously been reported (Gotoh et al. 1985). Since edema developed most rapidly during 12–24 hours after MCA occlusion, generation of these HETEs might be only a epiphenomenon succeeding the occurrence of cerebral infarction. However, the HPLC method used in this study has a certain limitation in its sensitivity and the result does not exclude the possibility that smaller amounts of these HETEs generated in some compartment within the brain were significantly increased in earlier periods. Studies using a more sensitive RIA method has shown a significant increase of leukotrienes immediately after ischemic insults (Kiwak *et al.* 1985). The above presumption is further reinforced by the study showing that the eicosanoid synthetic capacity of isolated brain microvessels was significantly enhanced following ischemia (Asano *et al.* 1985). In this study, brain microvessels were isolated from rat cerebral hemispheres rendered ischemic by permanent MCA occlusion. The conversion of exogenous radiolabeled arachidonic acid to each eicosanoid was studied using the radiochromatographic method. The results showed that the eicosanoid synthetic capacity of brain microvessels was significantly enhanced following ischemia, and that the production of hydroxyacids (HETEs) at 24 hours was greater than that at 72 hours after MCA occlusion (**Fig. IV-22**).

Thus, our results are in harmony with preceding *in vitro* studies which

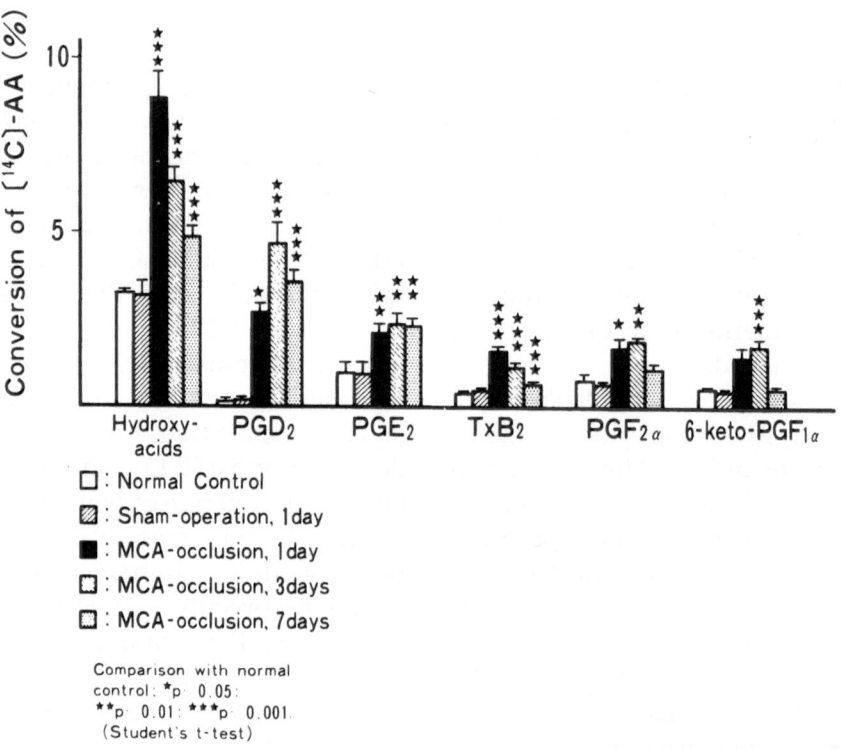

Fig. IV-22. Eicosanoid synthetic activities of brain microvessels in normal, sham-operated, and MCA-occluded rats

showed the capability of the brain tissue to synthesize 5-, 11-, 12-, and 15-HETEs (Wolfe and Pappius 1984, Adesuyi et al. 1985). It seems likely that different cells or tissues in the brain produce different lipoxygenase products and that formation of at least some of these products are increased following ischemia. Inasmuch as the involvement of lipoxygenase products in edema mechanism appears likely, it is needed to discriminate between those products which are truely responsible for the occurrence of ischemic brain edema.

c) Effects of Pharmacological Modulation of Eicosanoid Synthesis on CBF and Edema

Although the discrimination of the role of each eicosanoid in vivo is difficult, the use of cyclooxygenase inhibitors enables an evaluation of the integrative action of all the cyclooxygenase products. Most often used for this purpose is indomethacin, because the concentration of the drug required for cyclooxygenase inhibition is much lower than the concentrations which inhibit other enzymes (Flower 1974), and among other cyclooxygenase inhibitors such as aspirin, piroxicam and flufenamic acid, indomethacin proved to be the most potent inhibitor in the gerbil brain (Gaudet et al. 1980). Indomethacin (3 mg/kg) causes about 80% of prostaglandin synthesis inhibition in rat brain after one hour with a half life of 32 hours (Abdel-Halim et al. 1978). In the anesthetized but otherwise normal brain, indomethacin (10 mg/kg) significantly reduced CBF at normocapnea and severely impaired the response to hypercapnia (Pickard and MacKenzie 1973). Indomethacin did not affect the $CMRO_2$, mean arterial pressure, or the CBF response to hypocapnia (Pickard 1981). Also, the pronounced increase in cyclooxygenase products following recirculation has been effectively suppressed by pretreatment with indomethacin (Gaudet and Levine 1979, 1980). Thus, the effect of indomethacin on the alterations of CBF, neurological functions, and edema formation in various animal models has been studied to evaluate the role of cyclooxygenase products in toto.

So far, indomethacin has been shown to markedly improve cerebral circulation in the dog's brain after global ischemia (Furlow and Hallenbeck 1978), improve behavioral activity of gerbils exposed to transient carotid occlusion (Gaudet and Levine 1979), prevent arteriolar damage due to concussive brain injury in cats (Wei et al. 1981), improve CBF and cerebral metabolism (local cerebral glucose utilization) following focal cortical freezing lesion (Pappius and Wolfe 1983), and ameliorate hypoperfusion and edema formation following transient (2 hours) MCA occlusion in cats (Asano et al. 1984, Shigeno et al. 1985 a). In contrast, indomethacin exerts little effects on rCBF or edema formation in models of permanent ischemia (Johansson 1981, Ianotti et al. 1981). Further, Harris et al. (1982) showed using the MCA occlusion model in baboons that indomethacin significantly aggravated edema formation in ischemic foci, increasing the flow thresholds for changes in extracellular concentrations of K^+ and Ca^{2+} due to membrane depolarization.

That the effects of indomethacin on CBF and edema formation differs so much between continued ischemia and recirculation is well illustrated in the study of Johshita *et al.* (unpublished data). Using the cat MCA occlusion model, the effects of indomethacin (3 mg/kg) on l CBF and cortical specific

tween 20 and 25 ml/100 g/minute in both groups, and edema was clearly aggravated by recirculation.

Then, data were compiled according to the flow range of clip flow, *i.e.*, 0–15, 15–30, and more than 30 ml/100 g/minute. In the PI groups, the time course of l CBF in each flow range was

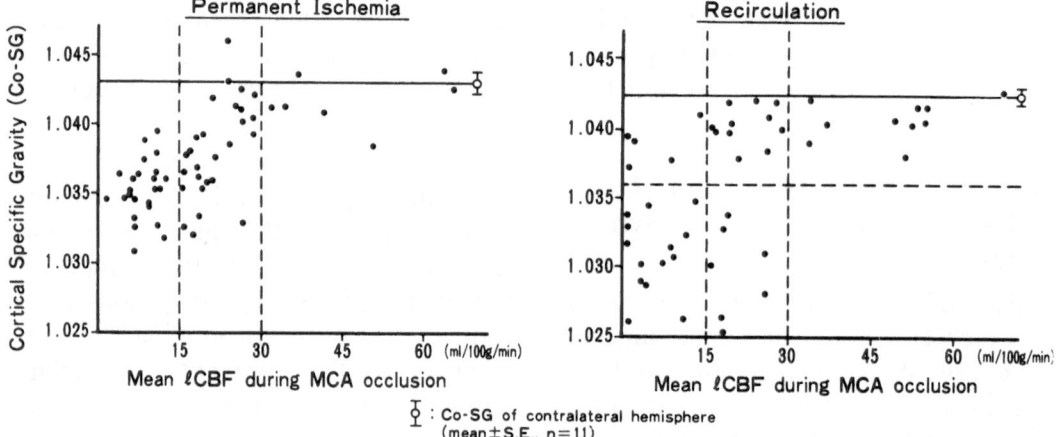

Fig. IV-23. Clip flow thresholds for edema development in permanent ischemia (4 hours MCA occlusion: left) and recirculation (2 hours MCA occlusion followed by 2 hours recirculation) in cats

gravity (coSG) were compared between the prolonged ischemia (4 hours MCA occlusion: PI) group and the recirculation (2 hours recirculation following 2 hours MCA occlusion: RC) group. The l CBF was measured at several locations over the ischemic cortical area using the hydrogen clearance technique. The mean l CBF value during MCA occlusion (clip flow) of each electrode site was correlated to the topographically corresponding coSG. As shown in **Fig. IV-23**, the l CBF threshold for edema formation lay be-

not affected by indomethacin. In contrast, the postischemic flow was significantly improved by indomethacin administration in the RC group (**Fig. IV-24**). In addition, indomethacin significantly ameliorated edema formation in the RC group, whereas it rather aggravated edema formation in the PI group (**Table IV-4**). The above results clearly indicate that the total effect of the cyclooxygenase products on CBF and edema formation in prolonged ischemia is different from that in recirculation. Such ambivalent effects of indometh-

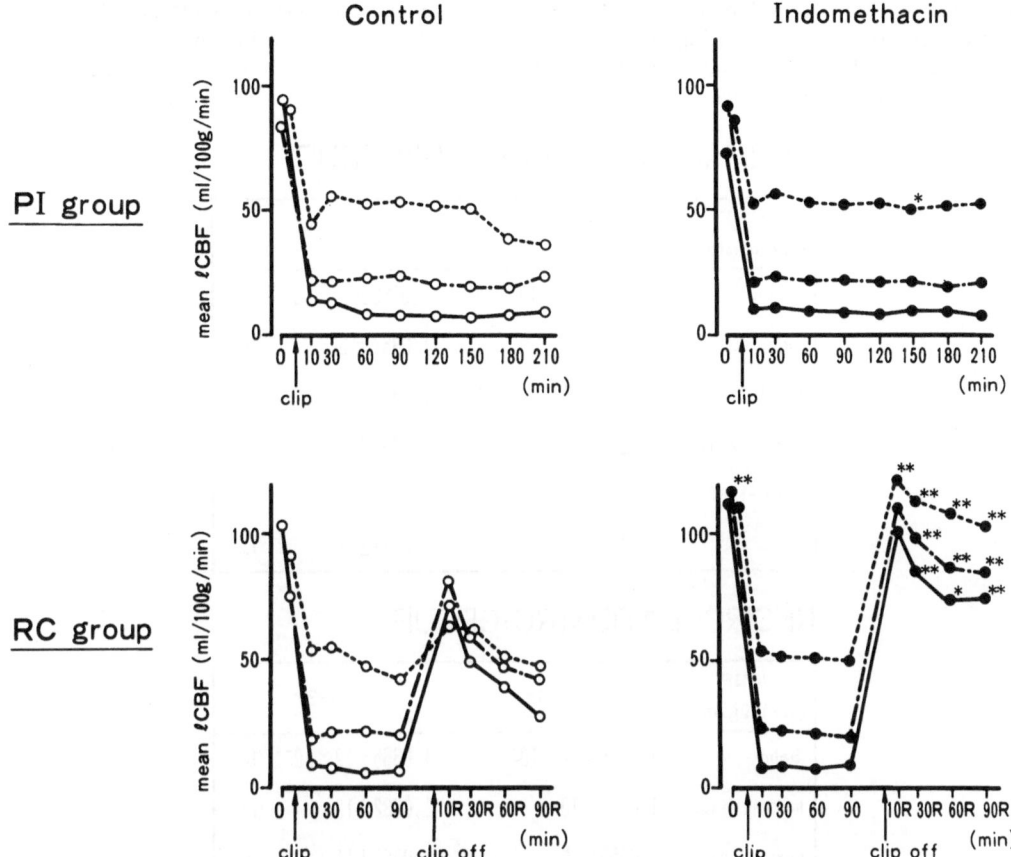

Fig. IV-24. The time course of lCBF following prolonged ischemia (upper figures) and recirculation (lower figures) and the effect of indomethacin administration. The electrode sites were assembled into three groups according to the flow ranges of 0–15, 15–30, and > 30 ml/100 g/minute. The lCBF following recirculation was significantly increased by indomethacin

acin may be ascribed to differences in the activities of cyclooxygenase, PG synthases, and lipoxygenases in each ischemia model. It has been known that suppression of cyclooxygenase leads to an increased substrate availability to the lipoxygenase system (Higgs and Vane 1983). In gerbil ischemia model, inhibition of PG synthesis by indomethacin was accompanied by a relative increase in the formation of LTB_4

(Dempsey *et al.* 1986). Thus it seems a likely event that in prolonged ischemia, edema was aggravated by indomethacin administration because of an increased synthesis of 5-lipoxygenase products. On the other hand, recirculation is followed by a temporary but explosive activation of the cyclooxygenase system. In addition to the possible effects of synthesized PGs, the oxygen free radical liberated from PGG_2 may

Table IV-4. Effects of indomethacin, ONO 3144, AVS, and a PGI_2 derivative, ONO 47241 on the cortical specific gravity following permanent ischemia (4 hours MCA occlusion: upper table) and recirculation (2 hours MCA occlusion followed by 2 hours recirculation: lower table). The cortical samples were assembled according to the clip-flow ranges of 0–15 and 15–30 ml/100 g/minute

PERMANENT ISCHEMIA(PI)GROUP

ICBF (ml/100g/min)	0–15	15–30
Saline	$1.0353 \pm 4 \times 10^{-4}$ (24)	$1.0382 \pm 7 \times 10^{-4}$ (26)
Indomethacin	1.0353 ± 5 (20)	1.0357 ± 8 (16)[a]
ONO-3144	1.0368 ± 7 (19)	1.0398 ± 8 (11)
ONO-47241	1.0373 ± 6 (13)[a]	1.0394 ± 6 (21)
OKY-1581		
AVS	1.0400 ± 4 (22)[c]	1.0433 ± 9 (13)[c]

RECIRCULATION(RC)GROUP

ICBF (ml/100g/min)	0–15	15–30
Saline	$1.0318 \pm 9 \times 10^{-4}$ (19)	$1.0365 \pm 12 \times 10^{-4}$ (18)
Indomethacin	1.0360 ± 10 (23)[b]	1.0382 ± 12 (20)
ONO-3144	1.0355 ± 7 (29)[b]	1.0381 ± 11 (16)
ONO-47241		
OKY-1581	1.0342 ± 13 (14)	1.0403 ± 10 (11)[a]
AVS	1.0346 ± 9 (19)[a]	1.0412 ± 1 (13)[b]

Statistical evaluation by Mann-Whitney's U-test.
a: $p < 0.05$, b: $p < 0.01$, c: $p < 0.001$.

exert deleterious influences on cerebral vasculature and parenchyma. In the same experimental condition using the cat MCA occlusion-recirculation model, a free radical scavenger, *i.e.*, AVS, significantly ameliorated post-ischemic hypoperfusion and edema formation (Asano *et al.* 1984 a, 1985 d). Further, a new antiinflammatory drug ONO-3144 which is known to stimulate PG hydroperoxidase activity presumably through quenching oxygen radicals liberated from PGG_2 (Aishita *et al.* 1983) also showed a similar beneficial effect (unpublished data). It is of interest that those three drugs which act on different sites of the arachidonate cascade **(Fig. IV-25)** showed similar

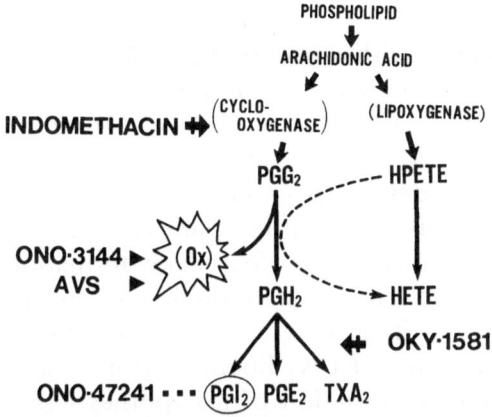

PHOSPHOLIPID

ARACHIDONIC ACID

INDOMETHACIN ➡ (CYCLO-OXYGENASE)   (LIPOXYGENASE)

PGG₂ --- HPETE

ONO-3144 ▶ (Ox)
AVS ▶   PGH₂ ----▶ HETE

⬦ OKY-1581

ONO-47241 ··· (PGI₂) PGE₂ TXA₂

Fig. IV-25. The site of action of indomethacin, ONO 3144, AVS, OKY 1581, or ONO 47241. Although each drug affects the cascade in different fashions, indomethacin, ONO 3144, and AVS have a common effect to decrease the level of oxidant (Ox) released from PGG_2. ⊣: inhibition of the enzyme activity; ▲: scavenging of free radicals

beneficial effects on postischemic CBF and edema formation (**Table IV-4**). The only action common to these drugs is to decrease the level of ambient oxygen free radical. So far as the condition following recirculation is concerned, therefore, it may be surmised that the oxygen free radical produced by the hydroperoxidase reaction from PGG_2 injure the cerebral microvasculature and parenchymal tissues, accelerating the development of brain edema. Nevertheless, the unbalanced formation of PGI_2 and TxA_2 may play some roles, since OKY-1581 (a TxA_2 synthase inhibitor), and ONO-47241 (a PGI_2 mimetic) exerted some beneficial effects on recirculation and permanent ischemia, respectively (**Table IV-4**).

3. Pathomechanisms Underlying Ischemic Brain Edema

The mechanisms involved in regulation of cellular volume and ion compositions in intra- and extracellular fluids can be represented as a "pump-leak" system in its simplest form. According to this paradigm, cell swelling as observed in ischemia is explained on the basis of two apparently independent mechanisms. The first mechanism involved in cell swelling is pump failure which is due to a decreased ATP synthesis in mitochondria or to a decreased activity of Na^+, K^+ ATPase. Since functions of both mitochondria and Na^+, K^+ ATPase depend on the integrity of biomembranes, lipid peroxidation primarily concerned with membrane damage is an attractive hypothetical mechanism in this respect. The second mechanism involved in cellular swel-

ling, *i.e.*, the sodium "leak", has remained elusive. Na^+/H^+ antiport system, the existence of which in neurons and glial cells have recently been demonstrated (Kimelberg and Bourke 1984), may provide such a leak pathway (Benos 1982, Siesjö 1985). Nevertheless, we shall not discuss on this subject here, since its relation to lipid metabolism has not so far been known.

In the final section of this chapter, we shall discuss on the possible third mechanism which may be involved not in glial swelling, but in the occurrence of brain edema. In the previous section, we showed the possibility that Na^+ transport across the BBB mediated by the capillary Na^+, K^+ ATPase might be responsible for edema formation due to regional ischemia. In this line of thinking, we have investigated on the underlying chemical mechanisms using brain microvessels. Hitherto obtained results indicate that regional ischemia is followed by an enhanced activity of arachidonate cascade in brain capillaries, which in turn is related to the increased activity of Na^+, K^+ ATPase. From our study emerged possible roles of oxygen free radicals and hydroperoxides (probably lipoxygenase products) as modulators of all the involved reactions. Pertinent findings as well as our current working hypothesis regarding the pathogenetic mechanism underlying ischemic brain edema will be described in some length.

a) Effect of Ischemia on Mitochondrial Function

Depletion of glucose and oxygen due to ischemia naturally causes cessation of mitochondrial function. When blood flow is restored, however, mitochondria may or may not resume function, the fate being dependent on the depth and duration of preceding ischemia. Since tissue viability is primarily determined by recovery of energy producing systems, the mechanism involved in mitochondrial damage pertains to the issue of reversibility of ischemic brain damage as well as to that of ischemic brain edema.

It has been known that FFAs act as uncouplers of oxidative phosphorylation in mitochondria (Lehninger and Remmert 1959, Wojtczak and Lehninger 1961, Borst et al. 1962), which is due to inhibition of the adenine nucleotide translocase (Wojtzcak 1976, Shrago

1978). Similar results have been obtained with mitochondria isolated from the brain (Ozawa et al. 1966, 1967, 1969, Sato et al. 1969, Lazarewicz et al. 1972). Observation of a significant breakdown of cerebral phospholipids following compression injury (Ishii et al. 1967), and enhanced in vitro edema formation in brain slices due to addition of "endogenous inhibitors", i.e., a crude preparation of FFAs liberated from brain, led to formation of a hypothesis that impairment of oxidative phosphorylation due to FFA accumulation may be associated with development of edema and other structural damages (Sato et al. 1969). In this theory, however, the particular actions of polyunsaturated fatty acids (Borst et al. 1962) were not discriminated from those of saturated fatty acids.

On the other hand, the deterioration of mitochondria under lipid peroxidation induced by ferrous ions, ascorbic acid, cystein, and glutathione was studied in Hunter's laboratory (Hunter et al. 1963, 1964 a, b, MacKnight and Hunter 1966). From these studies, the swelling and lysis of mitochondria was considered as the consequence of lipid peroxidation. Vladimirov et al. (1980) further showed that the uncoupling of oxidative phosphorylation is apparently a direct result of lipid peroxidation which may be explained by the increase in proton permeability of the membrane lipid bilayer and presumably also by the decrease of membrane electric stability. Treatment of isolated mitochondria with oxygen-radical-generating systems (hypoxanthine and xanthine oxidase in the presence of a suitable iron chelate) was shown to cause a severe inhibition of state 3 respiration (Hillered and Ernster 1983). In contrast to preceding reports, results of this experiment indicated that not lipid peroxidation but generation of hydroxyl radical is the mechanism responsible for mitochondrial damage.

Thus, regardless of the biochemical mechanism involved, there seems to be a good possibility that mitochondria are damaged following ischemia due to accumulation of PUFAs, possible generation of oxygen free radicals, and subsequent occurrence of lipid peroxidation. This possibility was supported by some studies which showed a pronounced reduction of state 3 respiration and RCR values in mitochondria isolated from brain exposed to transient ischemia of a brief duration (3–6 minutes) (Ozawa et al. 1967, Majewska et al. 1978). In contrast, more recent studies indicate that brain mitochondria may tolerate up to 60 minutes of *in vivo* ischemia without significant irreversible impairment of respiration (Schutz et al. 1973, Ginsberg et al. 1977, Rehncrona et al. 1979). This discrepancy has been attributed to the difference in techniques used to measure mitochondrial function *in vitro* (Siesjö 1981). Further it has been shown that mitochondria isolated after 30 minutes of recirculation following complete ischemia recovered and exhibited totally normal activities, whereas those isolated at the same period following incomplete ischemia showed further deterioration in their function (Rehncrona et al. 1979). These data were in good agreement with the previous finding that brain energy charge fully recovered after complete ischemia

but failed to recover after incomplete ischemia (Nordström et al. 1978). With respect to the possible effect of lactic acidosis in these experimental conditions, Hillered et al. (1985) showed that lactate levels reaching as much as 20 µmol/g does not adversely affect early postischemic recovery of mitochondrial function.

Although recent findings indicate that respiration of mitochondria is relatively resistant to irreversible impairment following cerebral ischemia, the markedly adverse influence of incomplete ischemia on mitochondrial function needs explanation. This problem, which was initially put forward by Siesjö (1978, 1981) and has not yet been resolved, still points to the possibility that the trickling oxygen supply may have a pathogenetic role in relation to free radical mechanisms.

In addition to the possible damage confined in the mitochondrial electron transport system, a relative substrate deficiency due to derangement in the TCA cycle has also been shown to occur following recirculation (Duckrow et al. 1981, Kogure et al. 1985). Such a substrate deficiency on recirculation causes an overoxidized state of the mitochondrial electron transport system, and may lead to generation of oxygen free radicals due to incomplete reduction of oxygen molecules (Kogure et al. 1985).

b) Membrane Damage Caused by Free Radical Injury: Disparity Between Results Obtained *in vitro* and *in vivo*

The presumption that lipid peroxidation injures respiratory function of

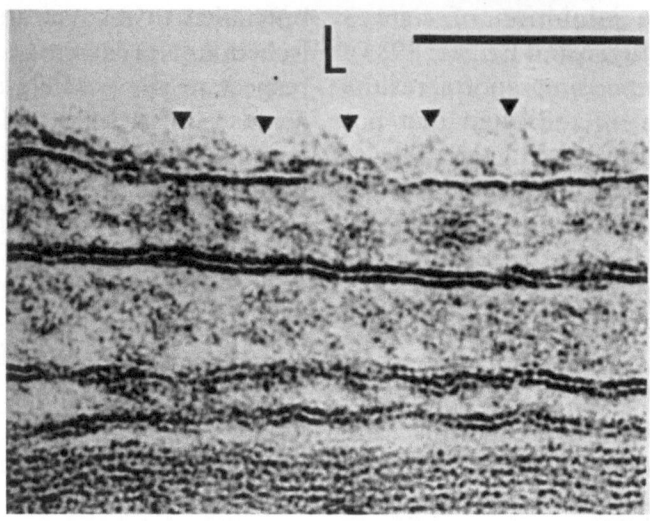

Fig. IV-26. Destruction of the outer leaflet of the endothelial membrane following intracerebral injection of arachidonic acid (indicated by arrows). The horizontal bar is $0.1\,\mu$. *L*: lumen

mitochondria has been well supported by *in vitro* experiments, but *in vivo* studies showed rather discrepant results. This situation also applies to the issue concerning membrane damage caused by lipid peroxidation.

So far as *in vitro* studies are concerned, that Na^+, K^+ ATPase activity in brain slices or synaptosomes is inhibited by lipid peroxidation has been well substantiated (Sun 1972, Kovachichi and Mishra 1980, 1981). Chan and Fishman have shown that addition of PUFAs in the incubation medium of rat brain slices enhances edema formation (1978), increases generation of O_2^- and TBAR (1980), and decreases the activity of Na^+, K^+ ATPase as well as the uptake of GABA and amino acids (Chan *et al.* 1983).

In this line of thinking, whether or not application of arachidonic acid or free radical generating system would cause edema has been examined. Intra-

cerebral injection of AA was shown to cause edema of vasogenic type (Wakai *et al.* 1982, Aritake *et al.* 1983, Chan *et al.* 1983). Chan *et al.* (1984) further showed that intracerebral injection of a superoxide-generating system (xanthine-xanthine oxidase system) cause brain injury characterized by a pronounced damage of endothelial cells. Electron microscopically, AA-induced endothelial damage had a peculiar feature that the outer leaflet of the luminal endothelial membrane was selectively destructed (**Fig. IV-26**: Wakai *et al.* 1982). Cristae of the mitochondria of the capillary endothelium in the central portion of brain edema were nearly lost. These pathological findings that AA and free radicals damage the capillary endothelium has been reinforced by results of *in vitro* studies using brain microvessels.

Levasseur *et al.* (1985) showed that AA and 15-HPETE in high concentrations

($200\,\mu g/ml$, $50\,\mu g/ml$ each) significantly reduced oxygen consumption in isolated brain microvessels. Challenge with the superoxide-generating system was shown to cause stimulation of phospholipase A_2 activity (Au et al. 1985). The very recent report of Lo and Betz (1986) is of particular interest. They showed that rubidium uptake of isolated rat brain capillaries incubated with a free radical-generating system was reduced 74%, whereas rubidium efflux, glucose transport, and capillary water space were unchanged. The reduction of rubidium uptake was prevented by catalase but not by other free radical scavengers such as SOD, ethanol, or mannitol. These results were interpreted as indicating that certain species of oxygen free radical specifically inhibited Na^+, K^+ ATPase, not through a generalized reduction of cell membrane integrity or transport systems.

Although the above studies showed that brain Na^+, K^+ ATPase activity is inhibited by AA or free radicals, results obtained with animal models of cerebral ischemia are not entirely consistent on this point. Schwarz et al. (1975) reported that in the gerbil cerebral cortex exposed to 1 hour unilateral ischemia, Na^+, K^+ ATPase activity did not change during ischemia but decreased on recirculation to 40–60% of that of control hemisphere by 5 hours. Mrsulja et al. (1980) showed that in gerbil forebrain exposed to 15 minutes transient ischemia due to bilateral carotid clamping, parenchymal Na^+, K^+ ATPase showed reduction only after recirculation, whereas capillary Na^+, K^+ ATPase

was significantly reduced at the end of ischemia and rapidly recovered on recirculation.

Using the same gerbil model of transient (15, 30, 60 minutes), incomplete, global ischemia, Enseleit et al. (1984) studied phospholipid content and Na^+, K^+ ATPase activity. Consistent to the results of Schwartz et al. (1975) and Mrsjulja et al. (1980), changes in the Na^+, K^+ ATPase activity of cerebral homogenates during ischemia were found small and, except for the 30 minute ischemic interval, statistically nonsignificant. The total cerebral phospholipid, phosphatidylcholine, phosphatidylserine, and phosphatidylethanolamine showed relatively small changes during ischemia. Phosphatidylinositol and phosphatidic acid markedly decreased. Recirculation resulted in a transient increase in cerebral phospholipid content ("rebound effect"), with the exception of phosphatidylserine. In parallel with the change in phospholipid content, reperfusion resulted in an increase to preischemic levels of cerebral Na^+, K^+ ATPase activity. This result is obviously discrepant from that of Mrsulja et al. (1980) which showed a transient reduction of the enzyme activity following recirculation.

Further, MacMillan (1982) reported that in the rat brain exposed to ischemia by four-vessel occlusion, the Na^+, K^+ ATPase activity showed a capability for enhanced activity (120–140% of control) during ischemia. Following recirculation, the enzyme activity showed phasic changes but never decreased below control. These authors further suggested that enhanced Na^+, K^+

ATPase activity was associated with hyperglycemia and hyperlactacidosis in the brain (MacMillan and Shankaran 1984).

In another brain ischemia model in rats, *i.e.*, bilateral carotid occlusion in spontaneously hypertensive rats, a time-dependent reduction of Na$^+$, K$^+$ activity was observed in both brain parenchyma and microvessels during ischemia (1, 3, 5 hours). Following recirculation, the enzyme activity in brain parenchyma showed no overshoot, but that in brain microvessels recovered rapidly and temporarily exceeded the control value (Fujita *et al.* 1985).

Thus, there are rather wide discrepancies between reported results concerning the ischemia-induced alterations of Na$^+$, K$^+$ ATPase activity in brain parenchyma or microvessels. A number of factors, such as the differences in species, depth and duration of ischemia, the kind of ischemia (regional or global), the method used to assay the enzyme activity, the kind of tissue fractions examined, and so on, would be responsible for these discrepancies. In addition, vasogenic edema induced by trauma (Demediuk *et al.* 1985) or a freezing lesion (Averet *et al.* 1984) has been more consistently accompanied by reduction of Na$^+$, K$^+$ ATPase activity. Although no generalization is possible, findings pertinent to lipid peroxidation and brain edema due to cerebral ischemia would merit consideration.

The finding common to all the above cited studies is that the Na$^+$, K$^+$ ATPase retains its activity in the early period of ischemia. Its reduction became significant only after several hours of ischemia (Palmer *et al.* 1985, Fujita *et al.* 1985). Whether the enzyme activity increases (Enseleit *et al.* 1984) or decreases (Schwartz *et al.* 1975, Mrsulja *et al.* 1980) following recirculation is the most contradictory point. As Enseleit *et al.* (1984) pointed out, changes in cerebral Na$^+$, K$^+$ ATPase activity may vary with different cerebral cell types. It seems possible that the increase in the activity of capillary Na$^+$, K$^+$ ATPase following recirculation (Mrsulja *et al.* 1980, Fujita *et al.* 1985) masked the decrease in that of parenchymal cells.

In any event, preservation or enhancement of the enzyme activity either during ischemia or following recirculation precludes the occurrence of membrane damage due to lipid peroxidation. The reported free radical damage of capillary endothelial cells *in vitro* (Lo and Betz 1986) and the increase in capillary Na$^+$, K$^+$ ATPase following recirculation (Mrsulja *et al.* 1980, Fujita *et al.* 1985) creates a most paradoxical situation. In all likelihood, generation of free radicals in the capillary endothelium would be enhanced on recirculation either through an increase in oxidation of hypoxanthine or in the activity of arachidonate cascade. Then, if the *in vitro* data of Lo and Betz (1986) apply to *in vivo* situation, a progressive reduction of capillary Na$^+$, K$^+$ ATPase activity should be observed. Obviously, this prediction is contrary to what was observed.

Regarding the relationship between the activity of Na$^+$, K$^+$ ATPase and brain edema *in vivo*, the report of MacMillan and Shankaran (1984) is most suggestive. They showed that upon recirculation following four vessel occlusion in rats, the cerebral Na^+, K^+ ATPase activity was increased in hypotensive hyperglycemic animals showing continuing massive lactacidosis, whereas it was normal or mildly suppressed in normotensive hyperglycemic animals showing progressive decreases · in lactate level. Hyperglycemic animals which show enhanced or normal ATPase activity show a high incidence of postischemic edema, whereas normoglycemic animals, which show a 30–40% decline in postischemic Na^+, K^+ ATPase activity, consistently show negligible or no evidence of edema on recirculation (Kalimo et al. 1981, Pulsinelli et al. 1982, MacMillan and Shankaran 1984). Recent in vitro study (Siesjö et al. 1985) indicates that acidosis triggers increased free radical formation. Thus the data of MacMillan and Shankaran (1984) are in apparent contradiction to the free radical hypothesis in dual meanings; in their study, acidosis led to an enhanced Na^+, K^+ ATPase activity, which was in turn associated with an increased edema formation.

By the same token, Mrsulja et al. (1985) recently showed a pronounced decrease of Na^+, K^+ ATPase activity (31% of control) in gerbil hippocampus 1 hour after 15 minutes of ischemia using the model of Kirino (1982). Since calculated brain swelling showed a slight increase, they interpreted the data as indicating the correlation between hypoactivity of Na^+, K^+ ATPase and ischemic brain edema. Nevertheless, in this model, the hippocampus which is determined to delayed neuronal death

lacks morphological evidence of any significant edema (Kirino, personal communication). Further in their result, the increase of brain water content in the hippocampus was not statistically significant (control $78.29 \pm 0.04 \rightarrow 78.99 \pm 0.04$).

Those paradoxical in vivo evidence as described above would demand a totally different interpretation of hitherto assumed relationships between lipid peroxidation, the activity of Na^+, K^+ ATPase, and ischemic brain edema.

c) Effect of a Lipid Hydroperoxide, 15-HPETE on the Eicosanoid Synthetic Capacity and Na^+, K^+ ATPase Activity of Brain Microvessels

The reasons why peroxidative mechanisms observed in vitro has not been verified by in vivo experiments seems to be that lipid peroxidation is especially liable to occur in incubated tissues (Barber and Bernheim 1967) and that very high concentrations of arachidonate or hydroperoxide (in millimolar concentrations) were employed to elicit significant changes in those experiments. Although generation of oxygen free radicals in living cells now seems unquestionable, multiple arrays of free radical scavenger systems prevent the in vivo occurrence of lipid peroxidation. In fact, there has not so far been any firm evidence showing the occurrence of lipid peroxidation in vivo in any pathological conditions (Halliwell and Gutteridge 1984 b). Further, the possible maximal concentrations of FFAs in the brain tissue exposed to ischemia scarcely reach the level of $10^{-6} M$

(Yoshida *et al.* 1980, Rehncrona *et al.* 1981). The tissue level of free AA is even much lower, since it is mostly bound to proteins or triglyceride as soon as it is liberated from membrane phospholipids. Therefore, the peroxidative damage observed in previous *in vitro* experiments regarding edema formation, superoxide or TBAR generation, or Na$^+$, K$^+$ ATPase activity, may be artefactual phenomena which are not likely to occur in *in vivo* conditions. As Halliwell and Gutteridge (1984 b) recently conjectured, it may be that the occurrence of lipid peroxidation represents not the cause but the result of irreversible tissue damage. If so, the implication of antioxidant therapy would be no more than to minimize the possible secondary adverse effects exerted from the necrotic focus. Parenthetically, lipid peroxidation in irreversibly damaged cells would evolve into termination of free radical reactions. The cell shrinkage with acidphilic cytoplasm observed in neuronal necrosis may represent such a process of termination reactions, *i.e.*, cross-linking, which may prevent intracellular molecules from breaking up into numerous small fragments and causing an enormous osmotic stress.

In contrast to lipid peroxidation, the enzymatic peroxidation of arachidonic acid, *i.e.*, the arachidonate cascade, represents a biologically controlled reactions. Presumably, the arachidonate cascade utilizes only free AA liberated in some particular compartments and its activity is regulated by a complex control system (Lands *et al.* 1984, Cutler 1984). Since it has been shown that liberation of AA and cerebral eico-

sanoid synthesis is enhanced following ischemia, investigation on the role of arachidonate cascade in distinction from nonenzymatic lipid peroxidation seems fully justified.

Regarding ischemic brain edema, we have already observed that the eicosanoid synthetic capacity of cerebral capillaries show a marked enhancement which parallels the development of brain edema following permanent MCA occlusion in rats (Gotoh *et al.* 1985, Asano *et al.* 1985, *cf.* **Fig. IV-22**). Oxygen radicals and hydroperoxides are known to enhance the activity of arachidonate cascade in general (Lands *et al.* 1984). Specifically in the canine basilar artery, 15-HPETE was shown to increase the release of HETEs (Koide *et al.* 1982, Shimizu *et al.* 1986). Therefore, we envisaged to investigate on the biochemical mechanisms involved in the alteration of eicosanoid synthesis in brain capillaries in relation to the pathogenetic mechanism underlying ischemic brain edema. As previously discussed, not cyclooxygenase but lipoxygenase products appear to be related to edema formation, so far as permanent ischemia is concerned. We have already shown that various HETEs increase in significant amounts in rat cerebral hemispheres exposed to permanent MCA occlusion (Usui *et al.* 1986, *cf.* **Table IV-3**).

Thus, we first examined the effect of 15-HPETE on the eicosanoid synthetic capacity of rat brain microvessels in comparison to the effects of monoamines and dibutyryl cyclic AMP (Asano *et al.* 1984 d, Koide *et al.* 1985). Serotonin, norepinephrine, dopamine, or dibutyryl cyclic AMP in high con-

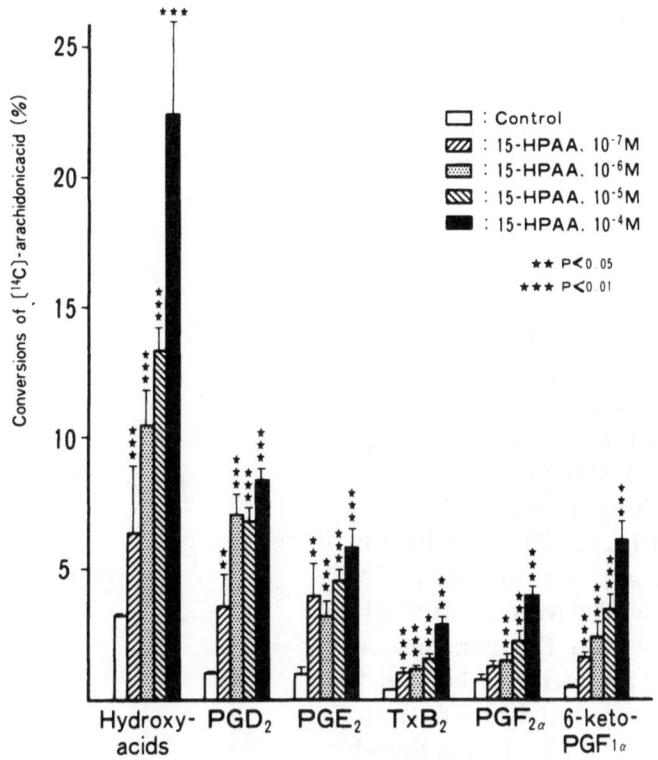

Fig. IV-27. 15-HPAA-induced stimulation of eicosanoid synthesis in rat brain microvessels (from Koide *et al.* 1985)

centrations $(10^{-3} \sim 10^{-4}\,\mathrm{M})$ did not show significant effects, except for a slight increase in PGD_2 synthesis. In contrast, 15-HPETE showed a remarkable stimulation of overall eicosanoid synthesis from low concentrations (**Fig. IV-27**). This enhanced eicosanoid synthesis by 15-HPETE was partially (only cyclooxygenase products) inhibited by indomethacin $(10^{-4}\,\mathrm{M})$, and totally inhibited by ETYA $(10^{-4}\,\mathrm{M})$. It was not influenced by SOD or mannitol, but was significantly suppressed by α-tocopherol, hydroquinone, or AVS. Thus it was surmised that singlet oxygen or hydroxyl radicals participate in the stimulatory effect of 15-HPETE.

Above results indicates that a trace amount of hydroperoxides or free radicals can stimulate the eicosanoid synthetic capacity of brain capillaries. The lowest concentration of 15-HPETE with a significant effect was $10^{-8}\,\mathrm{M}$ in this study, which is in the range of brain levels of HETEs observed in ischemic rat brain (*cf.* **Table IV-3**). Thus, our result is in conformity with the presumed role of ambient hydroperoxides in the regulation of the activity of arachidonate cascade (Lands *et al.* 1984), and indicates that lipid hydroperoxides or other free radicals generated in the vicinity of brain capillaries would greatly amplify the eicosanoid synthesis. It may be worthy of mentioning that a similar stimulatory effect of 15-HPETE on lipoxygenase activity was observed also in the canine basilar artery (Shimizu *et al.* 1986, **Fig. IV-14**). Then, what role would the increased eicosanoid synthesis play in edema formation?

Based on our previous finding that the unipolar transport of sodium across the BBB by capillary Na$^+$, K$^+$ ATPase is involved in edema formation (Asano *et al.* 1985 c, Shigeno *et al.* 1986), we next examined the effect of 15-HPETE on capillary and synaptosomal Na$^+$, K$^+$ ATPase activity (Koide *et al.* 1986). As shown in **Fig. IV-28**, 15-HPETE stimulated the activity of Na$^+$, K$^+$ ATPase from 1×10^{-8} M to 1×10^{-5} M. In synaptosomes, on the contrary, a dose-dependent inhibition of Na$^+$, K$^+$ ATPase was shown, consistent to findings of Chan *et al.* (1983). It was found that the stimulatory effect of 15-HPETE on capillary Na$^+$, K$^+$ ATPase was not due to a simple detergent-like action of the compound on the membranes. Further, this action of 15-HPETE was shown to be suppressed by a phospholipase A$_2$ inhibitor, quinacrine, lipoxygenase inhibitors, ETYA and caffeic acid, but not by indomethacin. Among free radical scavengers, α-tocopherol, hydroquinone, and AVS significantly reduced the stimulatory action of 15-HPETE, whereas mannitol, sodium benzoate, SOD, catalase, or SOD plus catalase did not. From these results, it was concluded that 15-HPETE stimulated capillary Na$^+$, K$^+$ ATPase through stimulation of phospholipase A$_2$ and lipoxygenase activities within the capillaries. It should be emphasized that relatively

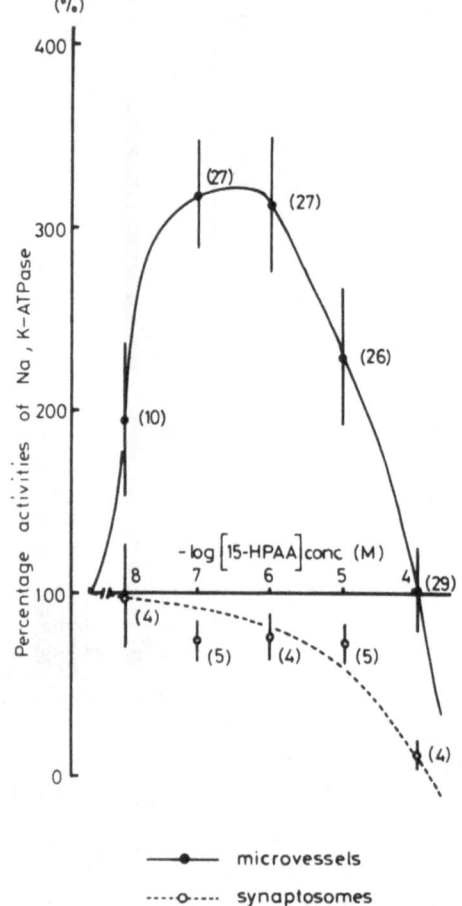

Fig. IV-28. Effect of 15-HPAA on the activity of Na$^+$, K$^+$ ATPase of brain microvessels and synaptosomes (from Koide *et al.* 1986)

low concentractions were required for the compound to stimulate the enzyme activity. In concentrations higher than 10^{-6} M, the once enhanced enzyme activity was inhibited.

In summary of this section, our studies show that hydroperoxides in low concentrations (10^{-8} M $\sim 10^{-6}$ M) cause a pronounced stimulation of the eicosanoid synthesis in brain capillaries, and that not the cyclooxygenase but the

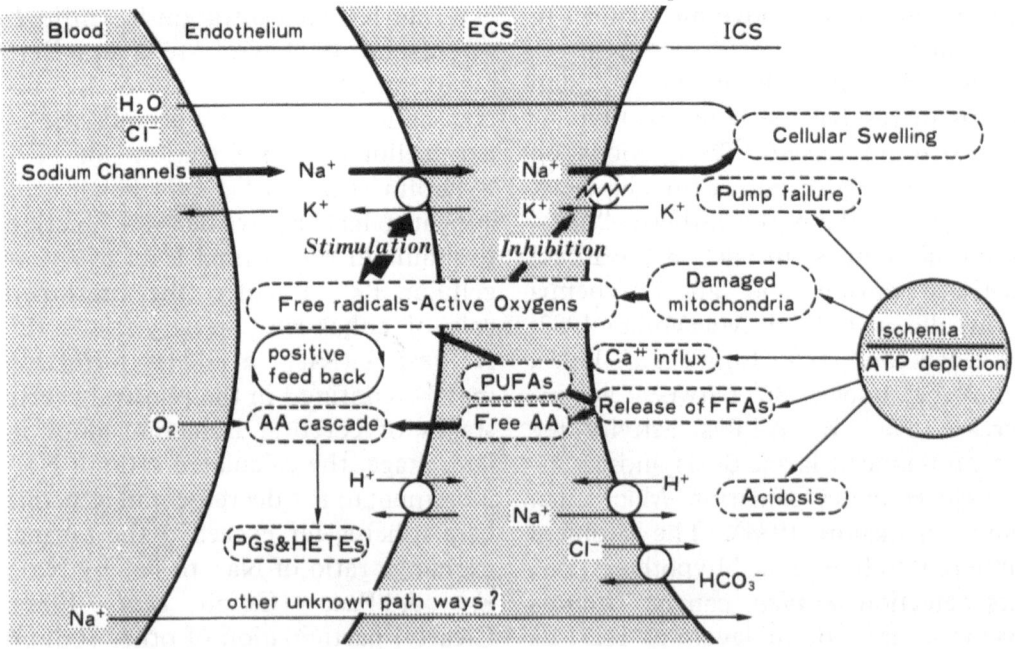

Fig. IV-29. Suggested mechanism of sodium influx in ischemic brain edema

lipoxygenase products enhance the activity of capillary Na$^+$, K$^+$ ATPase. The relevance of these mechanisms to edema formation *in vivo* is supported by the facts that the free radical scavenger AVS suppresses the stimulatory actions of 15-HPETE on the eicosanoid synthetic activity as well as on the Na$^+$, K$^+$ ATPase activity of brain microvessels *in vitro* (Koide *et al.* 1985, 1986), and that the drug suppresses the development of ischemic brain edema (Gotoh *et al.* 1984, Asano *et al.* 1984 a, Asano *et al.* 1985 d).

d) The Suggested Pathomechanism of Ischemic Brain Edema and Concluding Remarks

We may summarize our current knowledge about the effects of free radicals and eicosanoids on capillary endothelium by means of a paradigm as shown in **Fig. IV-29**. In essence, this paradigm which is presently meant to be a working hypothesis, is constructed form pieces of evidence showing that 1. endothelial cells have a better chance of survival than parenchymal cells in incomplete ischemia, 2. the unipolar sodium transport across the BBB participates in edema formation, and 3. stimulation of the lipoxygenase pathway by free radicals leads to enhanced activity of Na$^+$, K$^+$ ATPase of capillaries. Since our paradigm presupposes the viability of endothelial cells, it is only relevant to edema formation due to "incomplete" or "regional" ischemia, or to edema aggravation upon recirculation. In regional ischemia, the perifocal zone would provide the optimal condition for the above mechanism to operate.

As a working hypothesis, the paradigm helps to resolve the contradictions as revealed by recent investigations on lipid peroxidation and edema. Namely, it clearly explains the hitherto "paradoxical" increase in Na^+, K^+ ATPase activity during and after ischemia (Mrsulja et al. 1980, MacMillan 1982, Enseleit et al. 1984, Fujita et al. 1985), and the association between an increased Na^+, K^+ ATPase activity due to pronounced lactacidosis and an increase in edema formation (MacMillan and Shankaran 1984). The problem inherent to free radical hypothesis, i.e., its detection in vivo, can be circumvented, since brain levels of HETEs which are considered to reflect the reciprocal relationship between the ambient levels of free radicals and intrinsic free radical scavengers can be quantitated.

It should be emphasized that our paradigm is not in opposition to recently demonstrated ion exchange systems in neuronal and glial cell membranes. On the contrary, the concept of unipolar sodium transport would provide an essential link to the movement of sodium and water from blood to the extracellular fluid, and from there to the intracellular fluid (Siesjö 1985).

Insomuch as the stimulatory effect of hydroperoxides on Na^+, K^+ ATPase activity is highlighted in the diagram, it remains to be seen to what extent this mechanism is responsible for the enzyme activation. Since it has been shown that the ouabain-sensitive uptake of K^+ in isolated brain microvessel is half saturated when the external K^+ concentration is $3 \, mEq/L$ (Goldstein 1979, Goldstein and Betz 1983), and that the uptake of K^+ increases in parallel with an increase in extracellular concentration of K^+ (Chaplin et al. 1981), endothelial cells appear to have a physiological function to maintain the normal level of extracellular K^+. Further, the presumed role of enhanced capillary Na^+, K^+ ATPase activity in edema formation may be confined in the relatively early stages of cerebral ischemia, since in later stage, the calculated ratio of Na^+ increment to K^+ decrement in ischemic hemispheres far exceeded the normal exchange ratio of Na^+ to K^+ by Na^+, K^+ ATPase (Gotoh et al. 1985). Clearly, participation of other sodium channels must be sought.

It has been claimed that the occurrence of edema per se may represent an adaptive response of the brain to the ischemic insult, which provides a fluid medium for reparative activities of astrocytes and macrophages (Ikuta et al. 1983). The paradigm shown here is in apparent conformity with this contention. Like glial cells which act as "the silent partners of the working brain" (Plum 1986), endothelial cells may act as "the silent patrons" of the brain. Further, the adaptive or reparative responses of brain tissue to ischemia may not be confined to activities of glial and endothelial cells. A variety of biological responses which are inflammatory in nature, such as alterations in the permeability of the BBB to plasma proteins and biologically active compounds in the serum, blood flow changes due to vasoconstriction or vasodilation, adhesion and aggregation of platelets, and migration of leukocytes or monocytes, take place following an ischemic insult. Since some eicosanoids, particularly lipoxygenase products, are known to elicit or participate in those reactions, the fact that the activity of arachidonate cascade in the brain is stimulated by an increased level of ambient free radicals would assume particular significance from the therapeutic point of view.

Provided that all the reactions of the brain tissue to ischemic insults are reparative in nature, a complete suppression of the reactions, if possible at all, might be detrimental rather than beneficial. Obviously, the usefulness of a given drug which interferes with those reactions should be assessed from multifarious aspects, not from its effect on single experimental parameter. So far as animal experiments are concerned, the free radical scavenger, AVS significantly ameliorated cerebral vasospasm following SAH and ischemic brain edema. The ultimate judgement as to the validity and the benefit of such an approach, however, must be formed through future clinical trials. For the time being, we only expect that our past research efforts would provide certain breakthrough to unravel the pathogenetic mechanisms underlying cerebral vasospasm and edema, which are the major causes of AINDs and DINDs, and contribute to improve the outcome of SAH patients.

V. GRADING OF RISK, ANGIOGRAPHY AND COMPUTERIZED TOMOGRAPHY

A. Grading Systems

Despite the advances in investigation techniques, such as computerized tomography (CT), etc, the management of individual patients with a subarachnoid hemorrhage (SAH) due to an aneurysm rupture still depends greatly on the assessment of the patient's clinical condition. Several systems for grading patients with SAH have been proposed. Among them, the Botterell grading system (Botterell *et al.* 1956), the Nishioka grading system (Nishioka 1966), and the Hunt grading system (Hunt and Hess 1968, Hunt and Kosnik 1974) have been widely used in many neurosurgical centers of the world. Botterell *et al.* graded their cases as to operative risk from grade 1 to 5.

Botterell grading system (1956):

Grade 1 indicates a patient who was conscious with or without signs of blood in the subarachnoid space.

Grade 2 indicates a drowsy patient without significant neurological deficit.

Grade 3 represents the drowsy patient with a neurological deficit and probably an intracerebral clot.

Grade 4 includes patients with a major neurological deficit and deteriorating because of large intracerebral clots or older patients with less severe neuro-

logical deficit but preexisting degenerative cerebrovascular disease.

Grade 5 represents a moribund or near moribund patient with failing vital centers and extensor rigidity.

In the report on the cooperative study of intracranial aneurysms and SAH, Nishioka grading system was used (Nishioka 1966).

Nishioka grading system (1966):

Grade I—symptom-free: completely recovered from the effects of the last hemorrhage.

Grade II—minimally ill: complaining of headache but alert and responsive; no major neurologic deficit.

Grade III—moderately ill: a) lethargic with headache, neck stiffness but without hemispheric neurologic deficit; b) alert, recovered from the general effects of subarachnoid hemorrhage but having a hemispheric neurologic deficit.

Grade IV—seriously ill: a) severely obtunded without major neurologic deficit; b) lethargic or poorly responsive with hemispheric deficit (hemiparesis, dysphasia, mental confusion).

Grade V—moribund: decerebrate or unresponsive to all stimuli.

In the Nishioka grading system, which

is based upon the Botterell grading system, patients were graded according to the severity of illness. While it may differ in some details from the original Botterell system, this grading is more clearly defined.

Hunt and Hess (1968) described their modification of the Botterell classification. They graded patients according to the intensity of the meningeal inflammatory reaction, the severity of the neurological deficit, and the presence or absence of a significant associated disease.

Hunt and Hess system (1968):

Grade I—asymptomatic, or minimal headache and slight nuchal rigidity.

Grade II—moderate to severe headache, nuchal rigidity, no neurologic deficit other than cranial nerve palsy.

Grade III—drowsiness, confusion, or mild focal deficit.

Grade IV—stupor, moderate to severe hemiparesis, possibly early decerebrate rigidity and vegetative disturbances.

Grade V—deep coma, decerebrate rigidity, moribund appearance.

Note: Serious systemic disease such as hypertension, diabetes, severe arterioslcerosis, chronic pulmonary disease, and severe vasospasm seen on arteriography, result in placement of the patient in the next less favorable category.

In 1974, this grading was revised by Hunt or Hunt and Kosnik (Hunt and Kosnik 1974). In this revision the following grades 0 and I a were added, namely:

Grade 0—unruptured aneurysm.

Grade I a—no acute meningeal or brain reaction, but with fixed neurologic deficit.

Of these three grading systems, the Hunt grading system is probably the most popular. Recently, Lindsay et al. (1982 a, b, 1983) pointed out observer variability in the use of these grading systems. They suggested that the major source of variability in assessing patients with SAH is the difficulty in matching the features actually seen in clinical practice with the theoretical, idealized levels that constitute these grading systems. Namely, the cause of observer variability in grading might be vagueness of the level of consciousness, such as, drowsiness, confusion, stupor, etc.

The terms used to describe the level of consciousness in these systems were found to be significantly less consistent than the Glasgow coma scale (Jennett and Teasdale 1977, 1981, Teasdale et al. 1979), although Hunt pointed out that the Glasgow coma scale does not speak to memory loss or other focal deficits without "coma" (Hunt 1983). Using the Glasgow coma scale, the Glasgow group proposed the following grading (Teasdale et al. 1983, 1985).

Glasgow proposal:

Grade I—GCS score 15, without headache or stiff neck.

Grade II—GCS score 15, with headache or stiff neck.

Grade III—GCS score 13–14.

Grade IV—a) GCS 13–14, with severe neurological deficit. b) GCS 9–12.

Grade V—GCS score 3–8.

Note: A Glasgow coma scale (GCS) score of 15 indicates spontaneous eye-opening, the patient is oriented and obeys commands; 13–14 denotes eye-open and/or the patient is confused; and 8 indicates no eye-opening, no comprehensive words, and the patient does not obey commands.

Glasgow coma scale (GCS):

Eye-opening (E):

spontaneous	E 4
to speech	3
to pain	2
nil	1

Best motor response (M):

obeys	M 6
localizes	5
withdraws	4
abnormal flexion	3
extensor response	2
nil	1

Verbal response (V):

orientated	V 5
confused conversation	4
inappropriate words	3
incomprehensible sounds	2
nil	1

Coma score $(E + M + V) = 3$ to 15

Sano used a modified grading system based on the Hunt system using the Glasgow coma scale (GCS) (Jennett and Teasdale 1977, 1981) and the Japan coma scale (JCS) (Ohta et al. 1986) for grading of impaired consciousness.

Sano grading proposal:

Grade I—GCS score 15, (JCS score 0–1). Neurologically intact except for minimal headache or cranial nerve palsy.

Grade II—GCS score 15 (JCS score 0–1). With headache, neck stiffness, or both. Otherwise neurologically intact except for cranial nerve palsy.

Grade III—GCS score 13–14 (JCS score 2–10). a) without focal neurological deficit, b) with focal neurological deficit.

Grade IV—GCS score 7–12 (JCS score 20–100). With or without focal neurological deficit.

Grade V—GCS score 3–6 (JCS score 200–300). Unresponsive coma with or without abnormal posturing.

Note: Japan coma scale (JCS) for grading of impaired consciousness:

1-digit code: the patient is awake without any stimuli, and is;

1 almost fully conscious;

2 unable to recognize time, place, and person;

3 unable to recall name or date of birth.

2-digit code: the patient can be aroused (then reverts to previous state after cessation of stimulation);

10 easily, by being spoken to (or is responsive with purposeful movements, phrases, or words);

20 with loud voice or shaking of shoulders (or is almost always responsive to very simple words like yes or no, or to movements);

30 only by repeated mechanical stimuli.

3-digit code: the patient cannot be aroused with any forceful mechanical stimuli, and;

100 responds with movements to avoid the stimuli;

200 responds with slight movements including decerebrate and decorticate posture;

300 does not respond at all except for change of respiratory rhythm.

As the result of discussions with Japanese neurosurgeons and Professor Teasdale, the following proposal for grading patients with SAH due to aneurysm rupture was made by us (Sano and Tamura 1985).

Grade I—GCS score 15, without headache or neck stiffness (JCS score 0–1).

Grade II—GCS score 15, with headache, neck stiffness, or both (JCS score 0–1).
Grade III—GCS score 13–14 (JCS score 2–10).
Grade IV—GCS score 8–12 (JCS score 20–30).
Grade V—GCS score 3–7 (JCS score 100–300).
Note: GCS 13–14: eye open to sound and/or confused; GCS 3–7: no eye opening, no verbal response, does not obey commands.
However, the Sano grading system is more practical. Because, if the state in grade V means unresponsive coma or moribund, a GCS score of 7 should be assigned to grade IV.
In this proposed grading system, the Glasgow coma scale (GCS) score is used to express the level of consciousness because of its internationality.

Even though Japanese neurosurgeons are more familiar with the so-called the Japan coma scale (JCS), the JCS is well correlated with the GCS. **Tables V-1** and **V-2** show a comparison between the GCS score and the JCS score of patients with SAH at admission or just before an operation. Both scores are well correlated, although some differences are seen in grade IV patients because of speech disturbance. A problem of using the Glasgow coma scale is that in common practice three different features, such as eye opening, motor response, and verbal performance, are independently observed and summed, although the Glasgow group does not always recommend to use of such a summed score. With regard to this problem, to grade patients with SAH in an acute state, the Japan coma scale is the more useful.

Table V-1. Status on admission

Grade (JCS)		15	14	13	12	11	10	9	8	7	6	5	4	3	Total
	Grade	I, II	III		IV					V					
I, II	0	65	0	0	0	0	0	0	0	0	0	0	0	0	65 (23.7%)
	1	55	2	0	0	0	0	0	0	0	0	0	0	0	57 (20.8)
III	2	0	24	0	0	0	0	0	0	0	0	0	0	0	24 (8.8)
	3	0	10 III	3	1	0	1	0	0	0	0	0	0	0	15 (5.5)
	10	0	12	15	1	2	3	0	0	0	0	0	0	0	33 (12.0)
IV	20	0	0	3	2	3	0	1	0	0	0	0	0	0	9 (3.3)
	30	0	0	0	1	0	3	3	2	0	0	0	0	0	9 (3.3)
V	100	0	0	0	0	0	0	0	1	11	4	1	0	0	17 (6.2)
	200	0	0	0	0	0	0	0	1	2	6 V	7	5	0	21 (7.7)
	300	0	0	0	0	0	0	0	0	0	0	2	3	19	24 (8.8)
Total		120 (43.8%)	48 (17.5)	21 (7.7)	5 (1.8)	5 (1.8)	7 (2.6)	4 (1.5)	4 (1.5)	13 (4.7)	10 (3.6)	10 (3.6)	8 (2.9)	19 (6.9)	274

Table V-2. Preoperative status

		Glasgow Coma Scale (GCS)													Total
		15	14	13	12	11	10	9	8	7	6	5	4	3	
	Grade	I, II	III		IV					V					
Grade	(JCS)														
I, II	0	66	0	0	0	0	0	0	0	0	0	0	0	0	66 (30.8%)
	1	39	2	0	0	0	0	0	0	0	0	0	0	0	41 (19.2)
	2	0	19	1	0	0	0	0	0	0	0	0	0	0	20 (9.3)
III	3	0	11 III 2		1	1	1	0	0	0	0	0	0	0	16 (7.5)
	10	0	8	10	4	1	2	0	0	0	0	0	0	0	25 (11.7)
IV	20	0	0	3	2	1	2	3	1	0	0	0	0	0	12 (5.6)
	30	0	0	0	0	0 IV	1	1	3	0	0	0	0	0	5 (2.3)
	100	0	0	0	0	0	1	2	1	8	1	0	0	0	13 (6.1)
V	200	0	0	0	0	0	0	0	1	0	1 V 1		5	1	9 (4.2)
	300	0	0	0	0	0	0	0	0	0	0	2	0	5	7 (3.3)
Total		105 (49.1%)	40 (18.7)	16 (7.5)	7 (3.3)	3 (1.4)	7 (3.3)	6 (2.8)	6 (2.8)	8 (3.7)	2 (0.9)	3 (1.4)	5 (2.3)	6 (2.8)	214

B. Grading and Outcome

The operative results of our previous series (Sano 1985), which is classified according to Hunt's grading (grade I a is included in grade I), clearly show the lower the grade, the lower the operative mortality (*e.g.*, 1.9% in grade I) and the better the follow-up results (*e.g.*, 88.6% of grade I patients were able to work). As a whole, 75.9% of the aneurysm patients who underwent surgery were well and working a follow-up of from one to 14 years, 10.0% were able to care for themselves (*i.e.*, the patient needs no help to wash, eat, and go to the toilet, but is unable to work), and 6.6% were either bed-ridden or died of other diseases (see Chapter VI).

Since 1981, we have been using the latest proposed grading system as shown above which uses both the GCS score and the JCS score. The Glasgow outcome scale (GOS) is used to assess patients clinical results (Jenett and Bond 1975).

A total of 272 patients of SAH due to aneurysmal rupture were admitted to the Teikyo University Hospital since April, 1981 through December, 1985. In these patients, direct operation was performed in 214 cases (79%). More than half of the remaining nonoperated patients were admitted in a state of deep coma (grade V) and all of them died soon after admission. In 28% of non-surgical cases, fatal rebleeding occurred before surgery and 5% showed severe preoperative vasospasm. The relationship between the day of admission and the day of operation is shown in **Table V-3**. In our clinic, 67% of SAH patients were hospitalized within 3 days and early operation within one week

Table V-3

Day of operation	Day of admission									Total
	0	1	2	3	4	5	6	7–14	15–	
0	34	1	0	0	0	0	1	0	0	36 (16.8%)
1	32	11	0	1	0	0	1	0	0	45 (21.0%)
2	4	14	5	0	0	0	0	0	0	23 (10.7%)
3	2	3	4	0	0	0	0	0	0	9 (4.2%)
4	1	2	1	2	1	0	0	0	0	7 (3.3%)
5	0	0	0	0	1	1	0	0	0	2 (0.9%)
6	1	0	0	0	0	2	0	0	0	3 (1.4%)
7–14	4	4	2	4	2	2	1	10	0	29 (13.6%)
15–	8	5	5	3	0	2	1	18	18	60 (28.0%)
Total	86 (40.2%)	40 (18.7%)	17 (7.9%)	10 (4.7%)	4 (1.9%)	7 (3.3%)	4 (1.9%)	28 (13.1%)	18 (8.4%)	214

Table V-4. Grade on admission and outcome (6 months)

	Grade I GCS 15	Grade II GCS 15	Grade III GCS 13–14	Grade IV GCS 8–12	Grade V GCS 3–7	Total
Good recovery	20 (76.9%)	67 (71.2%)	37 (53.6%)	6 (24.0%)	2 (3.4%)	132 (48.5%)
Moderate disability	3 (11.5%)	9 (9.6%)	11 (15.9%)	4 (16.0%)	2 (3.4%)	29 (10.7%)
Severe disability	1 (3.8%)	5 (5.3%)	4 (5.8%)	7 (28.0%)	3 (5.2%)	20 (7.3%)
Vegetative state	1 (3.8%)	2 (2.1%)	3 (4.3%)	0	2 (3.4%)	8 (2.9%)
Dead	1 (3.8%)	11 (11.7%)	14 (20.3%)	8 (32.0%)	49 (84.5%)	83 (30.5%)
Total	26 (9.6%)	94 (34.6%)	69 (25.4%)	25 (9.2%)	58 (21.3%)	272

Table V-5. Preoperative grade and outcome (6 months)

	Grade I GCS 15	Grade II GCS 15	Grade III GCS 13–14	Grade IV GCS 8–12	Grade V GCS 3–7	Total
Good recovery	37 (86.0%)	47 (75.8%)	35 (62.5%)	5 (17.2%)	3 (12.5%)	127 (59.3%)
Moderate disability	4 (9.3%)	6 (9.7%)	8 (14.3%)	9 (31.0%)	2 (8.3%)	29 (13.6%)
Severe disability	1 (2.3%)	4 (6.5%)	5 (8.9%)	8 (27.6%)	2 (8.3%)	20 (9.3%)
Vegetative state	0	1 (1.6%)	1 (1.8%)	2 (6.9%)	2 (8.3%)	6 (2.8%)
Dead	1 (2.3%)	4 (6.5%)	7 (12.5%)	5 (17.2%)	15 (62.5%)	32 (15.0%)
Total	43 (20.1%)	62 (29.0%)	56 (26.2%)	29 (13.6%)	24 (11.2%)	214

Table V-6. Preoperative grade and outcome (6 months) in cases submitted to surgery in the first week (day 0–7)

	Grade I GCS 15	Grade II GCS 15	Grade III GCS 13–14	Grade IV GCS 8–12	Grade V GCS 3–7	Total
Good recovery	8 (100%)	36 (72.0%)	18 (56.3%)	2 (11.8%)	3 (13.0%)	67 (51.5%)
Moderate disability	0	6 (12.0%)	5 (15.6%)	6 (35.3%)	2 (8.7%)	19 (14.6%)
Severe disability	0	4 (8.0%)	3 (9.4%)	6 (35.3%)	2 (8.7%)	15 (11.5%)
Vegetative state	0	1 (2.0%)	1 (3.1%)	0	2 (8.7%)	4 (3.1%)
Dead	0	3 (6.0%)	5 (15.6%)	3 (17.6%)	14 (60.9%)	25 (19.2%)
Total	8 (6.2%)	50 (38.5%)	32 (24.6%)	17 (13.1%)	23 (17.7%)	130

Table V-7. Preoperative grade and outcome (6 months) in cases submitted to surgery in the second week (day 8–14)

	Grade I GCS 15	Grade II GCS 15	Grade III GCS 13–14	Grade IV GCS 8–12	Grade V GCS 3–7	Total
Good recovery	3 (60.0%)	7 (87.5%)	6 (85.7%)	2 (50.0%)	0	18 (75.0%)
Moderate disability	1 (20.0%)	0	0	0	0	1 (4.2%)
Severe disability	0	0	1 (14.3%)	1 (25.0%)	0	2 (8.3%)
Vegetative state	0	0	0	1 (25.0%)	0	1 (4.2%)
Dead	1 (20.0%)	1 (12.5%)	0	0	0	2 (8.3%)
Total	5 (20.8%)	8 (33.3%)	7 (29.2%)	4 (16.7%)	0	24

Table V-8. Preoperative grade and outcome (6 months) in cases submitted to surgery later than the second week (day 15–)

	Grade I GCS 15	Grade II GCS 15	Grade III GCS 13–14	Grade IV GCS 8–12	Grade V GCS 3–7	Total
Good recovery	26 (86.7%)	4 (100%)	11 (64.7%)	1 (12.5%)	0	42 (70.0%)
Moderate disability	3 (10.0%)	0	3 (17.6%)	3 (37.5%)	0	9 (15.0%)
Severe disability	1 (3.3%)	0	1 (5.9%)	1 (12.5%)	0	3 (5.0%)
Vegetative state	0	0	0	1 (12.5%)	0	1 (1.7%)
Dead	0	0	2 (11.8%)	2 (25.0%)	1 (100%)	5 (8.3%)
Total	30 (50.0%)	4 (6.7%)	17 (28.3%)	8 (13.3%)	1 (1.7%)	60

Table V-9. Age and outcome (6 months) in cases of Grades I and II submitted to surgery in the first week (days 0–7)

	Age < 60	Age ≥ 60	60–64	65–	Total
Good recovery	43 (86.0%)	1 (12.5%)	1	0	44 (75.9%)
Moderate disability	4 (8.0%)	2 (25.0%)	1	1	6 (10.3%)
Severe disability	2 (4.0%)	2 (25.0%)	0	2	4 (6.9%)
Vegetative state	0	1 (12.5%)	0	1	1 (1.7%)
Dead	1 (2.0%)	2 (25.0%)	1	1	3 (5.2%)
Total	50 (86.2%)	8 (13.7%)	3	5	58

Table V-10. Age and outcome (6 months) in cases of Grades III and IV submitted to surgery in the first week (days 0–7)

	Age < 60	Age ≥ 60	60–64	65–	Total
Good recovery	15 (41.7%)	5 (38.5%)	5	0	20 (40.8%)
Moderate disability	11 (30.6%)	0	0	0	11 (22.4%)
Severe disability	5 (13.9%)	4 (30.8%)	3	1	9 (18.4%)
Vegetative state	1 (2.8%)	0	0	0	1 (2.0%)
Dead	4 (11.1%)	4 (30.8%)	2	2	8 (16.3%)
Total	36 (73.5%)	13 (26.5%)	10	3	49

was done in 53.3%. In a total of 214 operated cases, rebleeding after admission occurred in 16 cases.

Table V-4 indicates the overall results according to the proposed grading system and **Table V-5** shows the postoperative results. In all tables, the proposed grading system shows a good correlation with the outcome of the patients. The correlation is better in the preoperative grading than in the grading on admission. This is understandable because the patient's condition may change in the course of hospitalization. This new grading system has the advantages that it is clear in its description of the patients condition, it results in less observer variability, and may be useful prognostically (Sano and Tamura 1985, Yasui et al. 1985 a).

Tables V-6 through **V-8** demonstrate the correlation between the preoperative grade and the outcome at 6 months after surgery, classified by the timing of operation. Grade I includes grade I a of Hunt, the outcomes in both grades I and II are favorable and are not different as regards the timing of operation. In patients of grade III, the

Table V-11. Age and outcome (6 months) in cases of Grades I and II submitted to surgery later than the first weeek (days 8–)

	Age < 60	Age ≥ 60	60–64	65–	Total
Good recovery	29 (87.9%)	11 (78.6%)	7	4	40 (85.1%)
Moderate disability	2 (6.1%)	2 (14.3%)	0	2	4 (8.5%)
Severe disability	0	1 (7.1%)	0	1	1 (2.1%)
Vegetative state	0	0	0	0	0
Dead	2 (6.1%)	0	0	0	2 (4.3%)
Total	33 (70.2%)	14 (29.8%)	7	7	47

Table V-12. Age and outcome (6 months) in cases of Grades III and IV submitted to surgery later than the first week (days 8–)

	Age < 60	Age ≥ 60	60–64	65–	Total
Good recovery	16 (66.7%)	4 (33.3%)	4	0	20 (55.6%)
Moderate disability	2 (8.3%)	4 (33.3%)	3	1	6 (16.7%)
Severe disability	2 (8.3%)	2 (16.7%)	0	2	4 (11.1%)
Vegetative state	1 (4.2%)	1 (8.3%)	0	1	2 (5.6%)
Dead	3 (12.5%)	1 (8.3%)	0	1	4 (11.1%)
Total	24 (66.6%)	12 (33.3%)	7	5	36

outcome of the delayed operation group shows a tendency to be better, but not significantly. An analysis of the causes of unfavorable outcomes (including severe disability, vegetative state, and death) in grades I to III cases indicates that 42% of the unfavorable results were caused by vasospasm and 21% were the result of operative failure, such as slipping out of clips, occlusion of arterial branches and/or intra-cerebral hematomas. Other causes are preoperative neurological deficit (that is grade I a), and general complications, such as pulmonary edema due to hyper-tensive therapy, pneumonia, and other infections.

As for the timing of surgery, age should also be considered. **Tables V-9** through **V-12** show the correlation between postoperative outcome and age, classified by timing of operation in each grade. In grade I and II cases of the early operation group, the outcome in older group, especially for patients older than 65, is worse than for the younger group. Therefore, delayed operation may be better for older patients, especially those in grades III or IV.

C. Angiography

1. Site of Aneurysm

In a total of the recent 272 cases of SAH due to aneurysmal rupture, angiography was done in 252 cases. The distribution of ruptured aneurysms in these 252 patients is shown in **Table V-13**. The outcome of each location of aneurysms according to clinical grades is indicated in **Tables V-14** through **V-16**. In internal carotid artery (IC) aneurysms, the outcome was better than anterior cerebral artery (AC) or middle cerebral artery (MC) aneurysms, because of better preoperative condition. One of the reasons of better preoperative condition in IC aneurysm cases may be lower incidence

of severe SAH and of severe intraventricular hematoma or intracerebral hematoma. The side of aneurysms may be another important factor affecting the outcome, especially functional recovery, because of the cerebral dominancy. In 63 IC aneurysms, 30 were on the right side and 33 on the left. There was no significant difference in the outcome between two sides (**Tables V-17, V-18**). The side difference existed in MC aneurysms. As shown in **Tables V-19** and **V-20**, the outcome in left MC aneurysms was significantly worse than that of the right. Mortality in each side group was not different. However, patients in good recovery group was less in the left side group. A half of patients with left MC aneurysms was unable to return to their previous work because of right hemiparesis, especially an affecting hand, and/or aphasia.

Multiple aneurysms were found by angiography in 48 cases (19%) out of our 252 cases. In multiple aneurysms, it is most important to decide which aneurysm has ruptured (Almaani and Richardson 1978, Nehls *et al.* 1985, Winn' *et al.* 1983). Usually, the ruptured one is angiographically diagnosed by size, shape, and/or local spasm etc. However, the combination of CT scan with angiography is more useful in deciding which aneurysm has bled (**Fig. V-1**). In our series, operation was done in 44 cases with multiple aneurysms and the diagnosis of the ruptured aneurysms was correct in 43 cases.

Table V-13. Site of ruptured aneurysm

Internal carotid artery (IC)	71 (28.2%)
Ophthalmic	5 (2.0%)
Posterior communicating	53 (21.0%)
Anterior choroidal	8 (3.2%)
Bifurcation	5 (2.0%)
Anteriro cerebral artery (AC)	97 (38.5%)
A¹ segment	3 (1.2%)
Anterior communicating	84 (33.3%)
A$_{2, 3 \text{ or } 4}$ segment	10 (4.0%)
Middle cerebral artery (MC)	59 (23.4%)
M$_1$	3 (1.2%)
Trifurcation (bifurcation)	56 (22.2%)
Vertebro-basilar artery (VB)	25 (9.9%)
Vertebral	2 (0.85)
PICA	7 (2.8%)
Vertebrobasilar junction	1 (0.4%)
AICA	2 (0.8%)
Superior cerebellar	3 (1.2%)
Basilar top	10 (4.0%)
Total	252

Table V-14. Preoperative grade and outcome (6 months). IC aneurysm

	Grade I	Grade II	Grade III	Grade IV	Grade V	Total
Good recovery	19(95.0%)	15(75.0%)	10(66.7%)	1(16.7%)	0	45(71.4%)
Moderate disability	0	3(15.0%)	3(20.0%)	1(16.7%)	0	7(11.1%)
Severe disability	1(5.0%)	1(5.0%)	0	3(50.0%)	0	5(7.9%)
Vegetative state	0	0	0	1(16.7%)	0	1(1.6%)
Dead	0	1(5.0%)	2(13.3%)	0	2(100%)	5(7.9%)
Total	20(31.7%)	20(31.7%)	15(23.8%)	6(9.5%)	2(3.2%)	63

Table V-15. Preoperative grade and outcome (6 months). AC aneurysm

	Grade I	Grade II	Grade III	Grade IV	Grade V	Total
Good recovery	9(69.2%)	17(70.8%)	15(71.4%)	3(20.0%)	0	44(55.0%)
Moderate disability	3(23.1%)	2(8.3%)	1(4.8%)	6(40.0%)	0	12(15.0%)
Severe disability	0	1(4.2%)	3(4.3%)	3(20.0%)	1(14.3%)	8(10.0%)
Vegetative state	0	1(4.2%)	1(4.8%)	0	1(14.3%)	3(3.8%)
Dead	1(7.7%)	3(12.5%)	1(4.8%)	3(20.0%)	5(71.4%)	13(16.3%)
Total	13(16.3%)	24(30.0%)	21(26.3%)	15(18.8%)	7(8.8%)	80

Table V-16. Preoperative grade and outcome (6 months). MC aneurysm

	Grade I	Grade II	Grade III	Grade IV	Grade V	Total
Good recovery	4 (80.0%)	13 (81.3%)	3 (30.0%)	1 (12.5%)	3 (20.0%)	24 (44.4%)
Moderate disability	1 (20.0%)	1 (6.3%)	3 (30.0%)	2 (25.0%)	2 (13.3%)	9 (16.7%)
Severe disability	0	2 (12.5%)	2 (20.0%)	2 (25.0%)	1 (6.7%)	7 (13.0%)
Vegetative state	0	0	0	1 (12.5%)	1 (6.7%)	2 (3.7%)
Dead	0	0	2 (20.0%)	2 (25.0%)	8 (53.3%)	12 (22.2%)
Total	5 (9.3%)	16 (29.6%)	10 (18.5%)	8 (14.8%)	15 (27.8%)	54

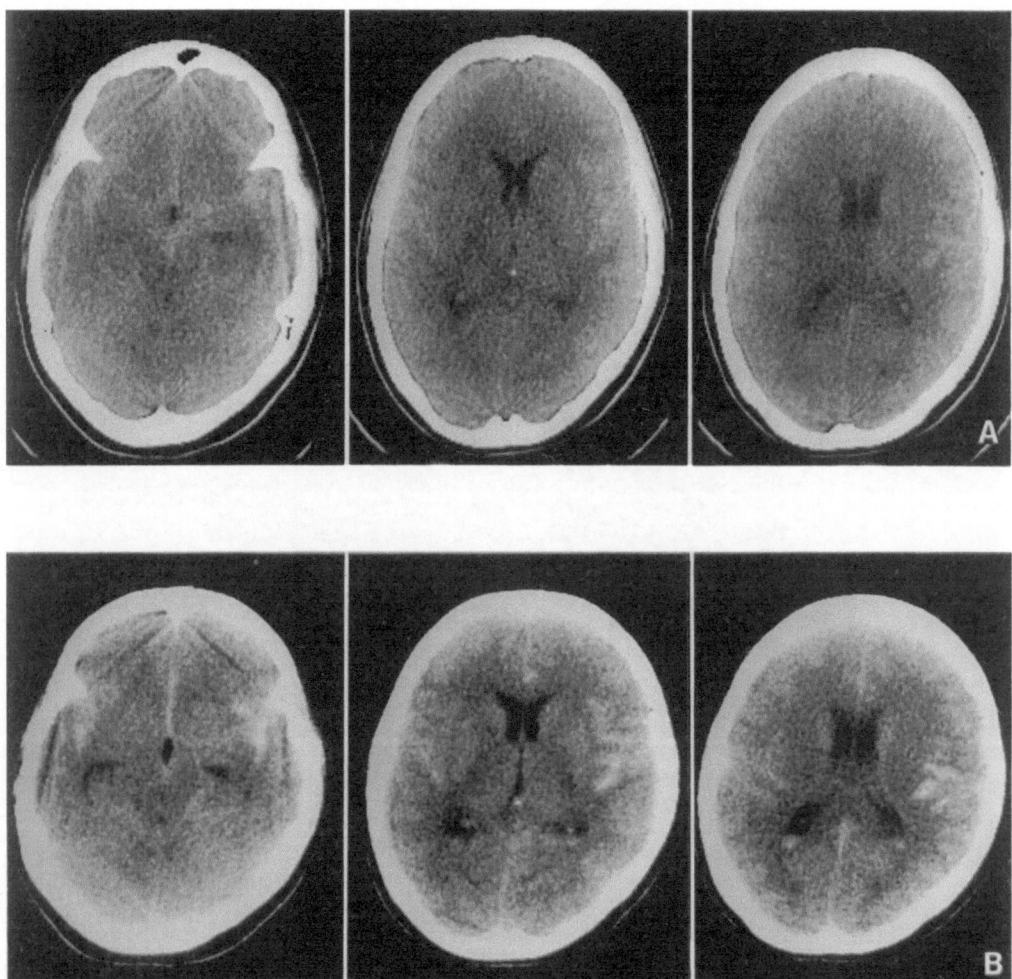

Fig. V-1. Case of multiple aneurysms. A) Thin subarachnoid blood is seen in the suprasellar cistern and both Sylvian fissures. There is no side difference. These initial tomograms were taken by a lower resolution CT scanner. B) Using a higher resolution CT scanner, subarachnoid blood in the right Sylvian fissure is more clear than previous tomograms. C) A left carotid angiogram with compression of the right carotid artery shows bilateral middle cerebral artery aneurysms. The right aneurysm operated on at first was diagnosed as ruptured by CT findings and angiographic criteria

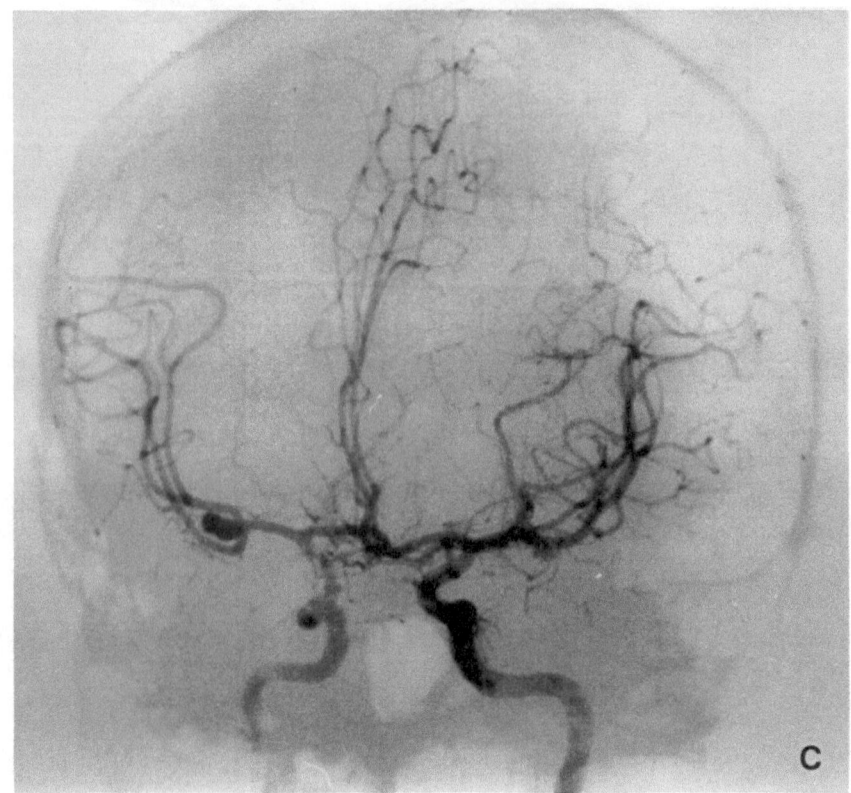

Fig. V-1C

2. Recurrent Hemorrhage During Angiography

The recurrent hemorrhage during angiography has been rarely reported (Tsementzis *et al.* 1984). However, Taneda *et al.* (1985) indicated that re-bleeding related to angiography is not so rare in an acute stage, especially within 10 hours after SAH. In a total of our 252 cases, rebleeding occurred in 5 cases (2.0%) during angiography. The rate of rebleeding during angiography

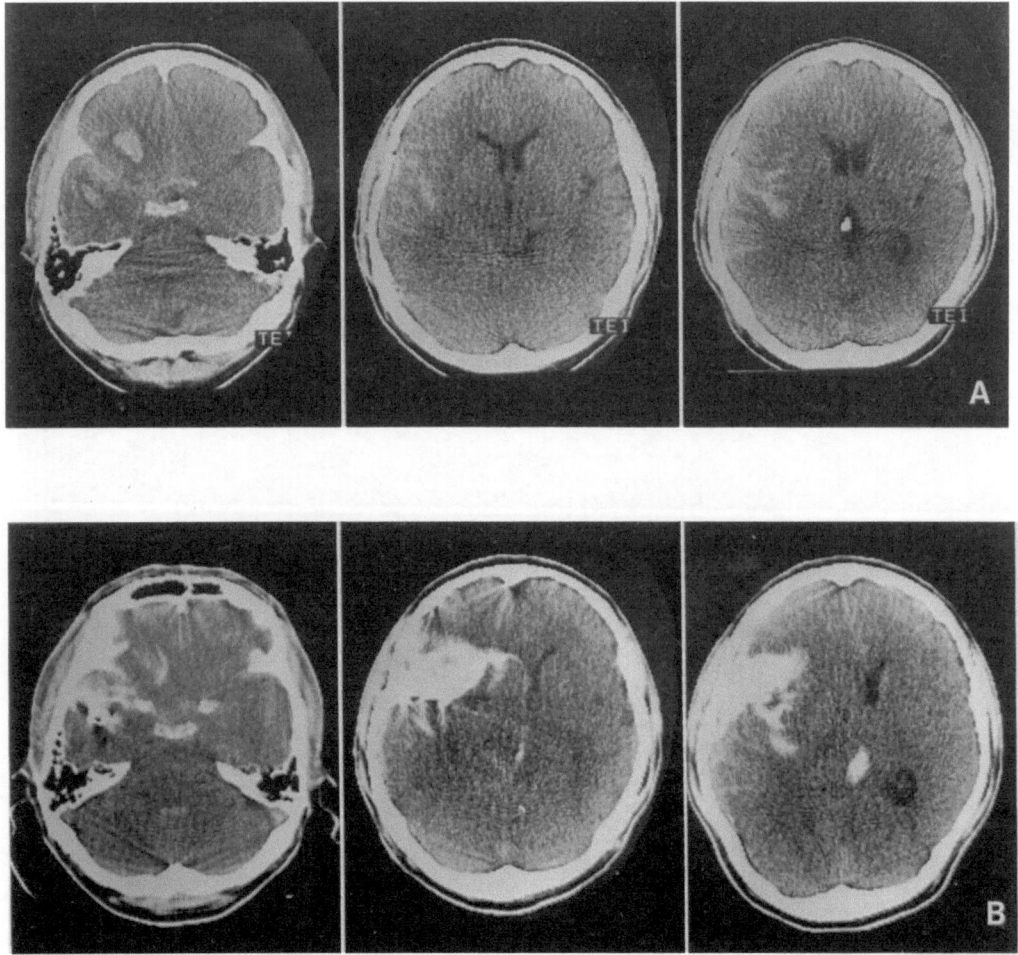

Fig. V-2. Patient with recurrent hemorrhage during angiography. A) CT scan at the time of admission. Subarachnoid blood is seen in the left Sylvian fissure. Small intracerebral hematomas are in the left frontal and temporal lobes. B) CT scan just after angiography. Marked high density lesion in the left Sylvian fissure and midline shift are seen. C), D) Angiograms indicate extravasation of contrast media (arrow)

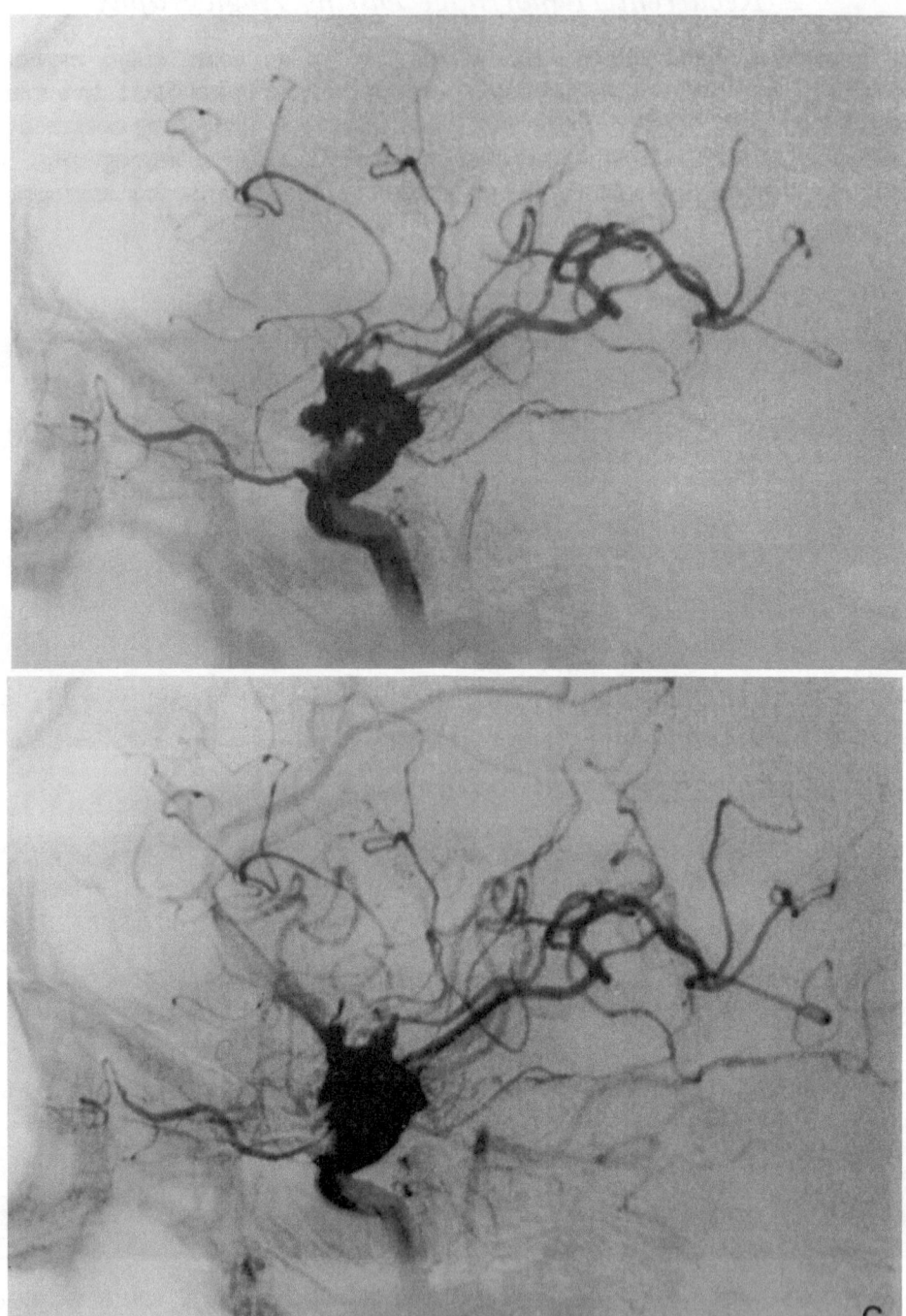

Fig. V-2C

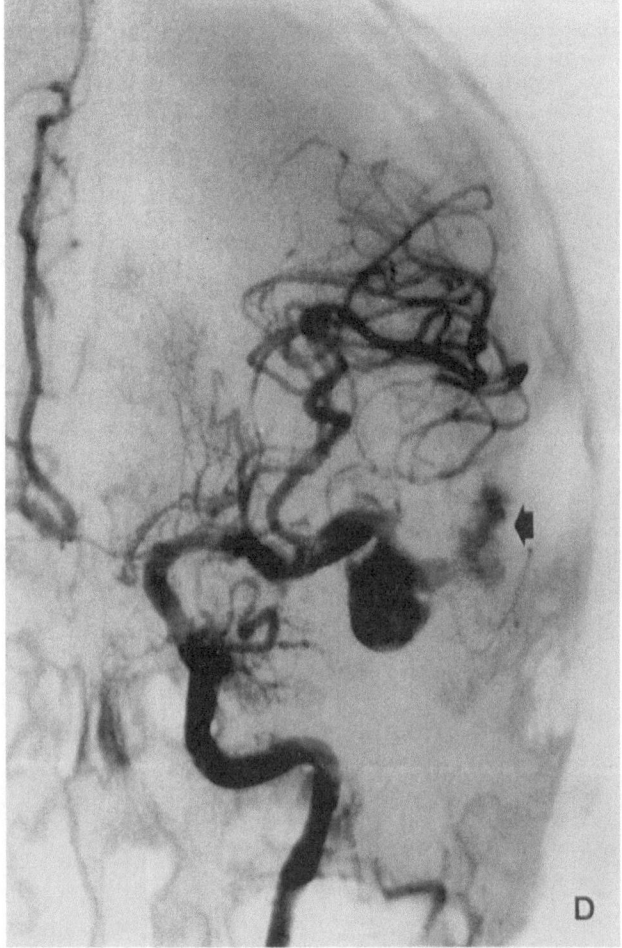

Fig. V-2D

was 5.6% in SAH patients who entered on day 0. In two cases, extravasation of contrast media was seen on angiograms (Figs. V-2, V-3). All of them resulted in a catastrophic outcome. Recently, the Cooperative Aneurysm Study has indicated that rebleeding occurred with the greatest frequency within 24 hours after SAH (see Chapter VII). Therefore, in our recent strategy, angiography has not been done within 6 hours after SAH.

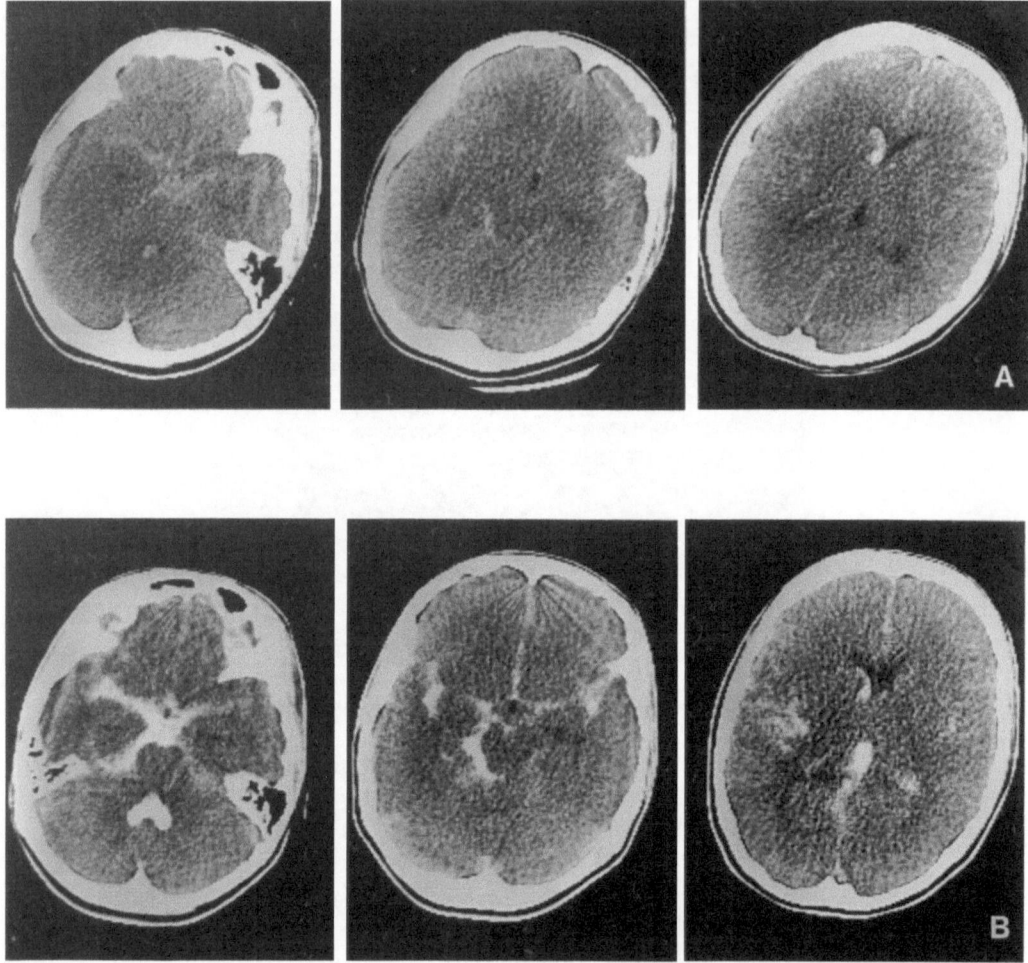

Fig. V-3. Another case with recurrent hemorrhage during angiography. A) CT scan at the time of admission. B) CT scan just after rebleeding during angiography. Marked high density lesions are seen in cisterns and the IVth ventricle. C) Angiograms show extravasated contrast media surrounding the left internal carotid artery-posterior communicating artery aneurysm and filling the IVth ventricle (arrows). D) Plain craniograms just after angiography. Extravasation of contrast media is seen in the IVth ventricle (arrows)

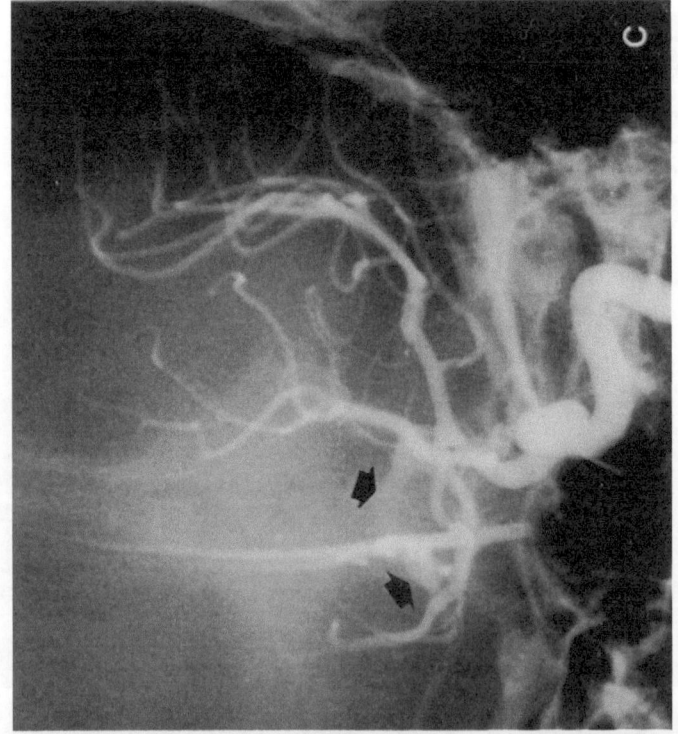

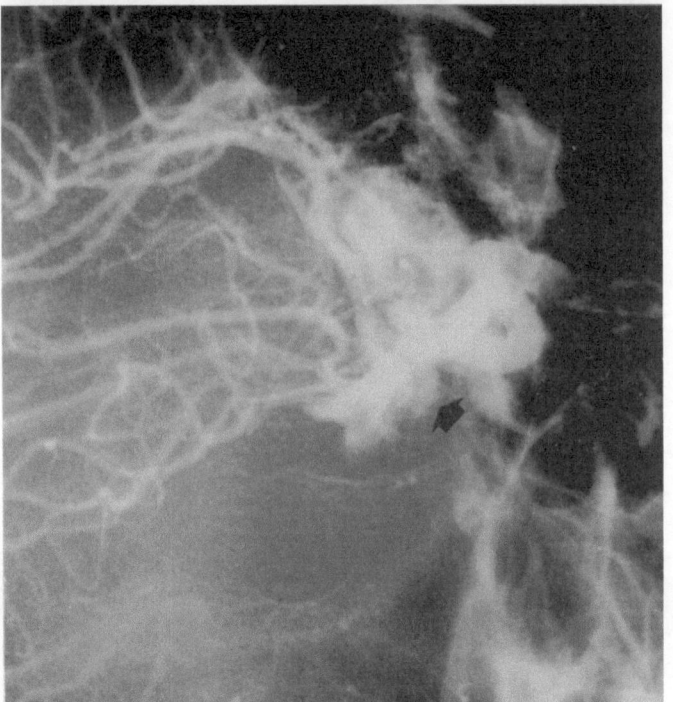

Fig. V-3C

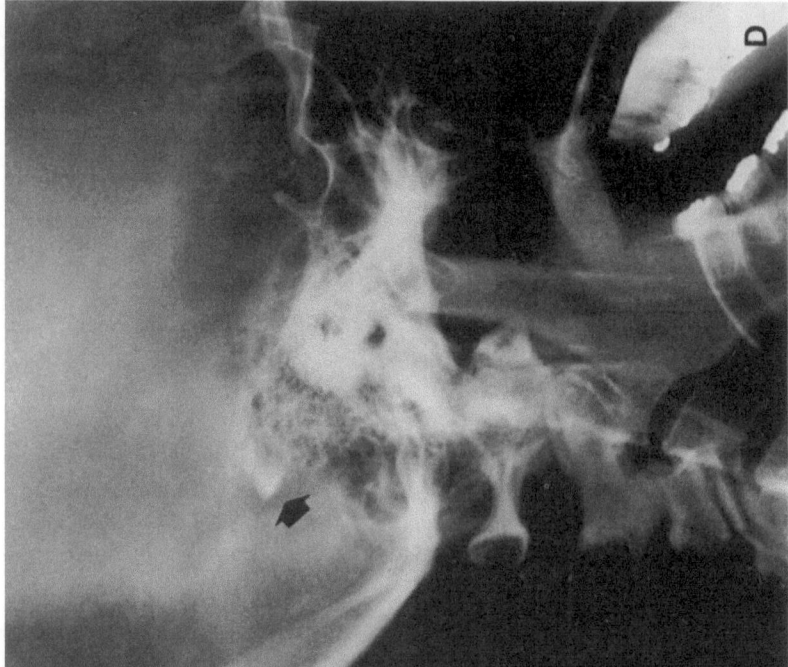

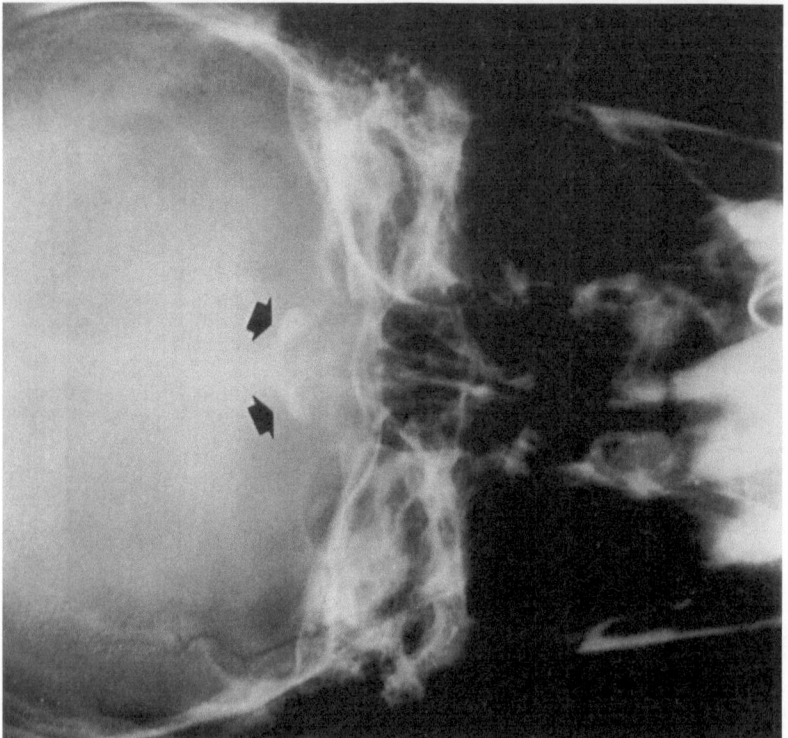

Fig. V-3D

D. Computerized Tomography

1. Subarachnoid Clot

Computerized tomography (CT) is the most important investigative tool, not only to diagnose SAH, but also to assist in predicting the occurrence of vasospasm by making an assessment of the volume or thickness of the subarachnoid clot. And also, the side of an aneurysm, or, in some cases, the site of an aneurysm is able to be predicted based on the distribution of subarachnoid blood (**Figs. V-4** through **V-10**). Nowadays, CT scan should be selected as the first study to diagnose SAH in the acute stage, if available.

Lumbar puncture may be the next desirable method for diagnosis, especially on day 0, because of its invasiveness, although removal of CSF may decrease intracranial pressure (ICP) and may relieve headache.

A close correlation between the development of vasospasm and subarachnoid clot is a well-known fact (Adams *et al.* 1983, Fisher *et al.* 1980, Fujita 1985, Gurusinghe and Richardson 1984, Kistler *et al.* 1983, Knuckey *et al.* 1985, Mizukami *et al.* 1980 b, Pasqualin *et al.* 1984, Suzuki *et al.* 1980, Takemae *et al.* 1978, Yamamoto *et al.* 1983) (**Fig. V-11**), although vasospasm does not always occur in all cases with thick subarachnoid blood (**Fig. V-12**) and does occur in some cases with thin subarachnoid blood (**Figs. V-13, V-14**). In 1977, Mizukami and others (Fisher *et al.* 1980, Takemae *et al.* 1978)

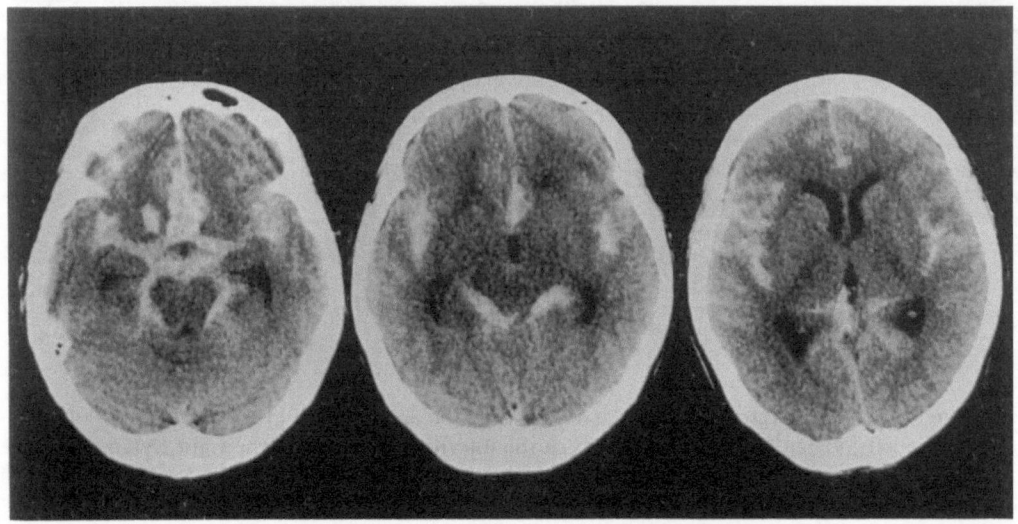

Fig. V-4. A typical SAH pattern of a ruptured anterior communicating artery aneurysm. Thick subarachnoid blood is seen in all cisterns and fissures. Especially thick clots are in the basal frontal interhemispheric fissure and a small intracerebral hematoma in the left frontal base

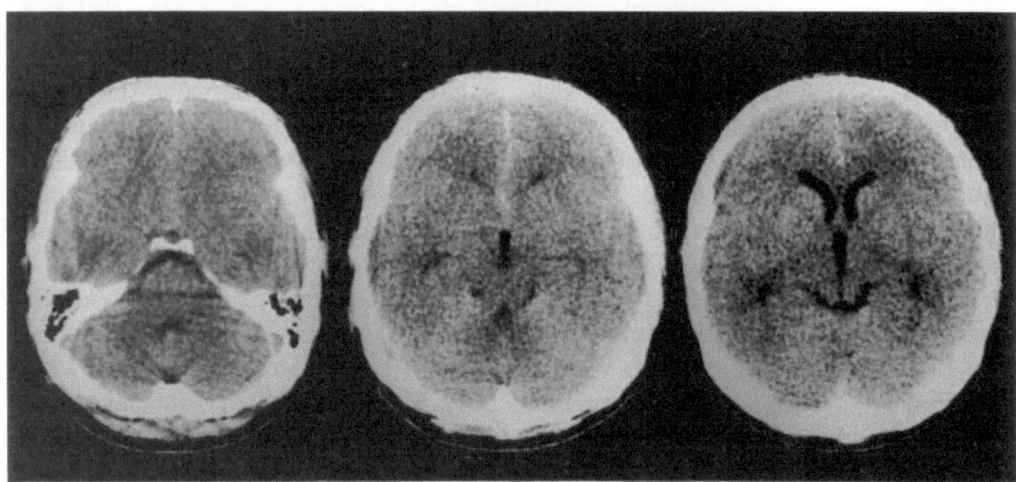

Fig. V-5. Thin subarachnoid blood is seen only in the interhemispheric fissure. CT scan of an anterior communicating artery aneurysm on day 0

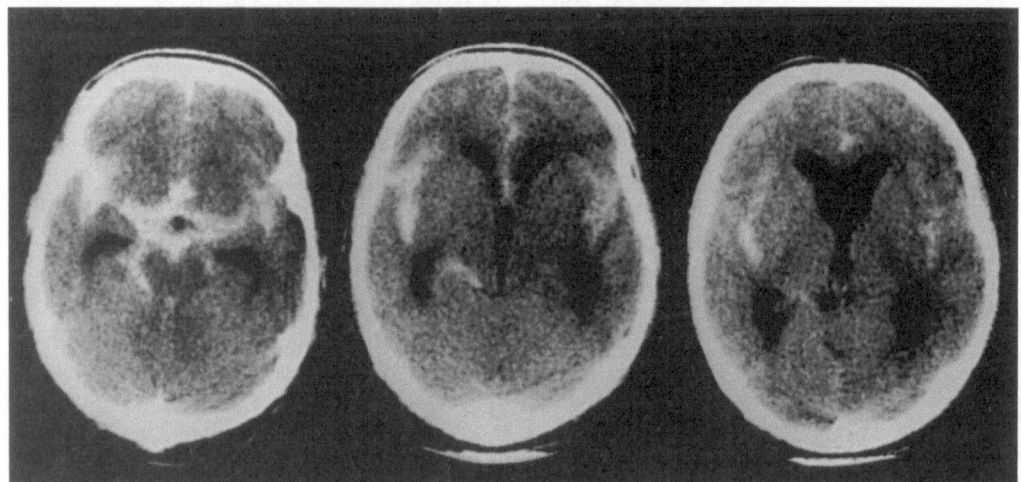

Fig. V-6. Case of a left internal carotid artery bifurcation aneurysm. Thick subarachnoid blood is seen in the left Sylvian fissure. Thin layers in the interhemispheric and the right Sylvian fissure. Laterality of subarachnoid blood is clearly demonstrated

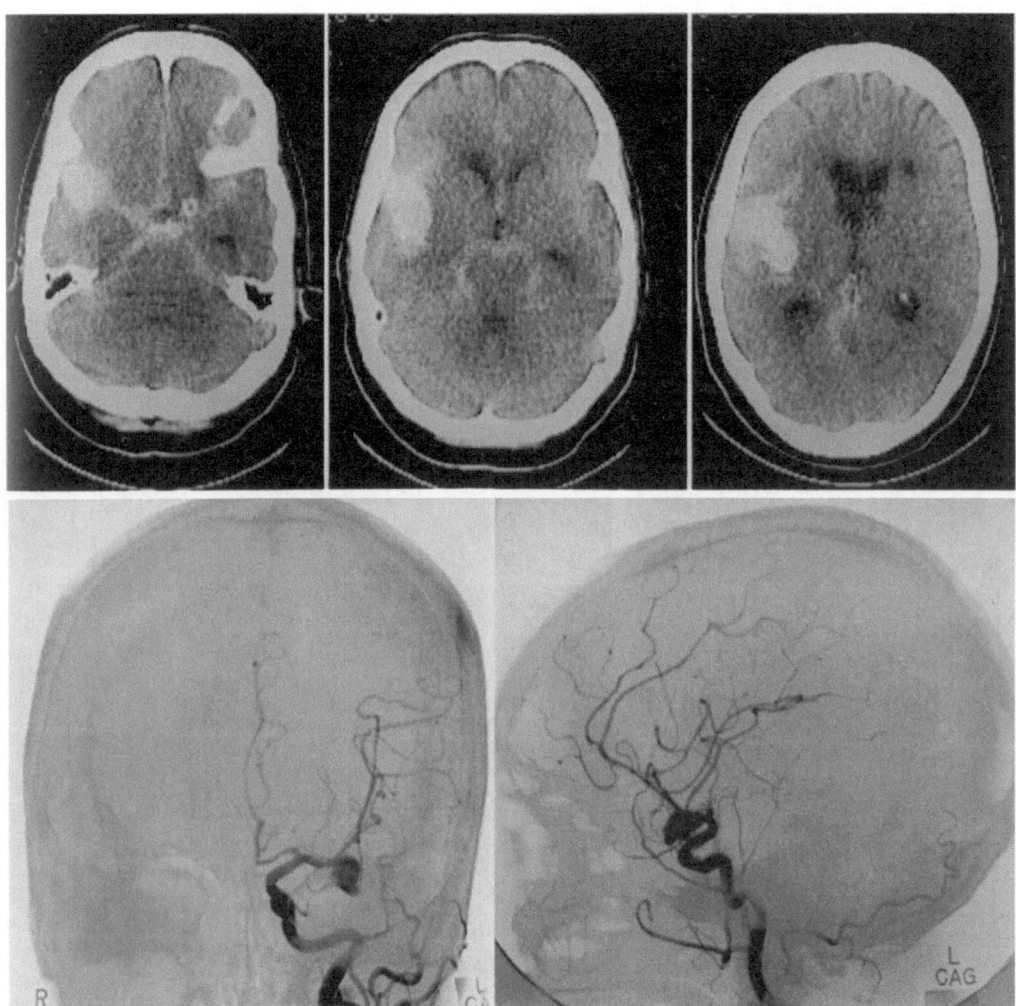

Fig. V-7. A typical pattern of SAH in a left middle cerebral artery aneurysm. CT scan shows a left Sylvian hematoma. There is no mass effect. Aneurysm arising from the bifurcation of the middle cerebral artery is seen in angiograms

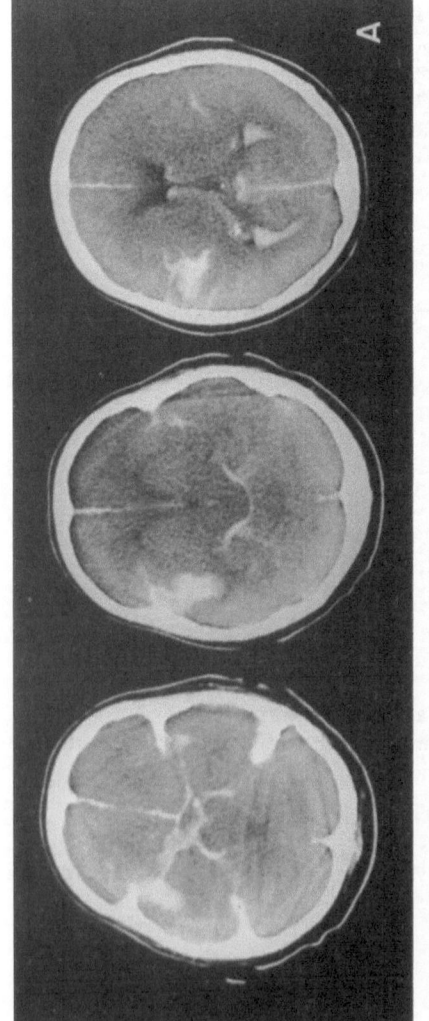

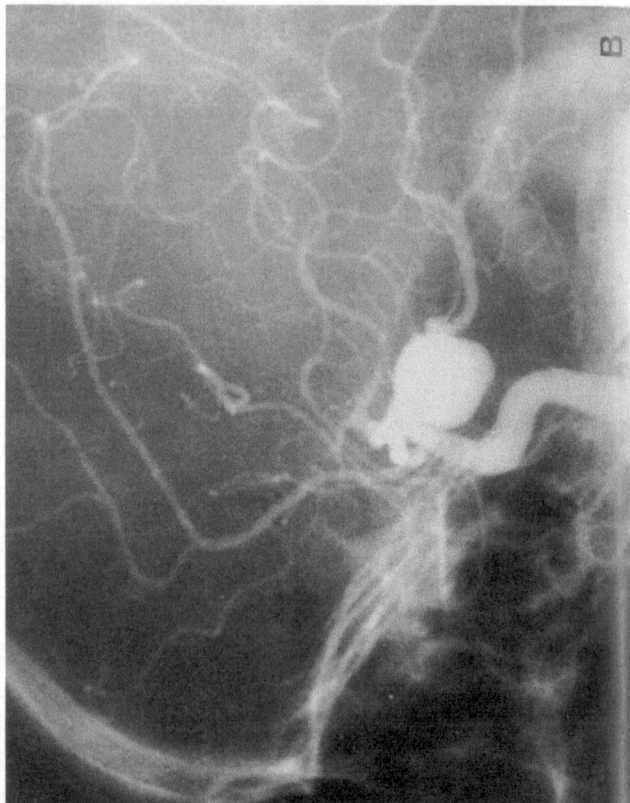

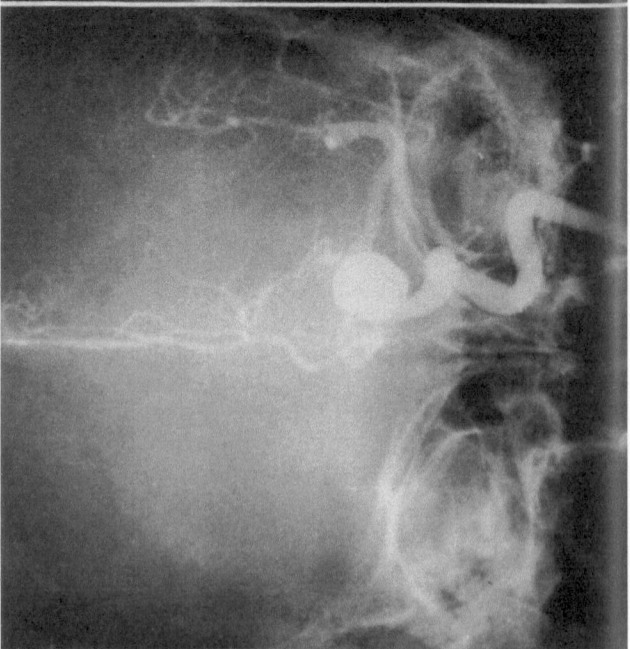

Fig. V-8. Case of a left internal carotid artery bifurcation aneurysm. A) CT scan shows a left Sylvian hematoma and thin subarachnoid blood in the suprasellar cistern, bilateral ambient and quadrigeminal cistern, and the interhemispheric fissure. A left middle cerebral artery aneurysm is highly suspected from this CT scan. B) Angiograms showing a large aneurysm of the left internal carotid artery

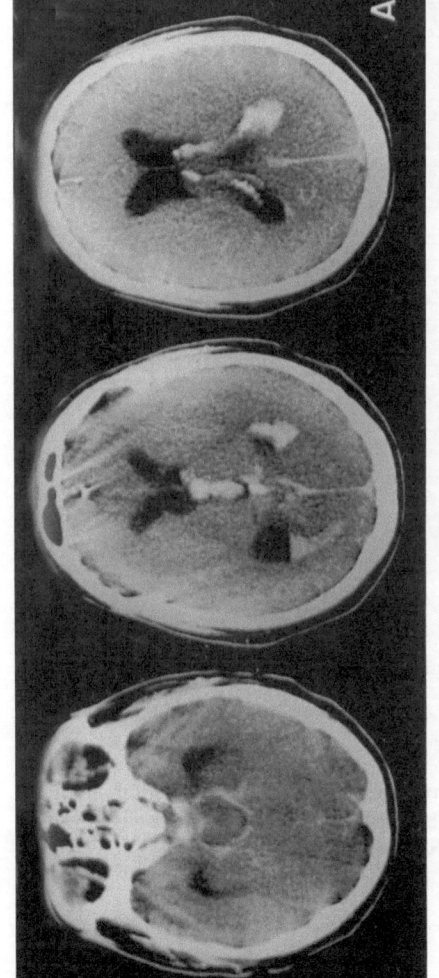

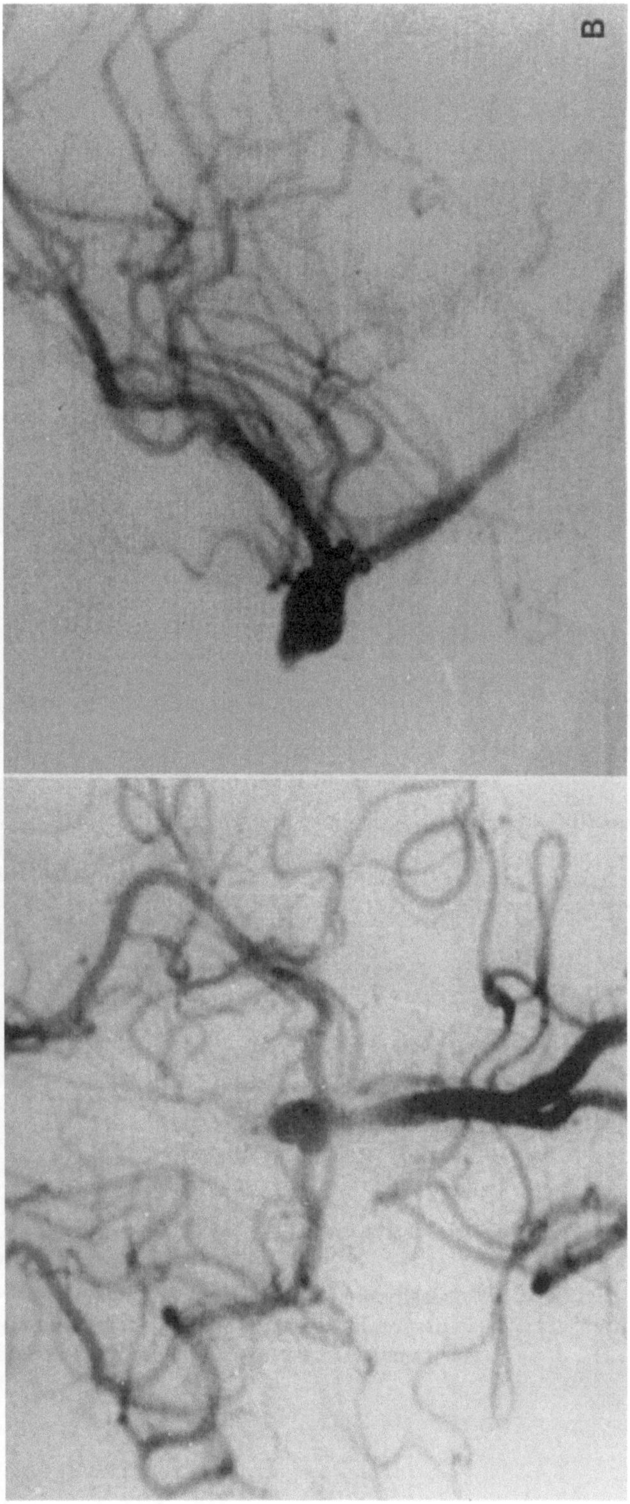

Fig. V-9. CT scan (A) and angiograms (B) of a basilar tip aneurysm. Intraventricular blood is seen in the IIIrd ventricle and both lateral ventricles. A basilar tip aneurysm sometimes ruptures directly into the third ventricle

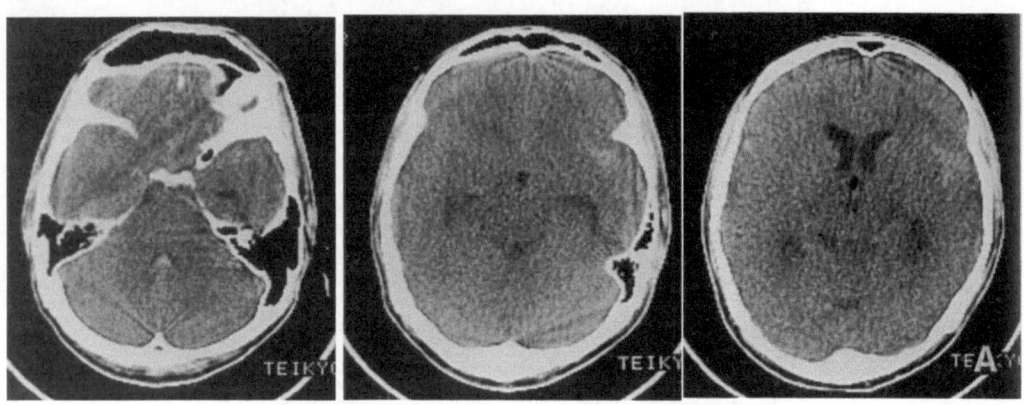

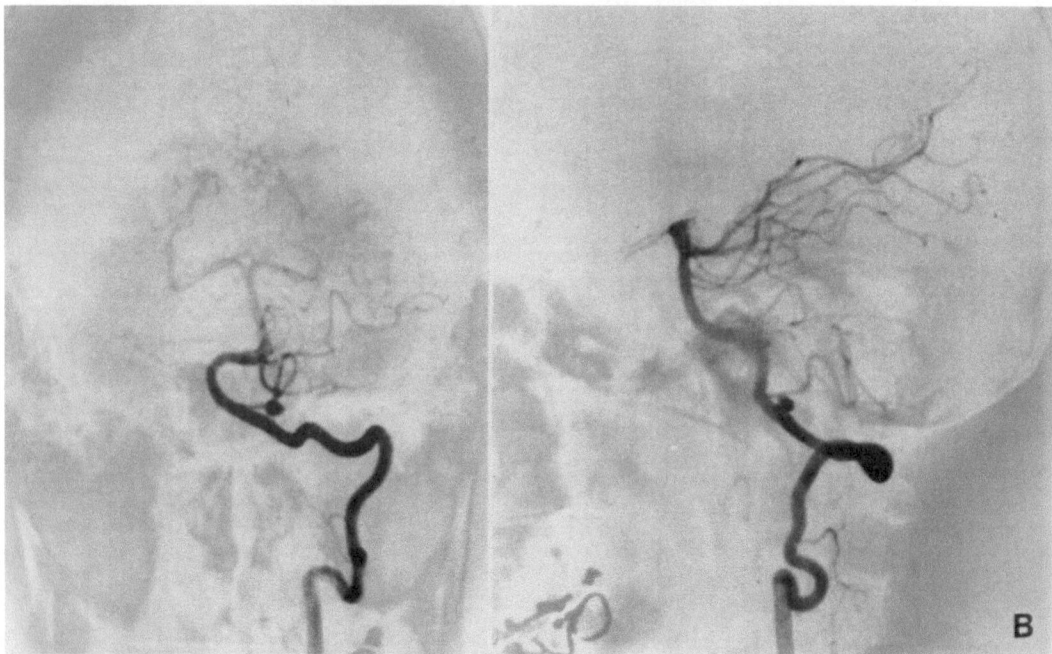

Fig. V-10. A) Small amount of blood clot is seen in the IVth ventricle and thin subarachnoid blood in the right Sylvian fissure. B) Angiograms show a left vertebral artery PICA aneurysm. In a vertebral artery aneurysm, reflux of blood into the IVth ventricle is usually seen

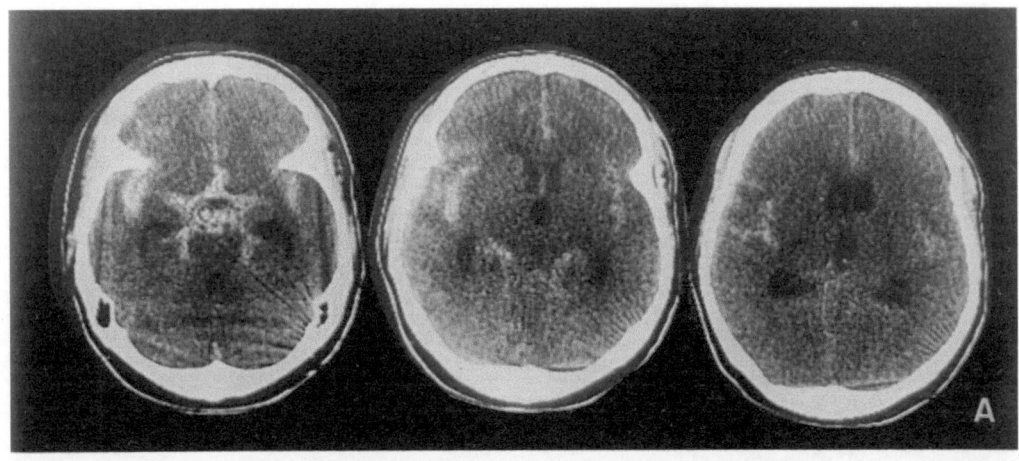

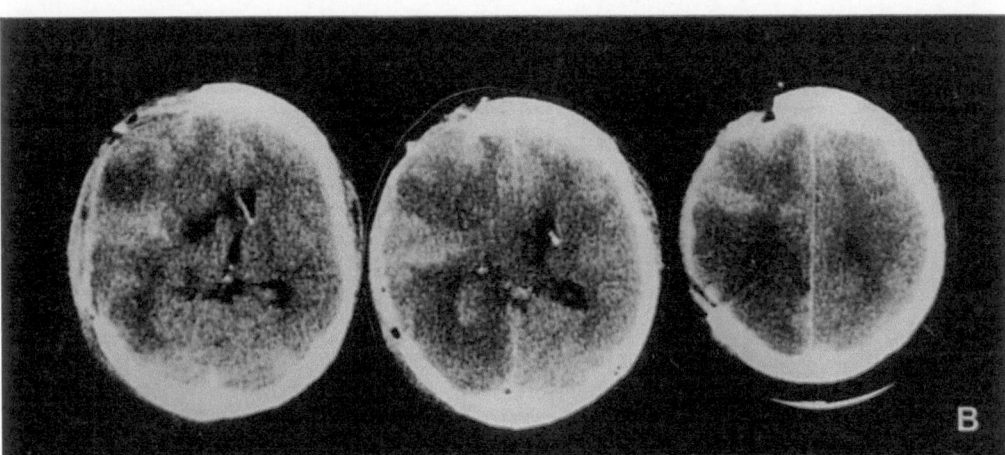

Fig. V-11. Case of a left internal carotid artery-anterior choroidal artery aneurysm. A) CT scan on day 0 shows thick subarachnoid blood in the left Sylvian fissure and the suprasellar cistern. B) Cerebral infarcts due to vasospasm in the left cerebral cortex are seen on CT scan on day 10

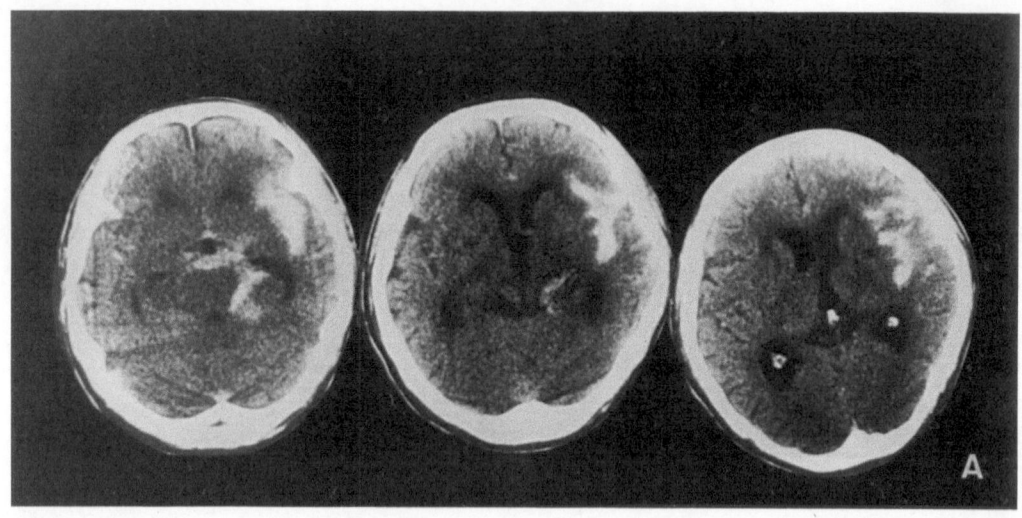

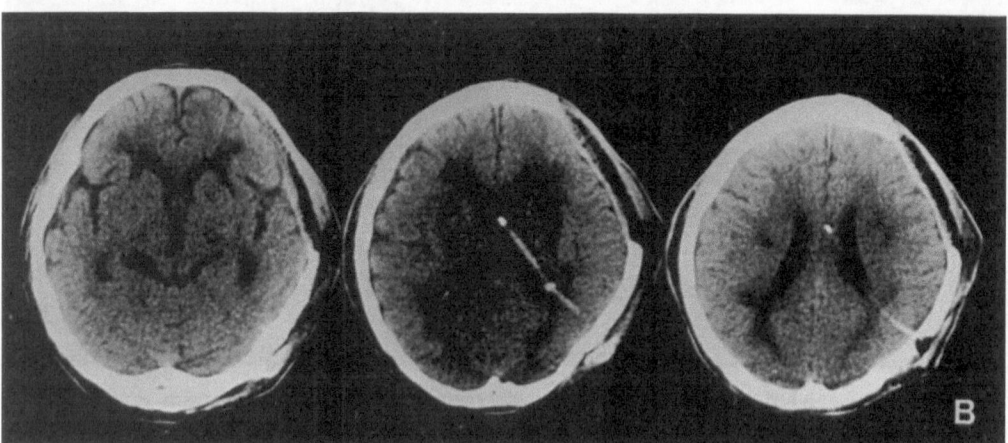

Fig. V-12. Case of a right middle cerebral artery aneurysm. A) Thick subarachnoid blood is seen in the right Sylvian fissure on CT scan on day 0. B) Follow-up CT scan does not show infarct due to vasospasm

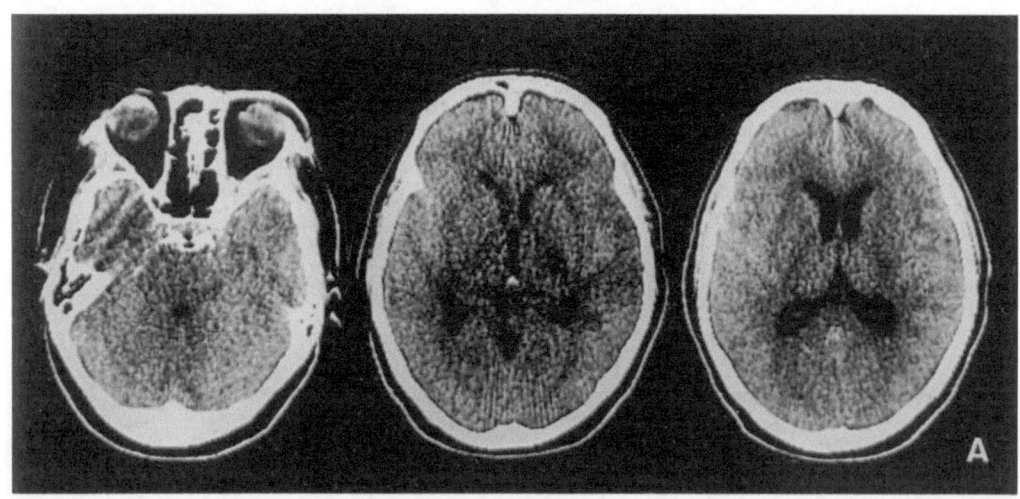

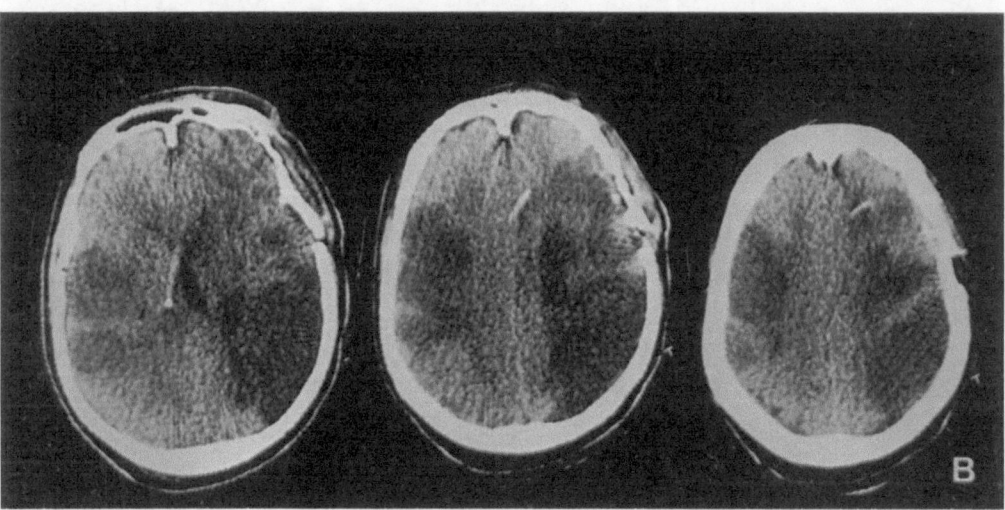

Fig. V-13. Case of a 64-year-old female with an anterior communicating artery aneurysm. A) CT scan on day 2 shows thin isodense subarachnoid blood in the interhemispheric fissure and bilateral Sylvian fissures. B) Infarcts due to vasospasm are seen in CT as low density areas of both hemispheres

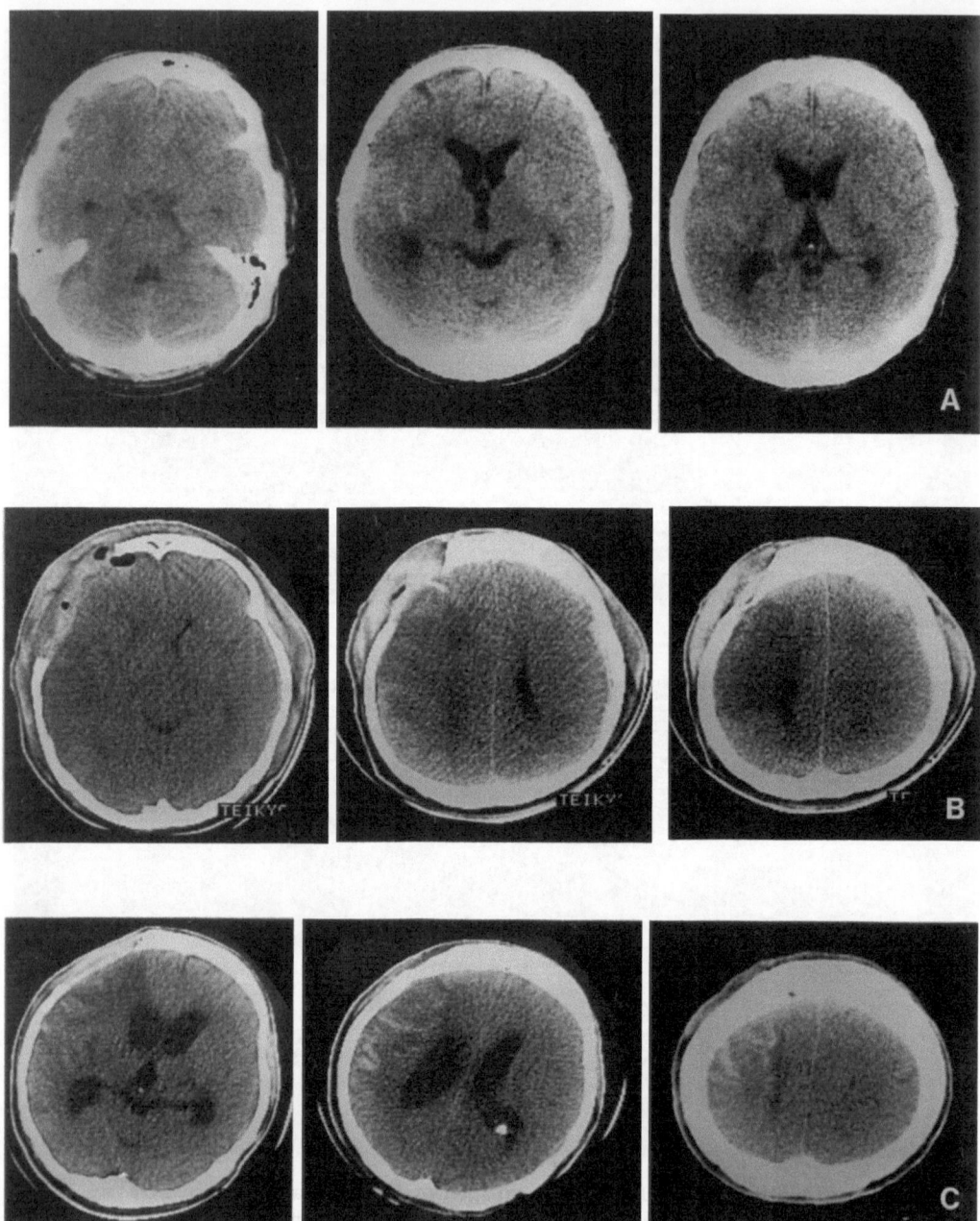

Fig. V-14. Case of a left internal carotid artery-posterior communicating artery aneurysm. A) Since the contour of bilateral Sylvian fissures is obscure, isodense subarachnoid blood is suspected from CT scan on day 5. B) One week later, low density area is seen in the territory of the left middle cerebral artery. C) Three weeks after the onset, patchy high density areas are scattered in the low density area. Hemorrhagic infarction is suspected to have occurred after the resolution of vasospasm. D) Two months later, ventricular dilatation and infarct in the territory of the left middle cerebral artery are shown

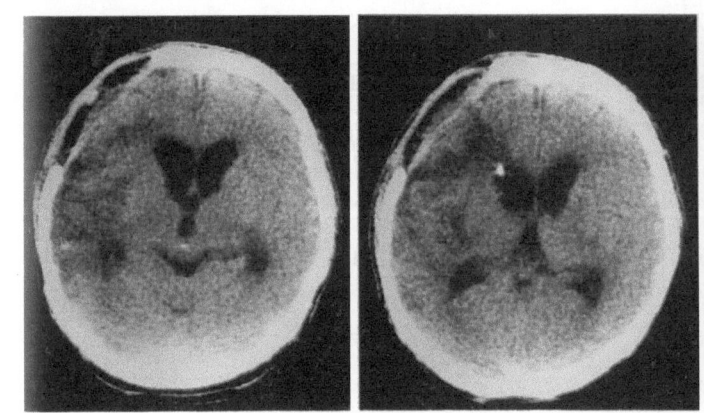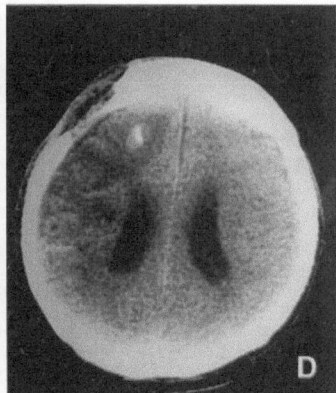

Fig. V-14D

have preliminarily reported the relationship between high density areas on CT and cerebral vasospasm. Fisher *et al.* (1980) classified subarachnoid blood on CT (EMI-1005) into four grades: group 1, no blood detected; group 2, a diffuse deposition or thin layer with all vertical layers of blood less than 1 mm thick; group 3, localized clots and/or vertical layers of blood 1 mm or greater in thickness; and group 4, diffuse or no subarachnoid blood, but with intracerebral or intraventricular clots. Using this classification, they have clearly demonstrated an excellent correlation between thick subarachnoid clot on CT scans within 5 days and the delayed clinical syndrome due to severe vasospasm. In their study, when the subarachnoid clot was not detected or blood was distributed diffusely, severe vasospasm was almost never encountered (1 of 18 cases). On the other hand, in the presence of subarachnoid blood clots larger than 5 × 3 mm (measured on the reproduced images) or layers of blood 1 mm or thicker in fissures and vertical cisterns, severe spasm followed almost invari-

ably (23 of 24 cases). Mizukami *et al.* (1980 b) have confirmed this relationship in 177 patients with ruptured aneurysm. The development of cerebral vasospasm was noted at the high rate of 84.6% in 26 cases where high density was demonstrated on CT scan within 4 days after SAH. Since then, many authors have reported similar results (Adams *et al.* 1983, Fujita 1985, Pasqualin *et al.* 1984, Suzuki *et al.* 1980).

However, subarachnoid clots are nowadays more clearly detected by modern higher resolution CT scanners. The initial types of CT scanners which were used by Fisher *et al.* were not capable of detecting small subarachnoid clots (**Fig. V-1**). Among a total of 128 patients who were admitted to the Teikyo University Hospital on the same day of SAH, that is day 0, none were classified as Fisher's group 1, that is "no blood detected".

In our department, 51 patients, who were admitted to the hospital within 4 days (day 0–3) after SAH due to aneurysmal rupture, were randomly selected and their CT films were ana-

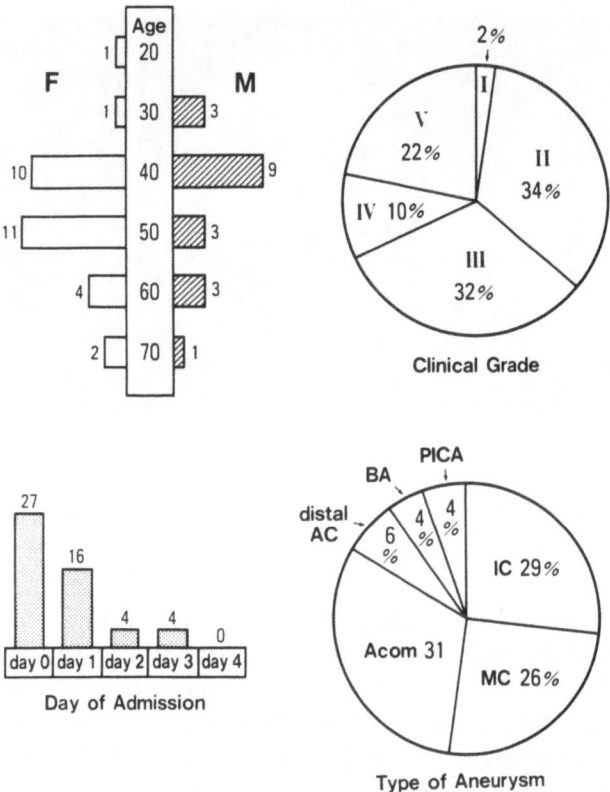

Fig. V-15. Basic characteristics of randomly selected 51 patients who were admitted within 4 days (days 0–3)

lysed prospectively. Basic characteristics, such as day of admission, grade at admission, site of aneurysm, and age distribution, are shown in **Fig. V-15**. According to the Fisher's classification on CT of these patients, no one belongs to group 1, that is "no blood detected". Incidences of groups 2, 3, and 4 were 12, 78, and 10, respectively **(Fig. V-16)**. Thus, more than three quarters of patients who were admitted at an early stage of SAH were classified as Fisher's group 3. Symptomatic vasospasm occurred in 59% of patient classified as Fisher's group 3. However, there were no patients with symptomatic vasospasm in group 2 or 4. It is clear that

symptomatic vasospasm mostly occurred in Fisher's group 3 patients. This classification, however, no longer remains useful, since using recent version of CT scanners most of SAH patients fall in only one group.

Several authors have reported modified classifications (Gurusinghe and Richardson 1984, Suzuki *et al.* 1980). However, still these classifications are either too complicated or in need of CT numbers. Therefore, we feel a more practical and simple classification should be developed.

For the purpose of examining whether the side and/or the site of aneurysms would be diagnosed by CT findings,

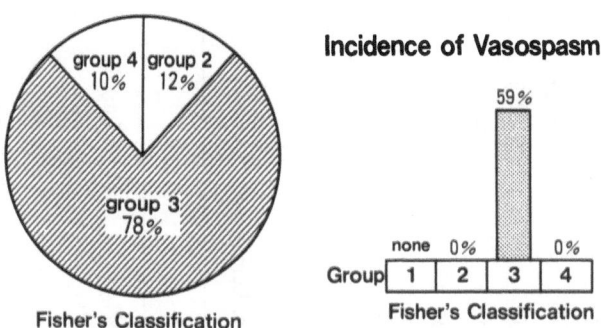

Fisher's Classification   Fisher's Classification

Fig. V-16. Fisher's classification on CT and incidence of vasospasm in these 51 patients

CT of these 51 patients were shown to 10 board-certified neurosurgeons in our department in a blind fashion. The sites of aneurysm was estimated accurately in 56% of the attempts. MC aneurysms or AC aneurysms were correctly located in 68 and 64% respectively. However, this figure was 42% in IC aneurysms and 35% in basilar aneurysms. Nevertheless, the usefulness of CT scan in the diagnosis of SAH, or in predicting the occurrence of vasospasm, and estimating the site of aneurysm is widely recognized (Adams et al. 1983, Gurusinghe and Richardson 1984, Kistler et al. 1983, Knuckey et al. 1985, Nehls et al. 1985, Pasqualin et al. 1984, Takayasu et al. 1985).

2. Intracerebral Hematoma

Intracerebral hematoma (ICH) was demonstrated in 59 cases of our 272 SAH patients. The incidence of ICH was 21.7% of the total cases and 19.2% of the operated cases. **Table V-21** shows the incidence of ICH in each site

Table V-21. Incidence of intracerebral hematoma

Type or aneurysm	Intracerebral hematoma			Total
	None	Moderate	Large	
IC	64 (90.1%)	5 (7.0%)	2 (2.8%)	71
AC	67 (69.1%)	17 (17.5%)	13 (13.4%)	97
MC	40 (67.8%)	4 (6.8%)	15 (25.4%)	59
VB	25 (100%)	0	0	25
Total	196 (77.8%)	26 (10.3%)	30 (11.9%)	252

IC: Internal Carotid Artery.
AC: Anterior Cerebral Artery.
MC: Middle Cerebral Artery.
VB: Vertebro-basilar Artery.

Table V-22. Preoperative grade and outcome (6 months) in SAH patients with intracerebral hematoma

	Grade I	Grade II	Grade III	Grade IV	Grade V	Total
Good recovery	2(66.7%)	1(25.0%)	5(50.0%)	1(7.1%)	1(10.0%)	10(24.4%)
Moderate disability	1(33.3%)	1(25.0%)	1(10.0%)	6(42.9%)	1(10.0%)	10(24.4%)
Severe disability	0	1(25.0%)	2(20.0%)	2(14.3%)	2(20.0%)	7(17.1%)
Vegetative state	0	1(25.0%)	0	1(7.1%)	1(10.0%)	3(7.3%)
Dead	0	0	2(20.0%)	4(28.6%)	5(50.0%)	11(26.8%)
Total	3(7.3%)	4(9.8%)	10(24.4%)	14(34.1%)	10(24.4%)	41

Table V-23. Preoperative grade and outcome (6 months) in SAH patients without intracerebral hematoma

	Grade I	Grade II	Grade III	Grade IV	Grade V	Total
Good recovery	35(87.5%)	46(79.3%)	30(65.2%)	4(26.7%)	2(14.3%)	117(67.6%)
Moderate disability	3(7.5%)	5(8.6%)	7(15.2%)	3(20.0%)	1(7.1%)	19(11.0%)
Severe disability	1(2.5%)	3(5.2%)	3(6.5%)	6(40.0%)	0	13(7.5%)
Vegetative state	0	0	1(2.2%)	1(6.7%)	1(7.1%)	3(1.7%)
Dead	1(2.5%)	4(6.9%)	5(10.9%)	1(6.7%)	10(71.4%)	21(12.1%)
Total	40(23.1%)	58(33.5%)	46(26.6%)	15(8.7%)	14(8.1%)	173

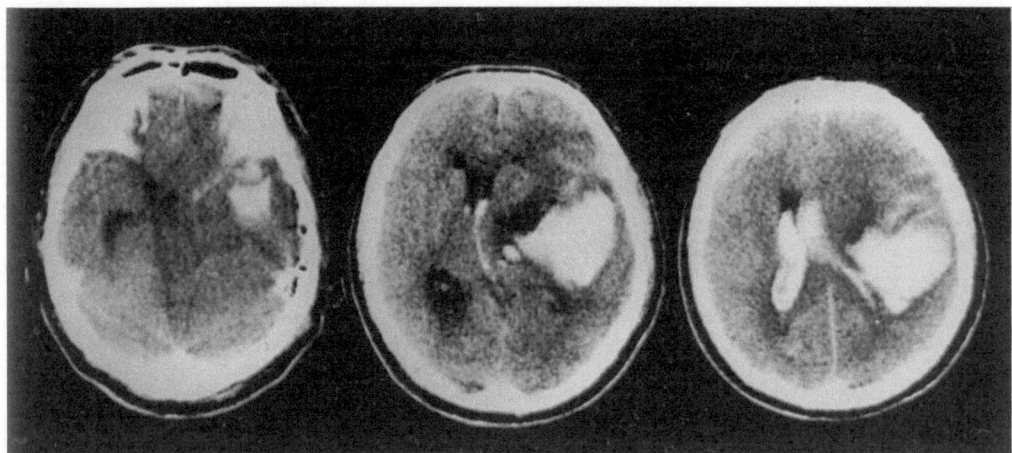

Fig. V-17. A large intracerebral hematoma in the right temporal lobe with intraventricular hematoma. A typical pattern of an intracerebral hematoma due to rupture of the right middle cerebral artery aneurysm

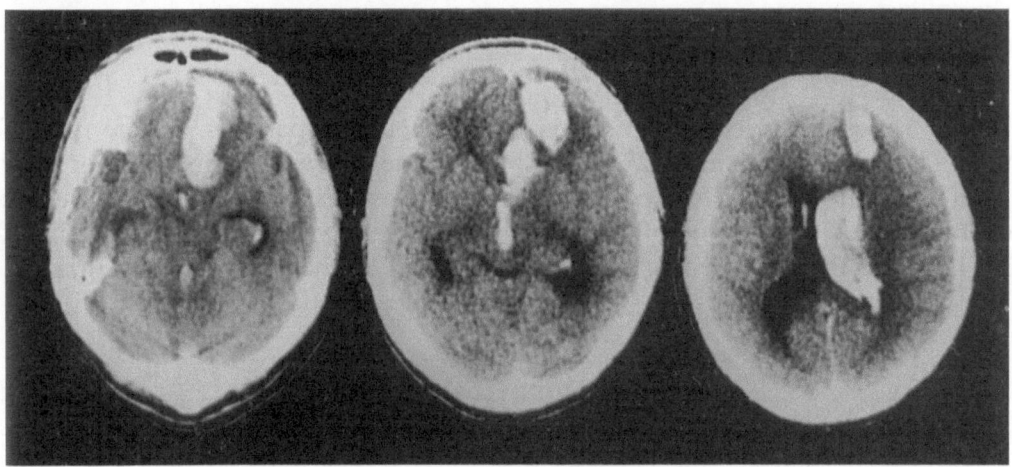

Fig. V-18. Case of an anterior communicating artery aneurysm. An intracerebral hematoma in the right frontal lobe with rupture into the right lateral ventricle and the IIIrd ventricle. This patient fully recovered after the operation

of aneurysm. In AC aneurysms and MC aneurysms, ICHs were found more frequently than in IC aneurysms. Especially, MC aneurysms produce large hematomas. The outcome in SAH patients with or without ICH is shown in **Tables V-22** and **V-23**. Operative results in ICH patients were unfavorable because of significantly high rate of disability. Typical hematomas in each site are shown in figures (**Figs. V-17** through **V-19**).

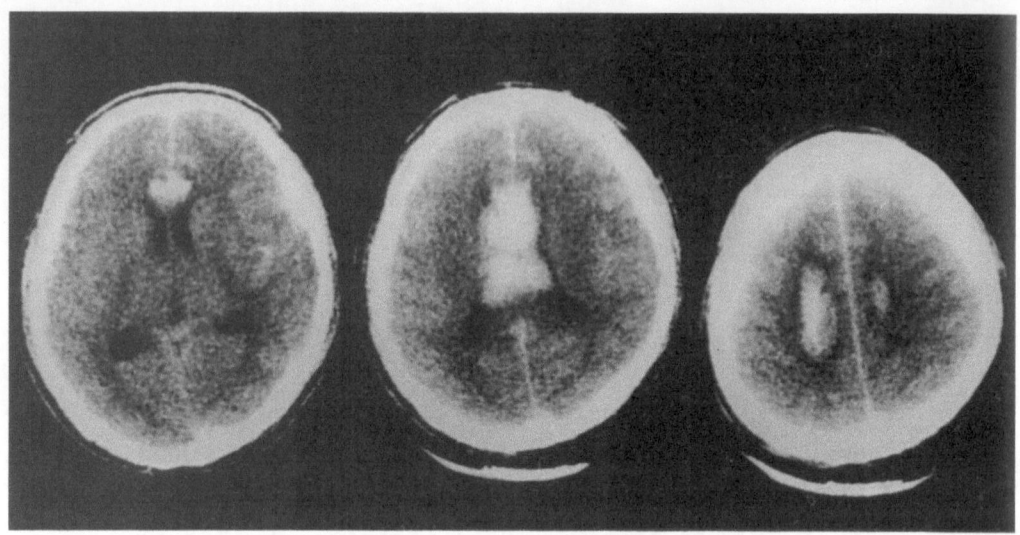

Fig. V-19. Case of an anterior cerebral artery-pericallosal artery aneurysm. CT scan shows an intracerebral hematoma in the corpus callosum

3. Intraventricular Hematoma

In our series, intraventricular hematoma (IVH) occurred in 23.2% of all SAH patients and 15.4% of the operated cases. As shown in **Table V-24**, IVH was found more frequently in AC aneurysm patients. Severe type of IVH means castings in three or four ventricles (**Figs. V-20, V-21**). Severe IVH usually resulted from direct rupture of aneurysms into ventricles.

Table V-24. Incidence of intraventricular hematoma

Type or aneurysm	Intraventricular hematoma			Total
	None	Moderate	Severe	
IC	62(87.3%)	7(9.9%)	2(2.8%)	71
AC	67(69.1%)	10(10.3%)	20(20.6%)	97
MC	49(83.1%)	7(11.9%)	3(5.1%)	59
VB	19(76.0%)	5(20.0)	1(4.0)	25
Total	197(78.2%)	29(11.5%)	26(10.3%)	252

IC: Internal Carotid Artery.
AC: Anterior Cerebral Artery.
MC: Middle Cerebral Artery.
VB: Vertebro-basilar Artery.

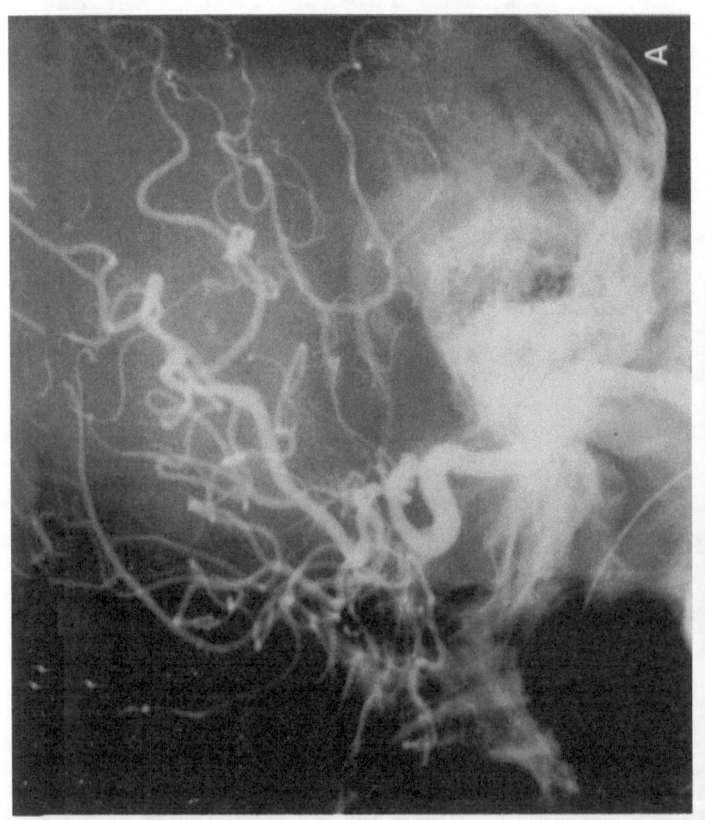

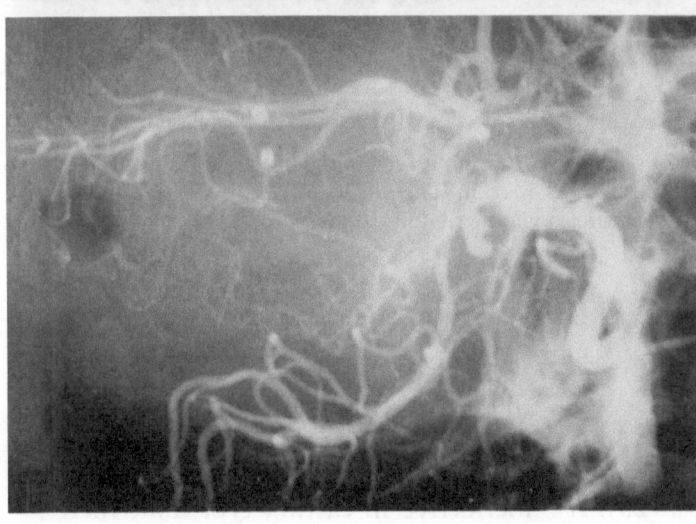

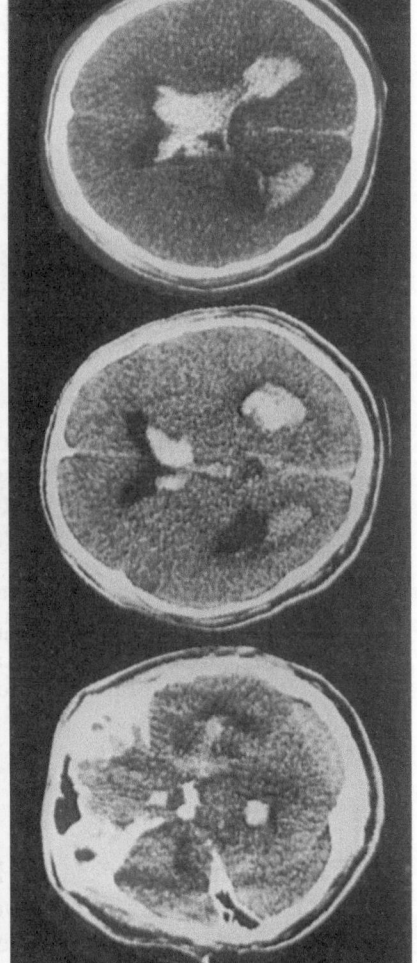

Fig. V-20. A) Angiograms indicate an internal carotid artery-posterior communicating artery aneurysm. The dome of the aneurysm points to the right temporal lobe. B) Intraventricular hematoma is seen in both lateral, IIIrd and IVth ventricles. The aneurysm seems to have ruptured into the right inferior horn

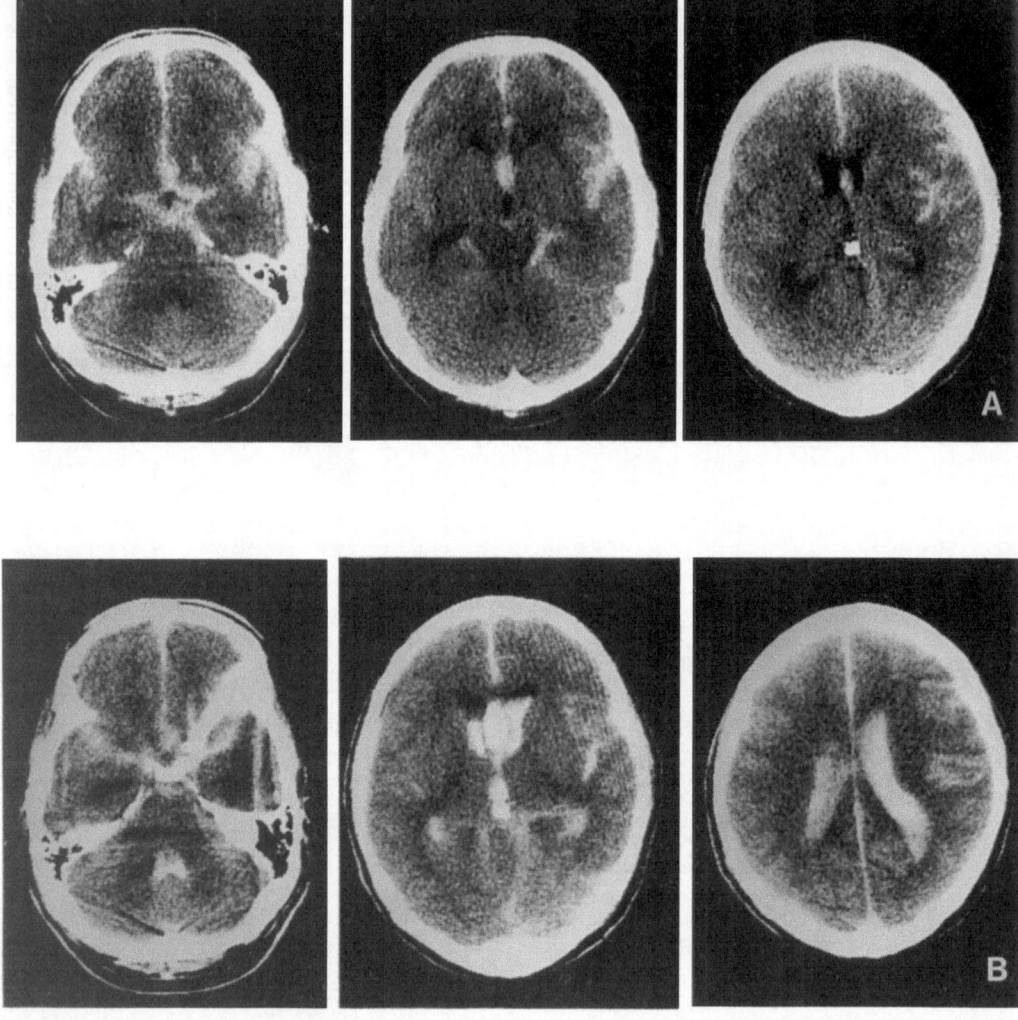

Fig. V-21. A) A small hematoma in the septum pellucidum is seen in the initial CT scan. B) After rebleeding on day 0, intraventricular hematoma is seen. The aneurysm seems to have directly ruptured into the ventricles via the septum pellucidum

The outcome of severe IVH cases was poor. In IVH group, however, the rate of disabled patients was lower than that of ICH group (**Tables V-25, V-26**). If secondary acute hydrocephalus is prevented by ventricular drainage or other therapy, the outcome in grade I to III patients was not unfavorable. However, ventricular drainage is usually ineffective in casting type and, therefore, the outcome in grades IV or V was miserable.

Table V-25. Preoperative grade and outcome (6 months) in SAH patients with intraventricular hematoma

	Grade I	Grade II	Grade III	Grade IV	Grade V	Total
Good recovery	4 (100%)	5 (100%)	8 (80.0%)	0	0	17 (51.6%)
Moderate disability	0	0	1 (10.0%)	3 (37.5%)	0	4 (12.1%)
Severe disability	0	0	0	1 (12.5%)	0	1 (3.0%)
Vegetative state	0	0	0	1 (12.5%)	1 (16.7%)	2 (6.1%)
Dead	0	0	1 (10.1%)	3 (37.5%)	5 (83.3%)	9 (27.3%)
Total	4 (12.1%)	5 (15.2%)	10 (30.3%)	8 (24.2%)	6 (18.9%)	33

Table V-26. Preoperative grade and outcome (6 months) in SAH patients without intraventricular hematoma

	Grade I	Grade II	Grade III	Grade IV	Grade V	Total
Good recovery	33 (84.6%)	42 (73.7%)	27 (58.7%)	5 (23.8%)	3 (16.7%)	110 (60.8%)
Moderate disability	4 (10.3%)	6 (10.5%)	7 (15.2%)	6 (28.6%)	2 (11.1%)	25 (13.8%)
Severe disability	1 (2.6%)	4 (7.0%)	5 (10.9%)	7 (33.3%)	2 (11.1%)	19 (10.5%)
Vegetative state	0	1 (1.8%)	1 (2.2%)	1 (4.8%)	1 (5.6%)	4 (2.2%)
Dead	1 (2.6%)	4 (7.0%)	6 (13.0%)	2 (9.5%)	10 (55.6%)	23 (12.7%)
Total	39 (21.7%)	57 (31.5%)	46 (25.4%)	21 (11.6%)	18 (9.9%)	181

4. Subdural Hematoma

Acute subdural hematoma due to aneurysmal rupture is rare. In our series, the subdural hematomas on CT were found in 7 cases (2.6%) **(Figs. V-22 through V-24)**. The outcomes of these cases were worse. Four of them died and two were disabled. Weir *et al.* (1984) reported that in these cases the ruptured aneurysms with subdural hematomas were mostly found on the

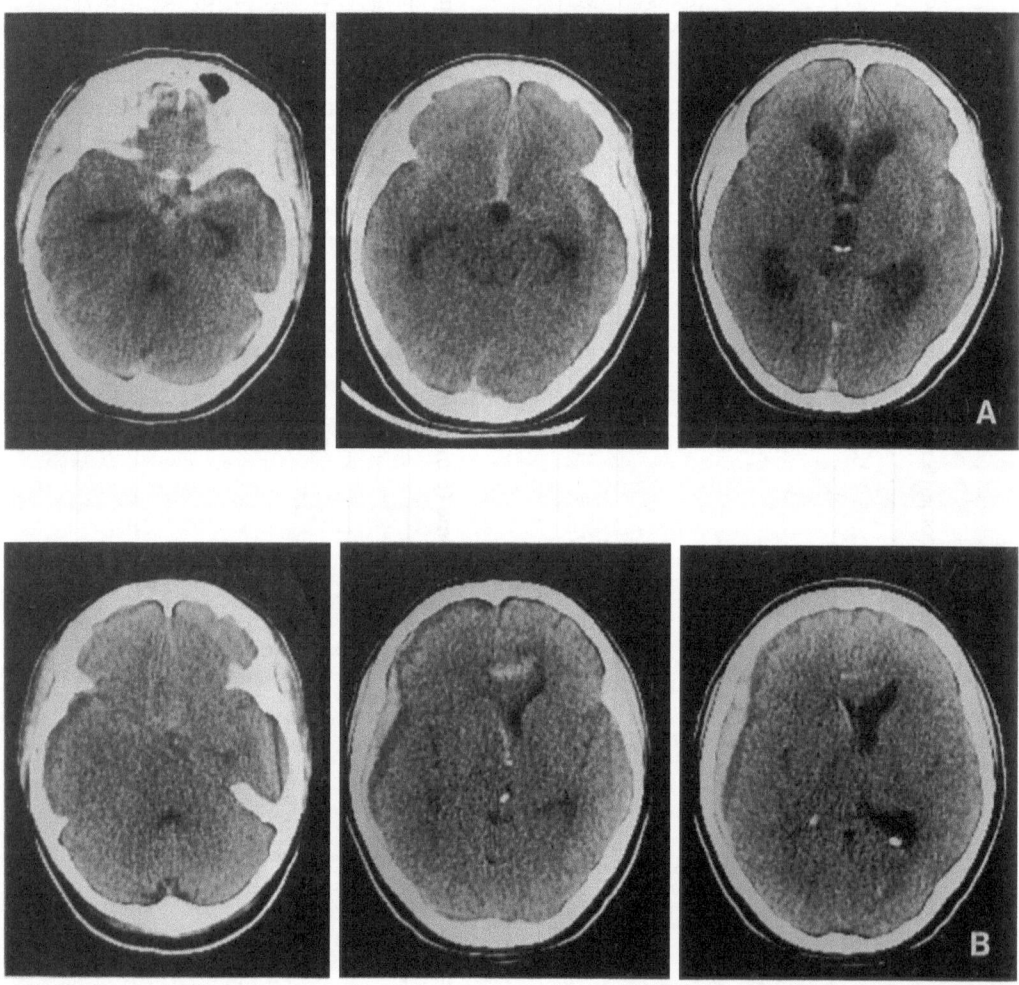

Fig. V-22. A) Subarachnoid blood is seen in the suprasellar cistern and the interhemispheric fissure. A typical SAH pattern of an anterior communicating artery aneurysm with symmetrical ventricular dilatation. CT scan on day 0. B) One week after the initial hemorrhage, rebleeding occurred. CT scan showed a left subdural hematoma with a midline shift. The outcome of this patient was "moderate disability"

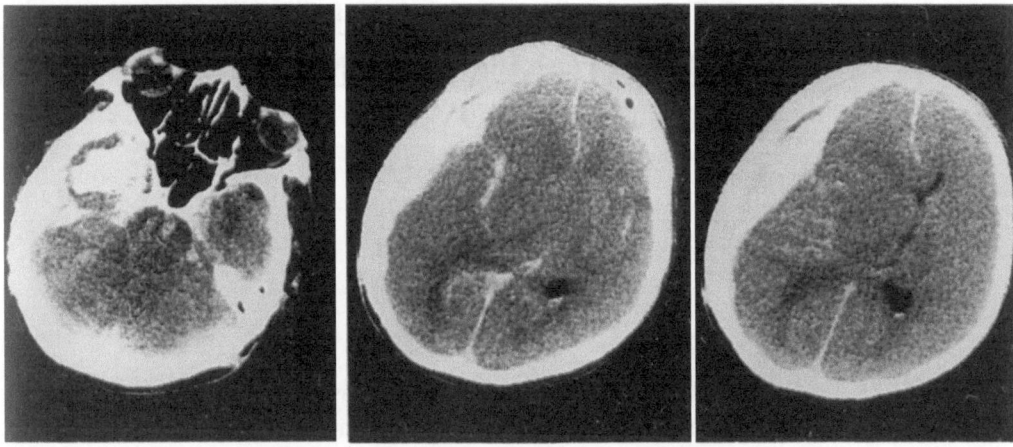

Fig. V-23. Case of a left middle cerebral artery aneurysm with a left subdural hematoma. CT scan on day 0. The status on admission was grade V with deep coma

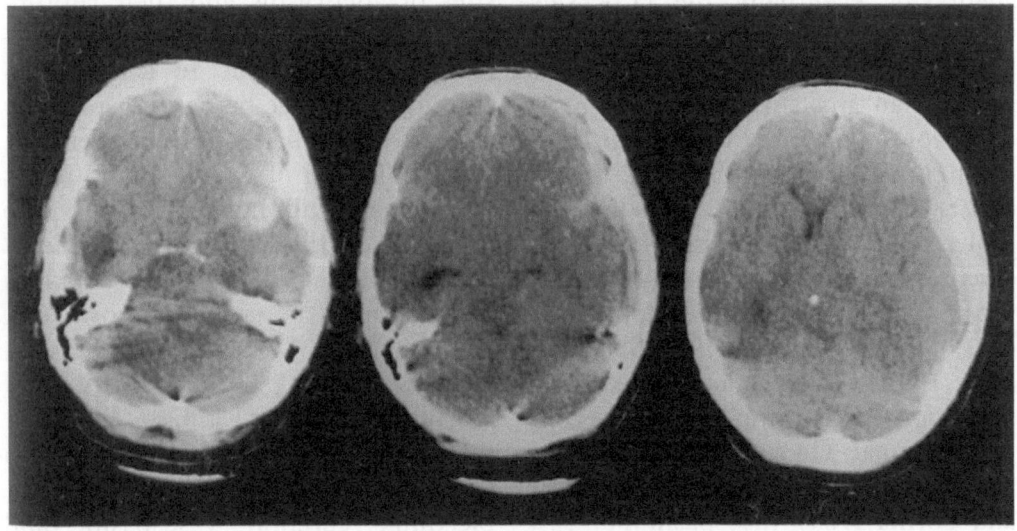

Fig. V-24. Case of a right middle cerebral artery aneurysm with a right subdural hematoma. A small hematoma in the right Sylvian stem is seen on CT scan on day 0. This patient fully recovered

internal carotid artery. However, in our series, only one case showed a ruptured aneurysm on IC, and three on MC and three on AC. The acute subdural hematoma from aneurysmal rupture tends to occur as a result of recurrent hemorrhage because of adhesion of subarachnoid space due to initial hemorrhage (Fig. V-22). Among our 7 cases, the subdural hematomas were seen after rebleeding in 5 cases.

VI. SURGICAL INDICATIONS AND DECISION MAKING

Introduction

Rupture of intracranial saccular aneurysms is the most frequent cause of subarachnoid hemorrhage (SAH). When conservatively treated, about 70% of the cases with ruptured aneurysms eventually die (Locksley 1966). The contention by some neurologists that conservative treatments are better than any surgical treatment is untenable. The three major causes of death in SAH patients are 1. direct effects of subarachnoid bleeding, 2. rebleeding of aneurysms, and 3. cerebral vasospasm, or cerebral arterial spasm, which may later develop in 40–60% of the cases (Odom 1975, Saito and Sano 1980) and may lead to cerebral infarction. According to the recent International Cooperative Study on timing of aneurysm surgery (Kassell and Torner 1984), vasospasm (33.5%), direct effects of SAH (25.5%), and rebleeding of aneurysms (17.3%) were important causes of disability and death in 1.272 patients. Direct effects of SAH or acute ischemic neurdogical deficits (AINDs) of the previous chapter involve the general condition of patients which can be expressed as grades (Botterell et al. 1956, Hunt and Kosnik 1974). Both the general condition and cerebral vasospasm are dependent upon the amount and distribution of subarachnoid blood (Fisher et al. 1980, Sano 1983). The same is true for both acute and chronic hydrocephalus after SAH (Black 1986). Furthermore, the sites of aneurysms and their rupture (e.g., the brain stem), increased intracranial pressure due to acute brain swelling (its cause being in dispute) or acute hydrocephalus, and intracerebral hematomas, all of which should affect the grades, will hold sway over the outcome of the patient. Other incidental pathological conditions, such as diabetes mellitus, atherosclerosis, etc. and the age of the patient may have an influence upon the mortality and morbidity.

Anyway, the most influential factors on the outcome of SAH patients are rebleeding, cerebral vasospasm, and direct effects of subarachnoid blood. Therefore, the most essential therapies of SAH patients are 1. prevention of rebleeding of aneurysms, and 2. prevention or treatment of cerebral vasospasm, and 3. if possible, to lessen the direct effects of SAH.

Theoretically, rebleeding can be prevented by clipping of aneurysms in the very early stage. Concurrent intracerebral hematomas can also be removed by early surgery. Acute hydrocephalus, if present, can be treated by ventricular drainage or shunting.

As was stated in the previous chapters, cerebral vasospasm is caused by free radical reactions initiated by clot lysis in the subarachnoid space (Asano et al. 1980, Sano et al. 1980). Therefore if all or most of subarachnoid blood clots can be removed by early surgery with irrigation and drainage of the subarachnoid space, cerebral vasospasm should be prevented by early surgery. This, however, is at variance with the reality. There may also be a possibility of decreasing the direct effects of SAH, at least to some extent, by washing out subarachnoid clots and by irrigating and draining the subarachnoid space (Saito et al. 1977, Sano and Saito 1980) and by the use of the so-called cerebral protectors such as barbiturates or more preferably nizofenone (Y-9179) (Sano 1983, Tamura et al. 1979).

According to the Cooperative Study in the United States from 1955 to 1966 (Locksley 1966), the rate of rebleeding is highest around 7 days after the onset of SAH. However, the so-called rebleeding in these statistics may partly include aggravation of the neurological status due to cerebral vasospasm.

According to Jane et al. (1977), for a patient with an anterior communicating aneurysm or a posterior communicating aneurysm seen on the first day, namely on the day of rupture, the chance he will rebleed within the first 6 months is about 50–60%. This rebleed rate rapidly diminishes, and on the 10th day the rate is approximately 35–50%. After 6 months this rate is about 3% a year; two thirds of these late hemorrhages, however, result in mortality.

In the Cooperative Study published in 1981 (Adams et al. 1981 a), rebleeding within 14 days after SAH occurred in 13% ($^{33}/_{249}$) of patients (9.6% occurred in good condition patients and 25% in poor condition patients), in spite of high dose antifibrinolytic therapy. According to the most recent International Cooperative Study (Kassell and Torner 1983, 1984), the peak of rebleeding occurred within the first 24 hours after the initial SAH; i.e., about 4.1% of the patients rebled. Then the rebleed rate dropped sharply until the end of 48 hours, when it was about 1.5% per day, declining gradually thereafter. The cumulative rebleed rate was 19% at 14 days.

In the same International Cooperative Study, antifibrinolytic agents significantly reduced the incidence of rebleeding; this effect, however, was offset by the higher incidence of vasospasm and focal ischemic deficit in the antifibrinolytic therapy group; consequently, the mortality at 14 days and at 30 days after SAH in this group was about the same as that in the no-antifibrinolytic therapy group (Kassell et al. 1984). Besides, 14% of the patients who received antifibrinolytic agents developed hydrocephalus compared to 7% of the patients without antifibrinolytic therapy and this difference was statistically significant. Furthermore, the rebleeding rates on days 0 and 1 in the groups with and without antifibrinolytic therapy were essentially the same. The reason is probably as follows. If antifibrinolytic agents are effective, there is usually a

Table VI-1. Summary of intracranial aneurysms experienced from December 1969 to December 1982 (1,047 cases)

20 cases of intracavernous aneurysms were treated by carotid ligation and/or trapping.
87 cases of unruptured (grade 0) aneurysms underwent microsurgery.

940 cases had subarachnoid hemorrhage (SAH) due to ruptured aneurysms.
a) 82 cases died before surgery.
b) 2 cases had spontaneous obliteration of aneurysms (Acom, M_1).
c) 4 cases did not want to have surgery.
d) 852 cases underwent microsurgery of aneurysms.

Timing of surgery	Cases	Op. mortality	Working in the follow-up
Within one week after SAH:	335	46 (13.7%)	219 (65.4%)
0*–2 day	222	31 (14.0%)	142 (64.0%)
3–7 day	113	15 (13.3%)	77 (68.1%)
Second week after SAH:	108	5 (4.6%)	95 (88.0%)
[within 2 weeks after SAH:	443	51 (11.5%)	314 (70.8%)]
Later than second week after SAH:	409	13 (3.2%)	333 (81.4%)

Note: 0* day is the day of SAH.

delay of 24 to 48 hours before adequate levels can be attained in the cerebrospinal fluid and perianeurysmal clot. Then, if the aim of early surgery of aneurysms is really to prevent rebleeding, one should do surgery as early as possible, may be on day 0 (the day of SAH). Logistical considerations will often make this approach difficult in many neurosurgical centers. Since antifibrinolytic agents are not effective within 48 hours after SAH, a practical means to decrease the probability of preoperative rebleeding in this early period may be induced systemic arterial hypotension as suggested by Kassel and Torner (1983).

In most neurosurgical centers of North America, delayed surgery, instead of early surgery, has been practiced because of the reportedly poor results of surgery done in the early stage of SAH (Drake 1981, Pia 1981). This trend, however, was recently criticized by Ausman et al. (1985).

From December 1969 to December 1982, we experienced 1,047 cases of intracranial aneurysms, among which 940 cases had SAH and 852 of these underwent microsurgical treatment as shown in **Table VI-1**. Review of this series shows that the operative mortality (hospital death) of surgery done within the first week of SAH (335 cases) was 13.7%, whereas that of surgery done later than the first week (517 cases) was as low as 3.5%; the total operative mortality in 852 cases was 7.5% (**Table VI-2**). Therefore, the operative mortality of surgery done

Table VI-2. Timing of microsurgery of ruptured aneurysm. Comparison of the group submitted to surgery within one week after SAH with the group submitted to surgery later than the first week after SAH

Location of Aneurysms	Cases	Within one week from the last SAH	Later than the 1st week	Operative Mortality
ICPcom	187	59(9)[†]	128 (3)[†]	12 = 6.4%
IC-bif	24	8(3)	16 (1)	4 = 16.7%
Ant. chor.	40	15(4)	25 (2)	6 = 15.0%
Acom	244	101 (14)	143 (7)	21 = 8.6%
A_1, A_2	43	17(3)	26 (1)	4 = 9.3%
MC	163	84(8)	79 (1)	9 = 5.5%
V-B	50	10(3)	40 (2)	5 = 10.0%
Multiple	101	41 (2)	60 (1)	3 = 3.0%
Total	852	335 (46=13.7%)	517 (18=3.5%)	64 = 7.5%

ICPcom : internal carotid-posterior communicating (here including ophthalmic).
IC-bif : internal carotid bifurcation, Ant. chor.: anterior choroidal.
Acom : anterior communicating, A_1, A_2 : A_1 and A_2 portions of the anterior cerebral artery, MC : middle cerebral, V-B : vertebro-basilar,
multiple : multiple aneurysms. ( )[†] : operative death.

within the first week was about 3.9 times as high as that of surgery done later than the first week (χ^2 test, $p < 0.1\%$). Two reasons account for this: one was the general condition of patients which can be expressed as higher grades (Hunt and Kosnik 1974), and the other was cerebral vasospasm which developed later after surgery.

A. Grading and Surgical Results

Grading of risk or classification of patients according to their general condition was done by Botterell, Hunt, and others (Botterell *et al.* 1956, Hunt and Kosnik 1974). Hunt has stressed the importance of the general condition of patients rather than the timing for surgery of aneurysms (Hunt and Miller 1977) and he established criteria for the grading of risk (Hunt and Kosnik 1974) which is now widely used. Although we are aware of observer variability in this grading (Lindsay *et al.* 1982a) and are proposing a new grading (Sano and Tamura 1985), Hunt's grading is used in this chapter because of its popularity.

Table VI-3 demonstrates the operative mortality (death during hospitalization) and morbidity of patients of our series in various preoperative conditions classified according to the grading of Hunt (grade I a was included in grade I). The lower the grade was, the lower was the operative mortality (*e.g.*, 1.9% in grade I) and the better the follow-up results were (*e.g.*, 88.6% of grade I patients were able to work). In the follow-up from two to 14 years, as a whole, 75.9% of the aneurysm cases that underwent surgery were well and working, 10.0% were able to care for themselves (namely, the patient needs no help to wash, to eat, to go to the toilet, but is unable to work), and 6.6% were either bed-ridden (4.8%) or died of other diseases after discharge (1.8%).

"Working" corresponds to "good recovery"; "caring for self" to "moderately disabled"; "bed-ridden"

patients and the cerebral vasospasm are closely related to the amount and distribution of subarachnoid bleeding. In this sense, the mortality and the morbidity of SAH patients may be said to already be determined to a considerable degree at the time of bleeding. In early surgery of ruptured aneurysms, all

Table VI-3. Operative mortality and morbidity of cases with ruptured aneurysms according to preoperative Hunt's grading

	Cases	Op. Mortality	Working	Caring for self	BD
Grade I	378	7 (1.9%)	335 (88.6%)	25 (6.6%)	11 (2.9%)
II	219	11 (5.0%)	188 (85.8%)	14 (6.4%)	6 (2.7%)
III	160	16 (10.0%)	99 (61.9%)	25 (15.6%)	20 (12.5%)
IV	78	29 (25.6%)	25 (32.1%)	19 (24.4%)	14 (17.9%)
V	17	10 (58.8%)	0	2 (11.8%)	5 (29.4%)
Total	852	64 (7.5%)	647 (75.9%)	85 (10.0%)	56 (6.6%)

[BD : bed-ridden (41 cases) or died of other diseases after discharge (15 cases).]

to "severely disabled" and "vegetative"; and "operative mortality" plus "died of other diseases after discharge" to "dead" of Glasgow outcome scale (Jennett and Bond 1975). In the International Cooperative Study (Kassell and Torner 1974) on patients with intracranial surgery, good recovery was 66.8%, moderately disabled was 9.6%; severely disabled 6.2%; vegetative survival, 2.2%; dead was 15.2%. Our results are comparable to or a little bit superior to these results.

In the early stage of SAH, patients are usually of higher grades, which seems to explain the higher operative mortality in early surgery. Both the grade of

these factors—especially the vasospasm which may later develop if subarachnoid blood is of a considerable quantity, regardless of surgical or conservative treatments—will influence the operative results—in addition to the surgeon's technique. In surgery of aneurysms in the later period, e.g., more than two weeks after SAH, the surgeon's technique is almost the sole factor influencing the results.

From the surgeon's point of view, he will prefer to perform surgery on good grade (I–II) patients, since postoperative outcome may be expected to be excellent in these patients.

B. Analysis of Timing of Surgery and the Surgical Results

The surgeon can prevent ruptured aneurysms from rebleeding by early surgery, but then he must encounter poor grade patients and also the problem of cerebral vasospasm which may develop later. Then when to do surgery? In order to solve the problem of timing of surgery, our own series of 852 cases of ruptured aneurysms submitted to microsurgery was analysed.

As far as the operative mortality (death during postoperative hospitalization) is concerned, χ^2 test shows that grade I cases are only a little bit superior to grade II cases ($p < 5\%$), however, definitely superior to grade III ($p < 0.1\%$), IV ($p < 0.1\%$), or V ($p < 0.1\%$) cases. Grade II cases are not significantly better than grade III cases, definitely better, however, than grades IV and V cases ($p < 0.1\%$).

Grade III cases are better than grade IV ($p < 0.5\%$) or V ($p < 0.1\%$), and IV cases are superior to grade V ($p < 1\%$).

It is concluded that the preoperative grading of patients' conditions is the definitive factor which influences the operative mortality. Since cerebral vasospasm, at least angiographical vasospasm, appears first on day 4 and not before (Saito et al. 1977), the first week after SAH was divided into 2 periods: from day 0 (the day of aneurysmal rupture) to 2 (48 hours after SAH), and from day 3 to 7. In the first week, of 335 cases that underwent microsurgery, grade I cases were only 39 (11.6%), grade II cases were 156 (46.6%), grade III cases were 81 (24.2%), grade IV cases were 44 (13.1%), and grade V cases were 15 (4.5%); i.e., grade III—V cases constituted 41.8% of the total.

The operative mortality in 222 cases that underwent surgery on day 0 through 2 (that is, within 48 hours after SAH) and the mortality in 113 cases submitted to surgery on day 3 through 7 were relatively high (14.0% and 13.3%, respectively) as compared with that in 108 cases submitted to surgery in the second week after SAH (4.6%) and that in 409 cases operated on later than the second week (3.2%), as seen in Table VI-4. If only grade I to III cases were considered, however, the operative mortality in cases submitted to surgery on day 0 through 2 was 8.0% and that in cases submitted to surgery on day 3 through 7 was 9.0%. That in cases operated on in the second week or later than the second week was 2.2 or 2.3%, as shown in the lowermost row of Table VI-4.

As far as the operative mortality is concerned, χ^2 test reveals that there was no statistically significant difference at any timing of surgery in each of grade I, II, IV or V cases. In grade III cases, however, surgery on day 3–7 resulted in a higher operative mortality as compared with surgery in the second week or later than the second week ($p < 1\%$). Surgery on grade III cases on day 0–2, however, showed no statistically significant difference in the operative mortality as compared with that in other timing of surgery.

Therefore, as for the operative mortality, it can be said that for grade III cases, surgery on day 3–7 is best avoided, as was suggested by us years

Table VI-4. Operative mortality and morbidity of cases with ruptured aneurysms according to timing of surgery and preoperative Hunt's grading

| | | 1st week after SAH | | | | | | | | | | IInd week after SAH | | | | | Later than IInd week | | | | |
| | | 0 (SAH day) - II day | | | | | III - VII day | | | | | | | | | | | | | | |
Grade	Cases	Op. Mt.	Wor-king	Caring for self	BD	Cases	Op. Mt.	Wor-king	Caring for self	BD	Cases	Op. Mt.	Wor-king	Caring for self	BD	Cases	Op. Mt.	Wor-king	Caring for self	BD
I	21	1	20	0	0	18	1	16	1	0	27	0	25	2	0	312	5	274	22	11
II	98	6	82	9	1	58	1	50	4	3	28	1	27	0	0	35	3	29	1	2
III	57	7	31	11	8	24	7	8	7	2	38	1	34	3	0	41	1	26	4	10
IV	34	12	9	7	6	10	3	3	3	1	14	2	9	1	2	20	3	4	8	5
V	12	5	0	2	5	3	3	0	0	0	1	1	0	0	0	1	1	0	0	0
Total	222	31	142	29	20	113	15	77	15	6	108	5	95	6	2	409	13	333	35	28
		‖	‖	‖	‖		‖	‖	‖	‖		‖	‖	‖	‖		‖	‖	‖	‖
		14.0%	64.0%	13.1%	9.0%		13.3%	68.1%	13.3%	5.3%		4.6%	88.0%	5.6%	1.9%		3.2%	81.4%	8.6%	6.8%

[Op. Mt.: operative mortality. BD: bed-ridden (41 cases) or died of other disease after discharge (15 cases).]

N.B.																				
Grade I - III Cases	176	14	133	20	9	100	9	74	12	5	93	2	86	5	0	388	9	329	27	23
		‖	‖				‖	‖				‖	‖				‖	‖		
		8.0%	75.6%				9.0%	74.0%				2.2%	92.5%				2.3%	84.8%		

ago (Sano and Saito 1978). This mortality is mostly due to cerebral vasospasm. Grade I and II cases showed good results at any timing of surgery. Therefore, early surgery is indicated for them. Grade IV and V cases exhibited poor results at any timing of surgery. Surgery, therefore, may better be postponed until they show better grades.

For timing of surgery, age should also be considered. In **Table VI-5**, grade I–IV cases (grade V cases were excluded because, as is universally recognized, surgery is not indicated for grade V cases) were divided into two groups; one below 60 years of age and one 60 years and over. There was a significant difference ($p < 0.1\%$) in the operative mortality between the group younger than 60 years (4.3%) and the group 60 years and over (13.6%). In the former group, there was no significant difference by the χ^2 test in the operative mortality at any timing of surgery, except that the operative mortality in surgery on day 0–2 was higher than that in surgery later than the second week ($p < 5\%$) and that in surgery on day 3–7 was even higher than that in surgery later than the second week ($p < 0.5\%$). In the group 60 years and over, the operative mortality in cases submitted to surgery on day 0 through 2 (26.2%) was significantly higher than that in the second week ($p < 5\%$) or that later than the second week ($p < 0.5\%$), but not significantly higher than that on day 3 through 7 (14.3%). This high mortality in early surgery in the higher age group

Table VI-5. Operative mortality and morbidity of grade I–IV cases with ruptured aneurysms according to age and timing

Age	Timing of Operation	Cases	Op. Mortality	Working	Caring for self	BD
	Ist week					
	0 (SAH) - II day	149	10 (6.7%)	118 (79.2%)	15 (10.1%)	6 (4.0%)
	III - VII day	82	8 (9.8%)	57 (69.5%)	12 (14.6%)	5 (6.1%)
	IInd week after SAH	83	3 (3.6%)	75 (90.4%)	4 (4.8%)	1 (1.2%)
	Later than IInd week	330	7 (2.1%)	276 (83.6%)	25 (7.6%)	22 (6.7%)
59 yrs.	Subtotal	644	28 (4.3%)	526 (81.7%)	56 (8.7%)	34 (5.3%)
60 yrs.	**Ist week**					
	0 (SAH) - II day	61	16 (26.2%)	24 (39.3%)	12 (19.7%)	9 (14.8%)
	III - VII day	28	4 (14.3%)	20 (71.4%)	3 (10.7%)	1 (3.6%)
	IInd week after SAH	24	1 (4.2%)	20 (83.3%)	2 (8.3%)	1 (4.2%)
	Later than IInd week	78	5 (6.4%)	57 (73.1%)	10 (12.8%)	6 (7.7%)
	Subtotal	191	26 (13.6%)	121 (63.4%)	27 (14.1%)	17 (8.9%)
	Grand Total	835	54 (6.5%)	647 (77.5%)	83 (9.9%)	51 (6.1%)

[BD : bed-ridden (38 cases) or died of other diseases after discharge (13 cases).]

was probably partly because these mortality cases belonged to the higher grades (III, IV) and partly because in the higher age group, the ability to recover from pathologic conditions may be lowered.

From these data, one may conclude that for cases 60 years of age or above when they are of grade III or of higher grades, surgery may better be postponed to the second week or later.

U test

To evaluate the mortality and morbidity together, the U test (Mann-Whitney) may be suitable. According to this test, there was no difference in the mortality and morbidity between grade I and II cases, at any timing of surgery. Both, grade I and II cases had significantly better mortality and morbidity than grade III (p < 0.1%) which was better than grade IV (p < 0.1%) which was, in turn, significantly better than grade V (p < 0.1%).

Grade IV cases showed no statistically significant difference in the mortality and morbidity for different timing of surgery. The same was true for grade V cases. The results in grade III cases, however, were different for different timing of surgery; e.g., the results were better in cases submitted to surgery in the second week than in later timing (p < 1%) or in the first week (p < 0.1%), and the results were slightly better in cases submitted to surgery on day 0 through 2 than on day 3 through 7 (p < 10%). In the group

Table VI-6. The rates of good recovery and postoperative death at 3 months postoperative. Comparison of the younger age and the older age groups, 1. with the borderline, 60 years of age and 2. 65 years of age

		Grade	I, II	III	IV	V
(1)	59 years or younger (109 cases)	Working (good recovery)	82.1%	69.0%	8.3%	0
		Death	3.6%	6.9%	25.0%	50.0%
	60 years or older (36 cases)	Working (good recovery)	58.3%	63.6%	0	0
		Death	8.3%	18.2%	25.0%	100%
(2)	64 years or younger (130 cases)	Working (good recovery)	81.0%	71.1%	6.7%	0
		Death	4.8%	10.5%	2.0%	57.1%
	65 years or older (15 cases)	Working (good recovery)	40.0%	0	0	0
		Death	0	0	40.0%	100%

younger than 60 years of age, cases that underwent surgery on day 3 through 7 showed poorer results as compared with cases operated on in later timing ($p < 0.1\%$), whereas, in the group aged 60 years or above, cases submitted to surgery on day 0 through 2 showed poorer results than the others ($p < 1-0.1\%$).

The borderline of 60 years of age was adopted here arbitrarily. We are not sure whether this borderline is valid or not. We are even more uncertain whether this borderline is valid for different grading or not. Therefore, we checked the latter in 146 recent cases of aneurysmal SAH that were submitted to surgery. The postoperative 3-month-results were as shown in Table VI-6. The results were not too different whether the borderline was 60 years or 65 years. And the younger age group (59 years and younger or 64 years and younger) showed good surgical results in grades I, II, and III. Considering the indiviudal biological variability, the borderline can be set between the ages of 60 and 65 years.

These differences in the operative mortality and postoperative morbidity in the different grading, timing, and age groups were mostly due to the amount of subarachnoid bleeding, its effects on the surrounding structures, and the consequential cerebral vasospasm, not to mention biological responses of each individual. Probably, surgery on aneurysms and removal of subarachnoid clots within 48 hours after SAH (if at all possible) can decrease the amount of subarachnoid bloot clots before their degradation occurs so as to decrease the extent of, or hopefully to prevent, development of the vasospasm. After day 3, degradation of clots may have already occurred to produce vasoactive substances or to initiate free radical reactions, so that surgery cannot prevent development of cerebral vasospasm. In the second week or later, if a case develops cerebral vasospasm, the grading becomes higher, and surgery may have to be postponed. Because of this inadvertent selection, the results of delayed surgery may seem to be better than those of early surgery. This observation is shared by other authors (Hugenholtz and Elgie 1982, Suzuki 1979, Suzuki et al. 1979).

One of the main reasons why many neurosurgeons hesitate to perform early surgery (which is the best means to prevent rebleeding of aneurysms) is fear of postoperative vasospasm, which is actually only remotely related to surgery itself. Therefore, if investigations yield effective drugs for prevention or cure of chronic cerebral vasospasm in the near future, the treatment of ruptured intracranial aneurysms will be changed; very early surgery to prevent rebleeding of aneurysms, combined with removal of as many subarachnoid clots as possible and fashioning of cisternal drainage or irrigation, will be performed—accompanied by ample administration (systemic or topical) of such drugs. Then the results of treatment of ruptured intracranial aneurysms will be greatly ameliorated.

C. Principle of Surgery of Aneurysms in the Acute Stage of SAH

Our principle for the surgical treatment of ruptured aneurysms in the acute stage is currently as follows:
1. During the first 3 days (48 hours) of SAH, namely on day 0 through 2, microsurgery of aneurysms with removal of subarachnoid blood clots is indicated for any case except for grades IV and V cases and probably for grade III cases 60 or 65 years and older. After clipping of aneurysms, treatment with induced hypertension and intravascular volume expansion (Kassell et al. 1982), and additional use of medicaments such as cerebral protectors, free radical scavengers, thromboxane A_2 inhibitors or antagonists (Sano 1983), calcium channel blockers, etc are recommended.

2. After day 3, surgery for cases in grades III, IV, and V should be postponed until they show improvement of neurological conditions by various conservative treatments. Microsurgery is indicated for cases in grades I and II at any time.
3. Any case showing neurological deterioration (to higher grades) should be submitted to computerized tomography (CT) and angiographical examinations to detect vasospasm, which is the most probable cause of the deterioration. If vasospasm is present, surgery should be postponed until vasospasm begins to subside and disturbance of consciousness begins to improve; however, this postponement cannot be for too long, because rebleed-

ing of aneurysms may occur when va-sospasm begins to subside (Sano 1983). The optimal timing of surgery for cases with preoperative vasospasm may be the second week after the onset of vasospasm.

4. If CT demonstrates an intracerebral hematoma caused by ruptured aneurysm, immediate surgery (clipping of the aneurysm and removal of the hematoma) is indicated for that case, because prognosis of the conservatively treated case is usually poor and, besides, surgery resulted in a useful social life for 48% of these cases in our series (Sano 1983).

VII. PERIOPERATIVE CARE

Subarachnoid hemorrhage (SAH) due to aneurysmal rupture is one of the most dramatic events in medicine. The Hisayama study in Japan (Omae *et al.* 1976) showed that death from SAH was found to be cause in 2.7% of all deaths, and 5.3% for those under 60 years of age. This means that the age-adjusted death rate for death from SAH in Japan per 100,000 people is about 4. Meanwhile, approximately 28,000 individuals in North America will experience aneurysmal rupture causing subarachnoid hemorrhage each year. Approximately 10,000 of these patients either die or are disabled as a result of the initial SAH insult. Out of these 10,000 fatal patients, 3,000 die rapidly without any warning. Therefore, the remaining 7,000 could be possibly saved by promoting public health or primary physician education, because causes of their death or disabled state are 1. ignored warning symptoms, 2. initial misdiagnosis, and/or 3. late referral. The other 18,000 patients could receive medical or surgical treatment. However, approximately 8,000 of these ultimately die or are disabled; 3,000 from rebleeding, 3,000 from vasospasm, 1,000 from medical complications, and 1,000 from surgical complications (Kassell and Drake 1982).

In fact, two greatest causes of death after SAH, which could be eliminated by medical or surgical treatment, are rebleeding of aneurysms and cerebral vasospasm (Kassell and Drake 1982). Therefore, the most essential therapies of SAH patients are 1. the preventive therapy against rebleeding of aneurysms and 2. the prevention or treatment of cerebral vasospasm which may lead to cerebral infarction (Ausman *et al.* 1985, Sano 1983, Sasaki *et al.* 1985 b).

A. Rebleeding

Rebleeding is the most influential factor on the outcome of SAH patients. In our series, rebleeding occurred in 74 cases (27.2%). **Table VII-1** shows the incidence of rebleeding by the site of aneurysm. In this table, results from 252 cases, in whome diagnosis was done by angiography, are shown. There was no site difference. In 34 cases (12.5%), rebleeding occurred after admission. Over all results with or without rebleeding after admission are shown in **Tables VII-2** and **VII-3**. The outcome of grades I to III in the rebled group is significantly worse than that of the nonrebled group. In the rebled group, mortality is 67.6% and favorable results including moderate disability are only seen in 33% of grade I or II patients.

Theoretically, the possibility of rebleeding should be eliminated by early surgery. However, the results from the Cooperative Aneurysm Study (Kassell

Table VII-1. Incidence of rebleeding

Type of aneurysm	Rebleeding
IC	$^{14}/_{71}$ (19.7%)
AC	$^{26}/_{97}$ (27.8%)
MC	$^{19}/_{59}$ (32.2%)
VB	$^{8}/_{25}$ (32.0%)
Total	$^{67}/_{252}$ (26.6%)

and Torner 1983) indicate that rebleeding occurs with the greatest frequency within 24 hours after the initial hemorrhage. In our data, rebleeding was also most frequently observed on day 0, especially within several hours after the intitial insult, ant this was sometimes fatal. Yasui et al. (1985 b) reported that, in a total of 420 SAH patients, 34.1% of them had more than two attacks. One third of these rebled patients (33.6%) had a second attack within 6 hours after the initial attack and about half of the rebleeding occurred within 24 hours. In their clinical results, the outcome of a patient with rebleeding was significantly bad. Taneda et al. (1985) indicated the risk of rebleeding during angiography. In their report, rebleeding occurred in 18.5% of 200 SAH

patients in whom a four vessel study by Seldinger's method was performed within 24 hours after SAH. Among them, angiography was related to rebleeding in 62% of the cases. Rebleeding in their series occurred in 92% of the patients within 10 hours after SAH. Therefore, therapy to prevent rebleeding should be started just after admission. From this point of view, invasive investigations, such as angiography, may better be avoided within the first 6 hours after SAH. Therefore, our strategy is as follows. In patients admitted to the hospital within a few hours after SAH, CT scanning is immediately performed but angiography is put off until several hours to half a day later. During this waiting period, the patients are sedated or anesthetized with intubation. After angiography, an immediate operation should be done according to the patient's grade, age, and general condition. Since preventive therapies against rebleeding, control of systemic arterial pressure, bed rest, and sedation, are in most cases in use.

Although, in some reports, the reduction in the patient's systemic arterial pressure has not significantly reduced the rate of rebleeding, a large number of physicians generally use antihypertensive drugs. Medical sedation is also used in general for the purpose of rest.

The value of antifibrinolytic therapy in the acute period following SAH is still controversial. In a report of the Cooperative Aneurysm Study, Adams et al. (1981 b) mentioned that antifibrinolytic therapy is a useful modality in the preoperative care of patients with SAH, although some minor and a few major

Table VII-2. Grade on admission and outcome (6 months) in SAH patients with rebleeding after admission

Grades	I	II	III	IV	V	Total
Good recovery	0	2	2	1	1	6 (17.6%)
Moderate disability	0	1	2	0	0	3 (8.8%)
Severe disability	0	1	0	0	0	1 (2.9%)
Vegetative state	1	0	0	0	0	1 (2.9%)
Dead	0	4	6	6	7	23 (67.6%)
Total	1 (2.9%)	8 (23.5%)	10 (29.4%)	7 (20.6%)	8 (23.5%)	34

Table VII-3. Grade on admission and outcome (6 months) in SAH patient without rebleeding after admission

Grades	I	II	III	IV	V	Total
Good recovery	20	65	35	5	1	126 (52.9%)
Moderate disability	3	8	9	4	2	26 (10.9%)
Severe disability	1	4	4	7	3	19 (8.0%)
Vegetative state	0	2	3	0	2	7 (2.9%)
Dead	1	7	8	2	42	60 (25.2%)
Total	25 (10.5%)	86 (36.6%)	59 (24.8%)	18 (7.6%)	50 (21.0%)	238

side effects occur. In their study, antifi-brinolytic therapy is associated with a 10% incidence of rebleeding, which is one half the rate among patients treated with bed rest alone.

Ramirez-Lassepas (1981) reviewed 25 published studies on the treatment with antifibrinolytic agents of SAH caused by ruptured intracranial aneurysms. In 13 controlled studies, a decrease in rebleeding was reported in 7 studies, although only 4 showed decreased mortality. Three studies showed no effect, and three reported a higher rate of rebleeding. He concluded that the data failed to demonstrate that antifi-brinolytic therapy alters the natural history of the disease.

Vermeulen et al. (1984) reported that they were unable to find beneficial results of antifibrinolytic treatment in SAH by a multicenter, randomized, double-blind, placebo-controlled trial. In a total of 479 patients, aneurysms were found in 130 of the tranexamic acid group and 155 of the placebo group. At three months there was no statistical difference between the out-comes in the tranexamic acid group and the control group. Rebleeding was sig-nificantly reduced in the tranexamic acid group ($p < 0.001$). The results were the same when only patients with an aneurysm demonstrated at angiog-raphy were considered. Cerebral in-farction occurred more frequently in the tranexamic acid group ($p < 0.01$). When only the results in patients with an aneurysm were considered, the find-ings were similar. Therefore, they con-cluded that until some method can be found to minimize ischemic complica-tions, tranexamic acid is of no benefit in patients with subarachnoid hemor-rhage.

These results have been verified by Kassell et al. (1984) in the preliminary observations from the International Cooperative Study on the Timing of Aneurysm Surgery. They found that the patients with antifibrinolytic therapy had a significantly lower re-bleeding rate (antifibrinolytic group: 11.7%, no-antifibrinolytic group: 19.4%), but higher rates of ischemic deficits (antifibrinolytic group: 32.4%, no-antifibrinolytic group: 22.7%) and hydrocephalus. Therefore, the net result was not different between the antifibrinolytic group and the control group in mortality in the 1st month following the initial SAH.

B. Vasospasm

1. Characteristics of Vasospasm

Late or chronic cerebral vasospasm, or more exactly, arterial spasm, has the following characteristics.

1. It usually appears after some time lag from the onset of SAH. In Saito and Sano's studies (Saito et al. 1977, Saito and Sano 1980) it appeared at the ear-liest on day 4 and, on average, on day 7.6 ± 2.5 in the natural course. Even postoperative vasospasm appeared with nearly the same average interval from the onset of SAH, namely, on day 6.7 ± 1.6.

2. There is a close correlation between

the development of vasospasm and subarachnoid clots, as pointed out by many investigators. One may say, without subarachnoid clots, no vasospasm.

These facts indicate that vasocontractile substances present in the extravasated blood are not by themselves the cause of vasospasm. Therefore, some secondary processes such as chemical transformation or interaction with the brain or blood vessels must be contemplated to explain the later occurrence of chronic vasospasm. In this regard, the idea that some vasocontractile substance is continuously released from the subarachnoid clot seems to be a useful notion. Even with this, however, no satisfactory explanation as to absence of vasospasm in the initial few days of SAH can be provided. Thus, the above two characteristics of vasospasm following SAH are considered to be the fundamental guidelines in the research and the therapy of vasospasm. In another view, the above two characteristics are not always disadvantageous to the therapy of vasospasm, because the occurrence of vasospasm can be frequently anticipated by CT films. Therefore, we have the time to begin the preventive therapy of vasospasm. The treatment for cerebral ischemia in clinical cases is usually started several hours later, at the earliest, after the onset of stroke, if the stroke patient is admitted to the hospital immediately after the onset. Experimental results show that preventive therapy is much more effective than treatment after the onset of stroke. Since most patients are admitted to the hospital after SAH, it is still possible to start drug administration before the onset of vasospasm.

2. Detection of Vasospasm

Angiographic vasospasm is diagnosed by demonstrating narrowing of the major cerebral arteries. The incidence of angiographic vasospasm in SAH patients has been reported to be about 40–70%, although there are some differences in definition. In a total of 272 cases of SAH who have been admitted to our clinic in the last few years, angiography was performed in 158 cases between days 4 and 21. The incidences of severe and moderate vasospasm in these patients were 34 and 26%, respectively. However, these figures of incidence are biased, because early postoperative angiography was not usually done in patients with severe fatal neurological deficit probably due to vasospasm and also in patients without any neurological deficit. CT is another valuable tool of investigation to disclose infarction due to vasospasm following SAH. In a incidence of 20.8% of our cases, low density areas of infarction due to vasospasm were identified.

Symptomatic vasospasm is the clinical consequence of cerebral ischemia due to vasospasm. The clinical diagnosis of vasospasm is based on the onset of neurological deficits, and the exclusion of other causes, such as, rebleeding, intracerebral hematoma, infarction or contusion due to operation, hydrocephalus, or hypoxia. The delayed ischemic neurological deficit (DIND) occurs a little bit later than appearance of angiographical narrowing of cerebral artery. While 60% of our patients developed severe or mild arterial narrowing in angiograms, only 27.9% developed clinical evidence of neurological deficits.

3. Treatment of Vasospasm

Prevention or treatment of vasospasm can be multifarious (Ausman et al. 1985, Heros et al. 1983, Sano and Saito 1980, Sasaki et al. 1985 b). The therapies which seem especially promising are listed in **Table VII-4** and are discussed briefly below.

1. Early surgery of aneurysms, removal of subarachnoid clots, and setting cisternal drainage for the purpose of continuously draining bloody CSF which may contain some vasocontractile substances are very important (Mizukami et al. 1982, Sano et al. 1980, Sano 1983, Suzuki et al. 1979, Taneda 1982). However, removal of all clots is, in many occasions, very difficult. This may be the reason why vasospasm may later develop, even in cases which undergo early surgery.

2. Hemodynamic improvement of cerebral circulation (by raising blood pressure, hypervolemia, etc) is, of course, necessary after aneurysm clipping (Farhat and Schneider 1967, Kosnik and Hunt 1976). In the distal parts of vasospastic arteries, perfusion pressure may be decreased, and, therefore, cerebral blood flow (CBF) is also decreased. The normal brain changes its cerebral blood flow with variations in $PaCO_2$, but not with physiological changes in the systemic arterial pressure. However, a loss of normal autoregulation is, more or less, observed in the acute phase after SAH (Heilbrun et al. 1972, Ishii 1979, Pickard et al. 1979, 1980, Terada 1985). Therefore, CBF is most likely dependent upon the systemic arterial pressure. The decreased CBF could be improved by raising blood pressure, lowering intracranial pressure, and/or decreasing blood viscosity (Grotta et al. 1982, Kee and Wood 1984). On the other hand, Maroon et al. (1979) showed that red blood cell mass and total blood volume were significantly decreased in patients with SAH. Solomon et al. (1984) reported that symptomatic vasospasm was clearly associated with depression of circulating blood volume. Because of these reasons, the hemodynamic therapy has been recommended by many authors. Several reports have demonstrated favorable results by the hemodynamic therapy (Kassell et al. 1982, Kosnik and Hunt 1976, Pritz et al. 1978, Tanabe et al. 1982). Kassell et al. (1982) reported that complete or partial resolution of neurological deficits occurred within 1 hour of commencement of the hypervolemic-hypertensive therapy in 81% of their 58 patients. Their most effective regimen consisted of intravascular volume expansion by whole blood or packed cell transfusions to maintain the hematocrit at approximately 40, blockade of the vagal depressor response by aropine, and administration of antidiuretic and vasopressor agents, such as, levarterenol,

Table VII-4. Possible modes of therapy against vasospasm

1. Removal of perivascular clot by early surgery and cisternal drainage of CSF

2. Hemodynamic improvement of impaired cerebral circulation
   elevation of systemic arterial pressure (dopamine, etc)
   hypervolemia
   dextran, mannitol, etc

3. Cerebral protection against ischemia by drugs
   barbiturates,
   nizofenone (Y-9179)

4. Use of free radical scavengers
   vitamin E
   barbiturates,
   nizofenone (Y-9179),
   1,2 bis (nicotinamido) propane (AVS),
   mannitol, etc

5. Suppression of platelet aggregation and pharmacological modification of prostaglandin synthesis
   ticlopidine,
   phosphodiesterase inhibitors
   (papaverine, phthalazinol, etc, see 9 below),
   cyclo-oxygenase inhibitors
   (indomethacin, aspirin, etc),
   lipoxygenase inhibitors,
   phospolipase A_2 suppression
   (steroids)

6. Inhibition of thromboxane-A_2 synthesis and/or stimulation of prostacyclin synthesis
   OKY-1581 (sodium (E)-3-(4-(3-pyridylmethyl)phenyl)-2-methyl-acrylate),
   OKY-046 (sodium(E)-3-(4(-1-imidazolimethyl)phenyl)-2-propenoate),
   nizofenone (Y-9179),
   trapidil,
   prostacyclin and its analogues

7. Calcium antagonists
   nimodipine,
   nifedipine,
   diltiazem,
   cinnarizine,
   verapamil,
   nicardipine

8. Adenyl cyclase stimulation
   beta$_2$-adrenergic drugs
   isoproterenol (beta$_1$ + beta$_2$) (+ lidocain)
   salbutamol (mainly beta$_2$)

9. Phosphodiesterase inhibition
   papaverine, methylxanthines, aminophylline, ascorboc acid, diazoxide, chlorpromazine, reserpine, hydralazine

10. Combination of 8 and 9
    salbutamol + aminophylline

11. Use of exogenous cyclic AMP or one of its congeners
    dibutyryl cyclic AMP

12. Guanyl cyclase inhibition
    phenoxybenzamine (blocking serotonin, alpha-adrenergic drugs and cholinergic drugs),
    phentolamine (alpha blocker),
    methysergide, reserpine, kanamycin (serotonin blocker),
    atropine (cholinergic blocker),
    prostaglandin-$F_{2\alpha}$ blocker (?),
    sodium nitroprusside (alpha blocker) +
    phenylephrine (alpha stimulator with little effect on the cerebral vessels)

13. Combination of 9 and 12
    aminophylline + nitropusside + dopamine

14. Angiotensin-converting enzyme inhitition
    teprotide

15. Topical application of local anesthetics
    lidocain

metaraminol, isoproterenol, dopamine, and dobutamine. Tanabe *et al.* (1982) have reported a beneficial effect of hyperdynamic therapy induced by administration of a large amount of human serum albumin without vasoactive drugs. As complications of these hemodynamic therapies, pulmonary hypoxia, congestive heart failure, pulmonary edema, or even myocardial infarction could occur. Therefore, monitoring of volume overload, such as Swan-Ganz catheter, should be needed (Kudo *et al.* 1981, Pritz 1984).

3. Protection of the brain against ischemia by barbiturates or nizofenone (Y-9179), is important. Nizofenone, a cerebral protector and a free radical scavenger, had less side effects than barbiturates. A recent cooperative double-blind clinical trial (Saito *et al.* 1983) showed that the group treated with this drug was superior to the placebo group (p < 0.05) in the "disability status scale" one month after development of vasospasm, although development of vasospasm was not prevented (administration of the drug was initiated before vasospasm developed and continued for 5 days and, if vasospasm developed, for an additional 5 days).

4. From the viewpoint of the pathogenesis of vasospasm, use of free radical scavengers such as vitamin E, barbiturates, nizofenone, AVS, mannitol etc. seems very important, especially in the early stage of SAH (Asano *et al.* 1980, Sano 1983).

5. Suppression of platelet aggregation and pharmacological modification of prostaglandin synthesis are also important. Phosphodiesterase inhibitors, which are thought to increase cyclic AMP are often used. Lipoxygenase inhibitors, if available, may be promising. Cyclooxygenase inhibitors and phospholipase A_2 inhibition, however, may aggravate vasospasm because they inhibit not only vasoconstrictive but also vasodilatatory prostaglandins.

6. Inhibition of thromboxane A_2 synthesis, by administration of OKY-1581 or OKY-046, or to a lesser extent by trapidil or Y-9179, is important. Use of prostacyclin or its analogues may also be useful.

7. Calcium can be considered as the final common pathway for muscle contraction. Therefore, use of the so-called calcium antagonists may be of value in the treatment of vasospasm. Better outcome of patients treated by calcium antagonists—as described, for example, by Allen—showed the usefulness of these drugs (Allen *et al.* 1983, Auer 1984, Brandt *et al.* 1985, van der Werf *et al.* 1985).

One of the main reasons why many neurosurgeons hesitate to perform early surgery (which is the best means to prevent rebleeding of aneurysms) is the fear of postoperative vasospasm, which is actually only remotely related to surgery (Ausman *et al.* 1985, Sano 1983). If some of these lines of investigation yield effective drugs for preventing or curing chronic cerebral vasospasm in the near future, the treatment of ruptured intracranial aneurysms will be changed so that very early surgery, to prevent rebleeding of aneurysms, combined with the removal of as much subarachnoid clot as possible and the fashioning of cisternal drainage will be performed, accompanied by ample systemic or topical administration of such drugs.

Recently, some well-designed double-blind controlled studies for the treatment of vasospasm were carried out and favorable results were reported. These studies are briefly mentioned below.

a) Calcium Channel Blocker

Recent studies both *in vitro* and *in vivo* suggest that vasoconstrictions in cerebrovascular smooth muscle can be effectively released by calcium channel blockers. Nimodipine is a lipid soluble calcium channel-blocking agent and, therefore, should cross the blood brain barrier. Its inhibitory effect is more selective against contraction of cerebral arteries than arteries of other parts of the body.

Allen *et al.* (1983) tried to use this calcium channel blocker against vasospasm due to subarachnoid hemorrhage because of its potent effect of blocking the influx of extracellular calcium, in a multiinstitutional prospective, double-blind, randomized placebo-controlled trial within 96 hours of subarachnoid hemorrhage. Patients received gelatin capsules containing either nimodipine or placebo. The initial dose was 0.7 mg per kilogram of body weight and, thereafter, 0.35 mg per kilogram was given every four hours for 21 full days. In their results, nimodipine significantly reduced the occurrence of several neurologic deficits, including death solely due to vasospasm. By the end of the 21-day-treatment period 8 of 60 patients in placebo group died or still had severe neurological deficits from vasospasm.

In contrast, only 1 of the 56 patients in nimodipine group died from vasospasm. Angiographic studies showed that the clinical efficacy of nimodipine might be the result of its inhibition of cerebral arterial spasm. Another explanation for the prevention of ischemic symptoms by nimodipine is that the infusion of nimodipine dilates predominantly the small resistance vessels and, therefore, the pial collateral circulation could be opened. Thus, nimodipine significantly reduced the occurrence of severe neurologic deficits due to vasospasm, although the total occurrence of neurologic symptoms due to spasm, including mild deficits that were transient and did not result in a persistent deficit, was similar in the two groups at the end of the treatment period.

b) Cerebral Protective Drug

Nizofenone (Y-9179), a new imidazole derivative, was found to possess a remarkable protective action against cerebral anoxia or ischemia in animal experiments (Ochiai *et al.* 1981, 1982, Tamura *et al.* 1979, Yasuda *et al.* 1981). Although little is known about the mechanism, the free radical scavenging action and the mild to moderate depressant action on the cerebral metabolic rate are considered especially pertinent to its cerebral protective action. Saito *et al.* (1983) reported favorable results of nizofenone against delayed ischemic neurological deficits following SAH due to aneurysmal rupture in a double-blind clinical study. In the study, patients with ruptured cerebral aneurysms admitted within day 9 after SAH received 30 mg of nizofenone or placebo (5% glucose) in a randomized fashion. Drug administration was immediately initiated after admission and continued for 5 days. When the occurrence of vasospasm was suspected by deterioration of neurological conditions, CT and angiography were carried out for exact diagnosis. If the diagnosis of vasospasm was made, drug administration was extended for an additional 5 days. Out of 42 cases of the nizofenone group and 48 of the placebo group, 25 (59.5%) and 29 (60.4%) developed vasospasm, respectively. Thus, nizofenone does not reduce the incidence of vasospasm. Since all the participating neurosurgeons had already adopted the principle of early operation advocated by Sano and Saito (1978), about half of the patients in each group were operated on within the first week after SAH. In comparing the disability status indices one month after admission and the neurological functions, such as motor and speech functions, between two groups, the nizofenone group with sufficient drug coverage had a significantly better outcome than the placebo group (p < 0.05).

In 1986, Ohta et al., after having repeated the same type of trial in a wider scale, also reported the effectiveness of nizofenone against delayed ischemic symptoms due to vasospasm. In their multicenter controlled double-blind clinical study, it was shown that treatment with nizofenone was significantly more effective (p < 0.05) than that with placebo based on functional recovery and other clinical findings. Their results show that nizofenone has favorable effects on the outcome of patients with SAH especially in those who are likely to suffer poor functional recovery such as patients with ischemic symptoms, those with moderately severe preoperative deficits, or those exhibiting diffuse high density areas (subarachnoid clots) on CT scans. These results are similar to the results of Saito's study. Thus, these double-blind clinical studies indicate that nizofenone has no effect against the occurrence of vasospasm itself, but has a favorable effect against ischemic neurological deficits due to vasospasm following SAH.

c) Platelet Aggregation Inhibitor

Ticlopidine is an inhibitor of platelet aggregation without any effect on prostacyclin generation. Ono et al. (1984) conducted a double-blind clinical study using ticlopidine and placebo, inorder to evaluate the effects of the agent drug against neurological deficits due to vasospasm following SAH. Patients with ruptured intracranial aneurysms operated on within three days after SAH were given a tablet of either 100 mg of ticlopidine or lactose as a placebo three times a day for four weeks. Neurological findings three months after the onset of SAH and the final outcome were compared. As a result, the incidence of cerebral vasospasm was not significantly different between both groups. However, the incidence of neurological deficit plus mortality at the time of discharge was 14% (ticlopidine group) and 29% (placebo group) and this difference was significant (p < 0.05). When both treatment groups were compared as a whole, no significant difference was observed in their clinical outcomes. However, by regrouping the categories into favorable and unfavorable outcome, it became apparent that incidence of favorable outcome was significantly higher (92 vs 79%) in the ticlopidine group ($p < 0.05$). The difference became more significant among the patients with angiographically proven vasospasm. Incidence of severe neurological deficit and death was 40% in the placebo group, compared with 18% in the ticlopidine group ($p < 0.05$). In their conclusion, ticlopidine has a clinical value in preventing neurological deficit and death due to vasospasm and this can be used after early surgery without increasing hemorrhagic complication.

d) Thromboxane A_2 Synthetase Inhibitor and Radical Scavenger

It has recently been suggested that the release of oxyhemoglobin (oxy Hb) from lysed erythrocytes and its conversion to methemoglobin might trigger cerebral vasospasm. This is of particular interest, because oxy Hb is contained in erythrocytes, i.e. subarachnoid clot, and therefore some period is required for its release into the subarachnoid space by hemolysis. This might partly explain the time lag in the occurrence of vasospasm. On the basis of this theory, Sano et al. stressed the role of free radical reactions leading to lipid peroxidation which is triggerd by hemolysis (Asano et al. 1980, Sano and Saito 1980). The tissue toxicity of free radical reactions has been well docu-

mented. Therefore, production of lipid peroxides or hydroperoxides by free radical reactions initiated by hemolysis seemed very likely to be a causative mechanism of chronic vasospasm after SAH. They found that thiobarbituric acid (TBA) reactive substances, which has been regarded as representing the presence of lipid peroxides, had increased especially in the CSF of SAH patient developing vasospasm (Asano et al. 1980, Sano and Saito 1980, Tanishima et al. 1979).

Using high performance liquid chromatography (HPLC) and gas chromatography-chemical ionization mass spectrometry (GC-MS), one of the TBA-reactive substances in the CSF of SAH patients with vasospasm on day 7 was identified as 5-hydroxyeicosatetraenoic acid (5-HETE), the metablite of 5-hydroperoxyeicosatetraenoic acid (5-HPETE) (Nakamura et al. 1982, Sasaki et al. 1982 c). Semiquantitative analysis of 5-HETE in the CSF showed a close correlation between the occurrence of cerebral vasospasm and the appearance of 5-HETE in the CSF. These facts, together with the rapid decrease of glutathione peroxidase and vitamin E in the CSF of SAH patients after day 3, and the decrease of prostacyclin in the cerebral artery in a few days after SAH, may well explain the time lag in the appearance of cerebral vasospasm (Sano 1983, Sasaki et al. 1981 b).

The toxic effects of lipid peroxides or free radicals on tissues, especially on the endothelial membrane, are well documented. This endothelial damage must cause impairement of prostacyclin (PGI_2) synthesis in the endothelium. In addition, lipid hydroperoxides are known to inhibit synthesis of PGI_2 from endoperoxides. In the presence of endothelial damage, the artery, being devoid of or deficient in PGI_2 synthesis, would be exposed to the unopposed vasocontractile action of thromboxane A_2 and of other vasocontractile substances released from the aggregated platelets. Peroxidation of polyunsaturated fatty acids (PUFAs) is increased in SAH, as shown by an increase of TBA-reactive substances, which means an increase in the endoperoxides and hydroperoxides of PUFAs. Endoperoxides become sources of prostaglandin synthesis and from these, vasoconstrictive thromboxane A_2 (TXA_2), PGF_2, D_2, E_2, etc will be produced. Therefore one reliable treatment would be the correction of this imbalance between PGI_2 and vasoconstrictive prostaglandins, especially TXA_2.

On the basis of this hypothesis, OKY-1581 or OKY-046, a thromboxane A_2 synthetase inhibitor have been tried for the purpose of preventing vasospasm in experimental and clinical studies. The preventive action of OKY-1581 or OKY-046 have been reported on experimental vasospasm and also on clinical vasospasm following SAH (Naito et al. 1983, Sasaki et al. 1982 a, Shikinami et al. 1985, Suzuki et al. 1985, Tani et al. 1984, Uyama et al. 1985). A double-blind, randomized placebo-controlled trial was done recently in Japan. Patients in whom surgical treatments were completed within 96 hours after subarachnoid hemorrhage received OKY-046, 400 mg/day, or 80 mg/day, or placebo by continuous

infusion for 10 to 14 days following operation. In the results, the incidence of angiographical cerebral vasospasm and the incidence of low density on the CT scan on day 21 or later were significantly suppressed in the OKY-046-treated group. Analysis of activities of daily life (ADL) at one month later indicated that patients in OKY-046-treated group with preoperative GCS at 14 or less, which corresponded to grade III or worse, were significantly better in ADL than that in the placebo group.

It has been concluded that the early postoperative administration of OKY-046, a TXA_2 synthetase inhibitor, significantly reduces the occurrence of cerebral vasospasm and low density on the CT scan. It has also been shown that the OKY-046 administration produces a significantly favorable prognosis in patients with higher severity of symptoms before surgery.

Recently, Sano et al. tried to use AVS, one of free radical scavengers, against vasospasm due to subarachnoid hemorrhage from the viewpoint of the pathogenesis of vasospasm (see Chapter I). A randomized, double-blind placebo-controlled study was done in Japan by Sano et al.. Patients with ruptured cerebral aneurysms received AVS, 4g/day, 2g/day or placebo by intravenous drip infusion for 14 days. The drug administration was initiated by day 4.

This study showed a statistically significant effect against delayed ischemic neurological deficits, such as hemiparesis, due to vasospasm. And also, a significant decrease of the incidence of low density was shown on the CT scan.

e) Steroid

The effect of steroid against cerebral vasospasm or cerebral ischemia following SAH is still controversial. Hashi *et al.* (1986) have shown that the large doses of hydrocortisone dilates the arterial narrowing. The preventive effect of high-dose hydrocortisone against ischemic symptoms is thought to be immediate, but transient in our trial. They carried out a randomized, double blind controlled study of high-dose hydrocortisone in 140 SAH patients (Hashi 1986). A dosage of 3,000 mg of hydrocortisone or a placebo was intravenously infused every 12 hours for three days beginning from just after the onset of ischemic symptoms. In the hydrocortisone group, the improvement of ischemic symptoms was significant four days after the treatment started and also one month later. They concluded that high-dose hydrocortisone is effective against ischemic symptomes due to vasospasm following SAH, although systemic hypotension and hyperglicemia were frequently present.

VIII. SURGICAL TECHNIQUES

Fig. VIII-1 shows the distribution of aneurysms experienced in our clinic until December, 1982. The distribution is similar to the larger series of Suzuki, and to all Japan and USA series (Suzuki and Yoshimoto 1979). Specifically, 89.9% of aneurysms are located in the anterior half of the circle of Willis or its branches, 2.3% are located in the intracavernous portion of the internal carotid artery, and 7.8% in the vertebral-basilar system.

As Rhoton (1980) and others pointed out, these saccular aneurysms arise at a branching site on the parent artery, or they arise at a turn or curvature in the artery. The aneurysmal dome, or fundus, points in the direction of the maximal hemodynamic thrust in the preaneurysmal (proximal) segment of the parent artery. This knowledge is useful during an operation when a surgeon searches for the aneurysm which is buried in blood clots or a dense arachnoid adhesion (**Fig. VIII-2**). If he identifies the preaneurysmal and post-aneurysmal (distal) segments of the parent artery, the aneurysm must be present on the extension line of the preaneurysmal segment.

There are certain sophisticated methods to occlude aneurysms, such as stereotaxic aneurysm thrombosis (Alksne *et al.* 1966, Alksne and Smith 1977 b) or the detachable balloon catheter technique (Serbinenko 1974). These are, however, still in the experimental stage. The surest way to occlude most intracranial aneurysms is intracranial dissection and clipping or ligation of the aneurysmal neck under the operating microscope.

The wall of an aneurysm, especially of a ruptured one, is so thin and fragile that even the slightest tension may cause intraoperative rupture. In order to minimize stress to the aneurysmal wall, meticulous sharp dissection (not blunt dissection) of the aneurysmal neck is necessary. The dome of the aneurysm should be judiciously avoided, at least at the onset of the dissection. Once the neck of the aneurysm is freed, it can be clipped or ligated, depending on the preference of the surgeon. Most neurosurgeons prefer replaceable spring clips, because if a clip applied to the neck disturbs the blood stream of the parent artery or its branches, it can be removed and applied again so that vital blood supply can be preserved. There are many types of spring clips and their appliers. For most cases, Sugita clips or Yaşargil clips are useful. For deep aneurysms, the slender type III clip applier of Sano (Mizuho Ika Co.) may be convenient. For ordinary aneurysms, the multipurpose all-angle clip

LOCATION(1480 ANEURYSMS/1284 CASES)

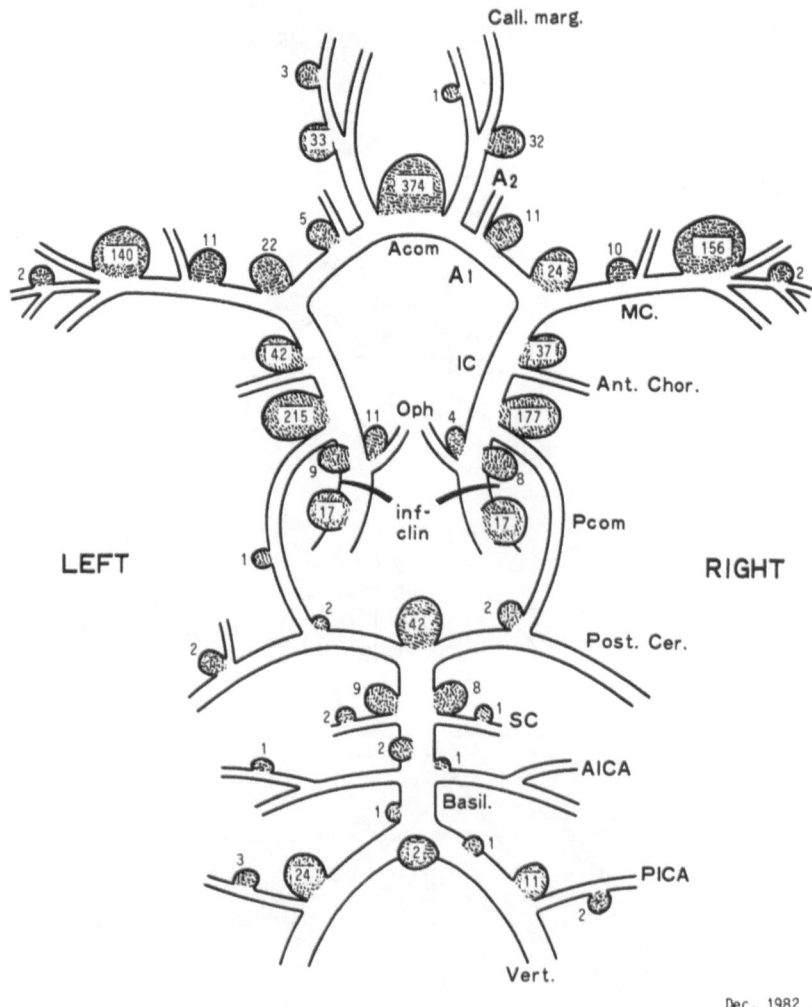

Dec. 1982

Fig. VIII-1. Distribution of intracranial aneurysms. *Acom* anterior communicating artery, A_1 A_1 portion of the anterior cerebral artery, A_2 A_2 portion of the anterior cerebral artery, *AICA* anterior inferior cerebellar artery, *Ant. Chor.* anterior choroidal artery, *Basil.* basilar artery, *Call. marg.* callosomarginal artery, *IC* internal carotid artery, *inf. clin.* infraclinoid portion of the internal carotid artery, *MC* middle cerebral artery, *Oph* ophthalmic artery, *Pcom* posterior communicating artery, *PICA* posterior inferior cerebellar artery, *Post. Cer.* posterior cerebral artery, *SC* superior cerebellar artery, *Vert.* vertebral artery

applier of Sano (Mizuho Ika Co.) (Sano 1980 a) may be of great use.
Bipolar coagulation, microscissors, self-retaining retractors, and re- placeable spring clips have greatly improved precision and safety in intracranial aneurysm surgery. Electrocoagulation of the aneurysmal neck

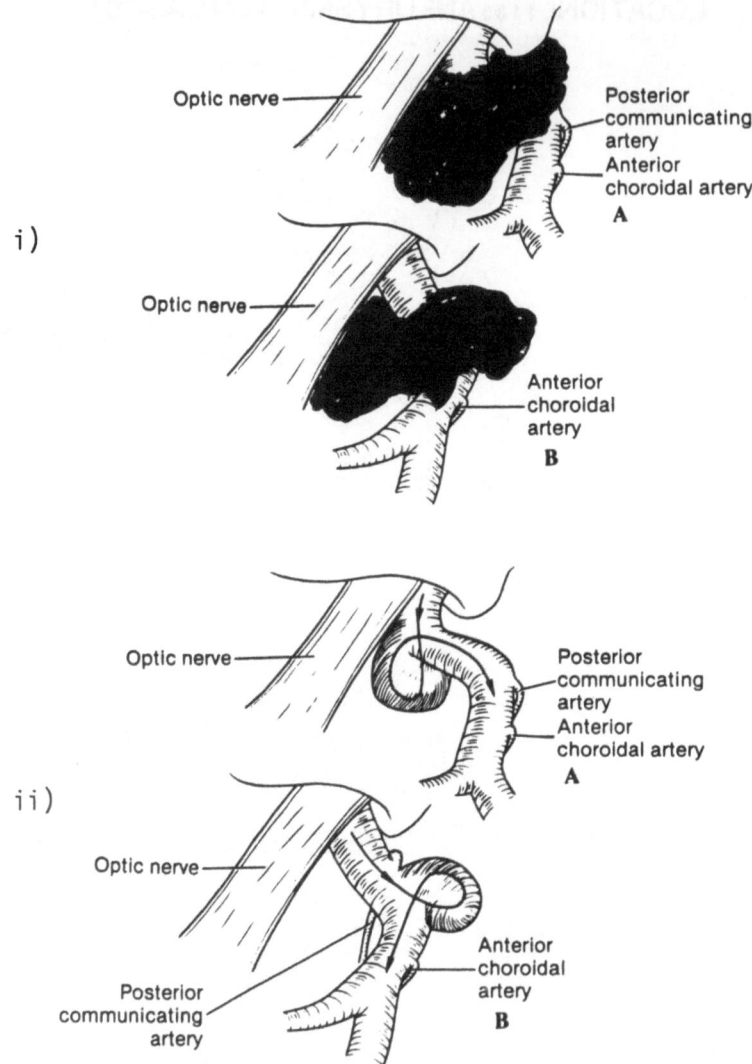

i)

ii)

Fig. VIII-2. Aneurysm and blood stream. i) Before removal of blood clots. ii) After removal of blood clots and before clipping the aneurysm. *A* Paraophthalmic aneurysm. *B* Internal carotid-posterior communicating aneurysm (modified from Sano: Cerebral aneurysms. In: Gendai Gekagaku Taikei **4**, 129–171 (1982). Tokyo: Nakayama Shoten)

with bipolar cautery set at low levels of current contrived by Yaşargil (Yaşargil and Fox 1975), is a very useful technique. This not only causes the neck to shrink and its wall to toughen, but also gives the often irregularly shaped neck a more regular surface, allowing better application of the clip. During the whole maneuver, 10x to 16x magnification of the operative field is recommended.

A. Technique of Washout of Clots

When explored in the early stage of subarachnoid hemorrhage (SAH), the ruptured aneurysm is usually found to be buried in subarachnoid blood clots. The clots around the aneurysm are gently suctioned out and the aneurysm is exposed. After clipping the neck of the aneurysm, as much as possible of the subarachnoid clot are removed by suction and irrigation with saline.

In cases with aneurysms of the anterior circle of Willis, the Sylvian fissure, the chiasmatic cistern, the interpeduncular cistern, and often the interhemispheric fissure should be opened in order to remove the clots.

A silicone rubber tube, 2.4 mm in outer diameter, with a radiopaque tip, is inserted into the basal cistern, mostly through the chiasmatic cistern and the Liliequist membrane, and the other end of the tube is brought out through the dural closure to a small incision of the skin where it is fixed by a stitch (Saito et al. 1977, Sano and Saito 1980). This tube is connected with a rubber or plastic bag and serves as continuous cisternal drainage. In cases of severe SAH, or in cases with already existent vasospasm, continuous ventricular drainage is also fashioned in order to control intracranial pressure. For removal of oozing epidural blood, epidural drainage is also set up (**Fig. VIII-3**). Some neurosurgeons prefer to insert 2 cisternal drains and irrigate the subarachnoid space with saline with or without urokinase (48-192 IU/ml) between 2 drains or between ventricular and cisternal drains.

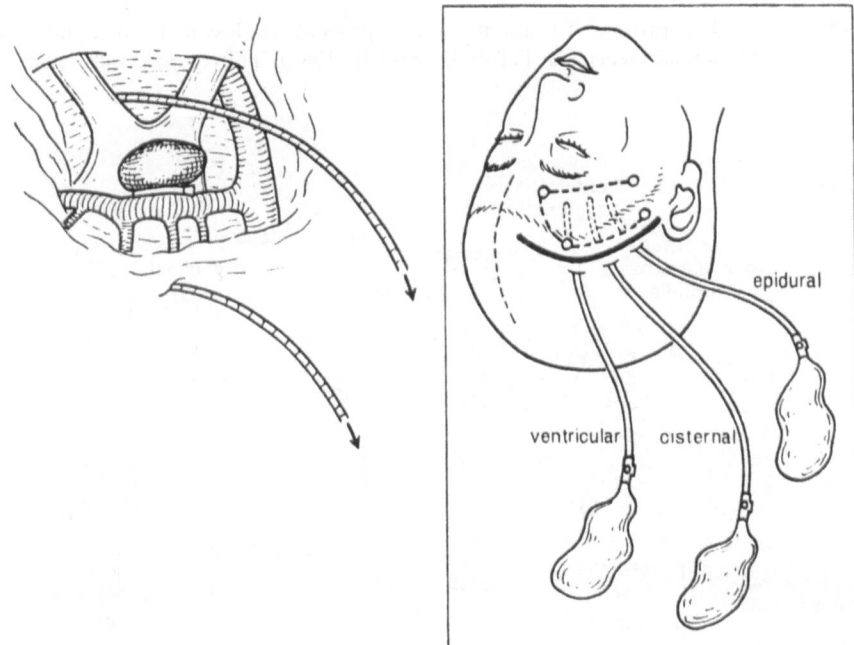

epidural

ventricular   cisternal

Fig. VIII-3. Cisternal drainage and other drainages. The illustration is a case of anterior communicating aneurysm

B. Internal Carotid-Posterior Communicating Aneurysm and Internal Carotid-Anterior Choroidal Aneurysm

The internal carotid-posterior communicating (ICPcom) aneurysm, along with the anterior communicating and the middle cerebral ones, are the most frequent of all intracranial aneurysms.

Among 1,480 aneurysms which we experienced until the end of 1982, 392 (26.5%) belonged to this type. In addition, the internal carotid-anterior choroidal (Ach) aneurysm was found in

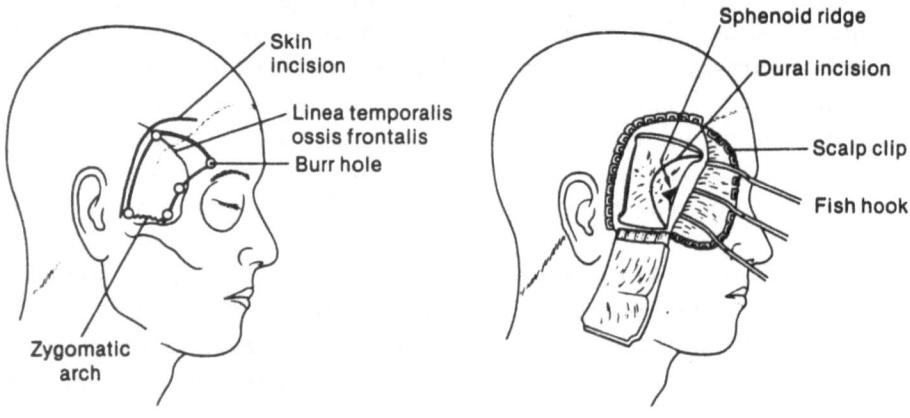

Fig. VIII-4. Pterional approach (frontotemporal approach) (redrawn from Sano: Cerebral aneurysms. In: Gendai Gekagaku Taikei **4**, 129–171 (1982). Tokyo: Nakayama Shoten)

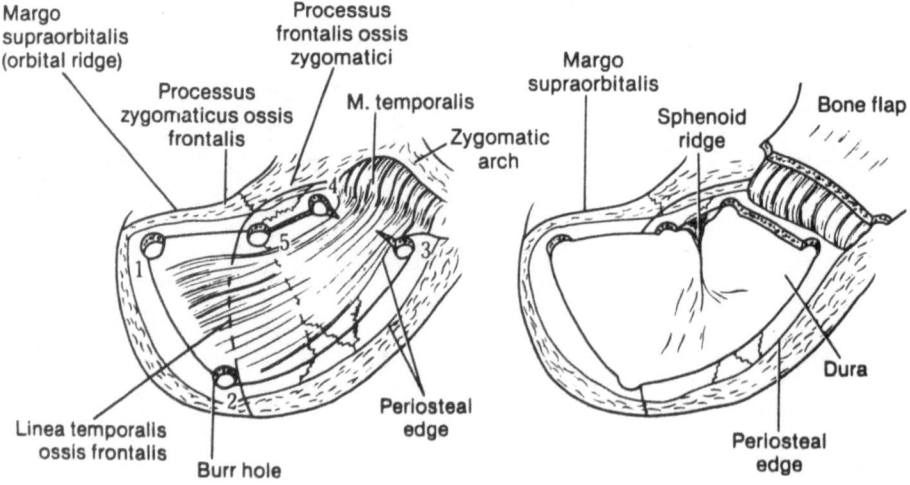

Fig. VIII-5. Pterional approach. Surgeon's view (Figs. 5 to 11) (redrawn from Sano: Neurol. Surgery **10**, 21–28 (1982))

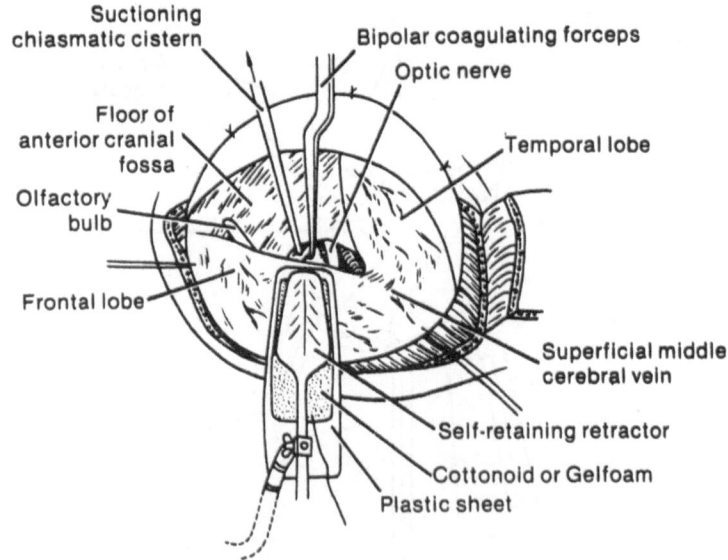

Fig. VIII-6. Retraction of the frontal lobe (redrawn from Sano: Cerebral aneurysms. In: Gendai Gekagaku Taikei **4**, 129–171 (1982). Tokyo: Nakayama Shoten)

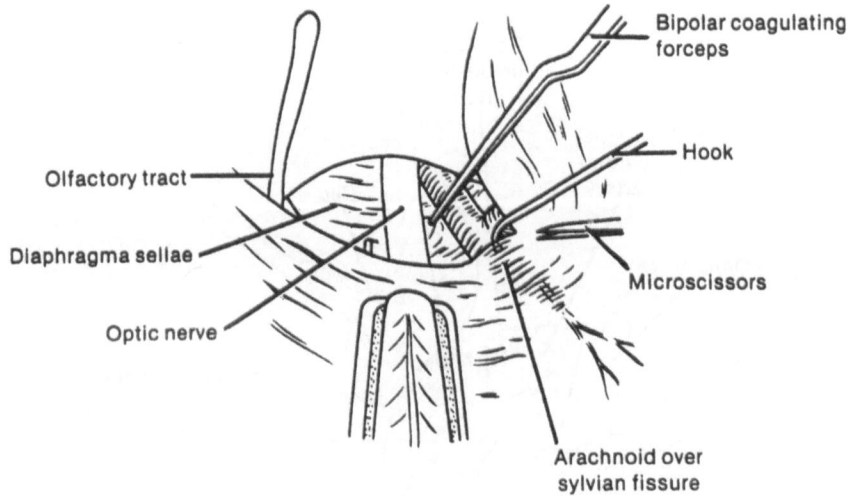

Fig. VIII-7. Dissection of the arachnoid (redrawn from Sano: Cerebral aneurysms. In: Gendai Gekagaku Taikei **4**, 129–171 (1982). Tokyo: Nakayama Shoten)

5.3% (79 in number) and what Pia (1979) called the paraophthalmic aneurysm, that is, the one located between the branchings of the ophthalmic and the posterior communicating artery, was observed in 1.1% (17 in number). If all added, 33% of all aneurysms were located in this region. ICPcom aneurysms arise on the internal carotid, proximally or distally to the

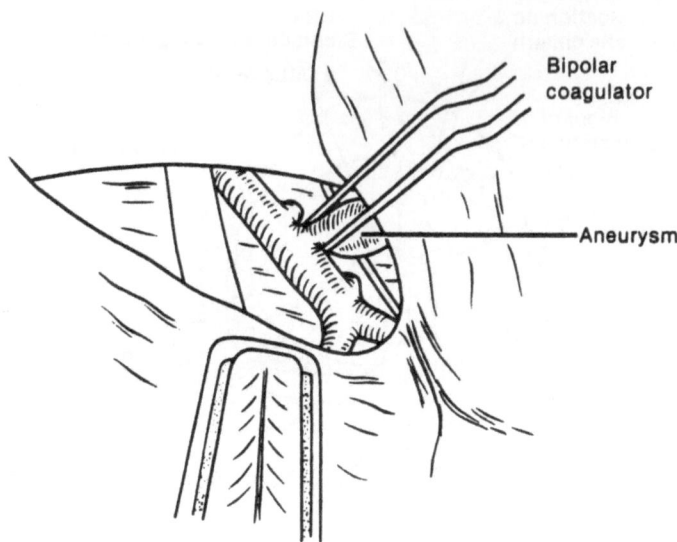

Fig. VIII-8. Electrocoagulation of the aneurysmal neck (redrawn from Sano: Cerebral aneurysms. In: Gendai Gekagaku Taikei **4**, 129–171 (1982). Tokyo: Nakayama Shoten)

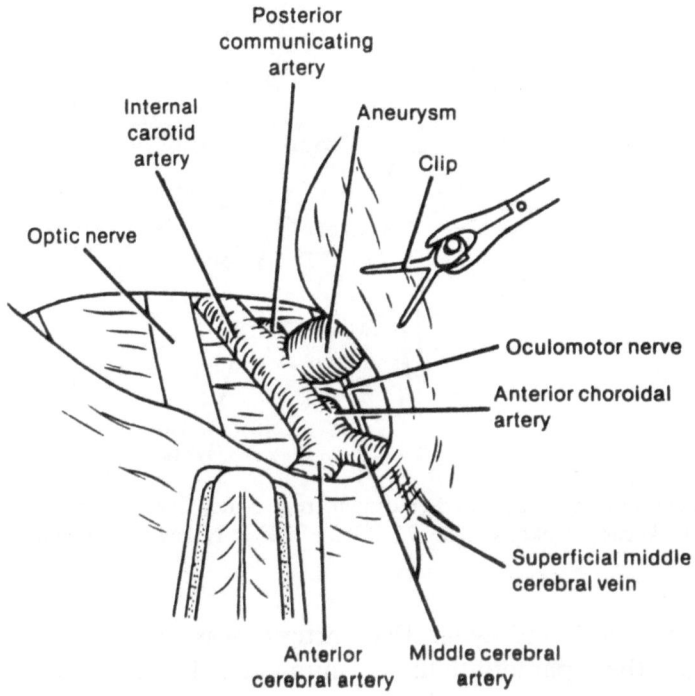

Fig. VIII-9. Aneurysm neck clipping (1) (redrawn from Sano: Cerebral aneurysms. In: Gendai Gekagaku Taikei **4**, 129–171 (1982). Tokyo: Nakayama Shoten)

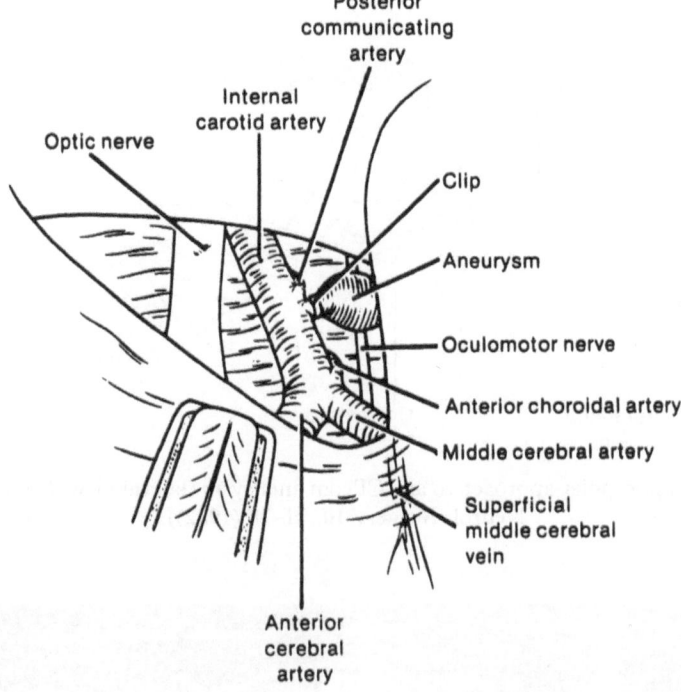

Fig. VIII-10. Aneurysm neck clipping (2) (redrawn from Sano: Cerebral aneurysms. In: Gendai Gekagaku Taikei **4**, 129–171 (1982). Tokyo: Nakayama Shoten)

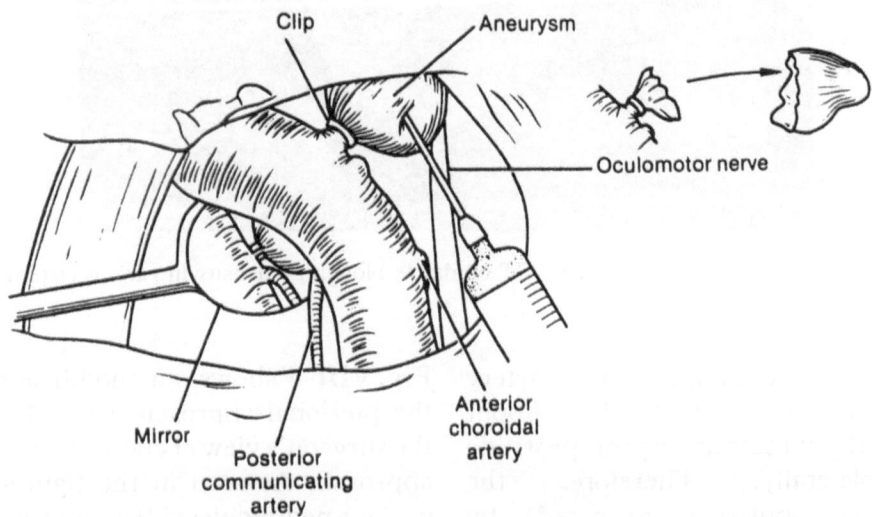

Fig. VIII-11. Use of a small mirror to check the position of a clip (redrawn from Sano: Cerebral aneurysms. In: Gendai Gekagaku Taikei **4**, 129–171 (1982). Tokyo: Nakayama Shoten)

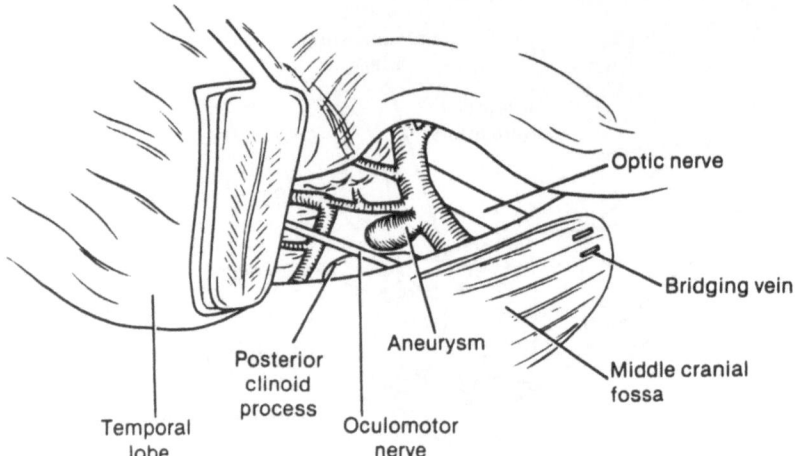

Fig. VIII-12. Temporopolar approach to an ICPcom aneurysm (lateral view) (redrawn from Sano: Neurol. Surgery **10**, 21–28 (1982))

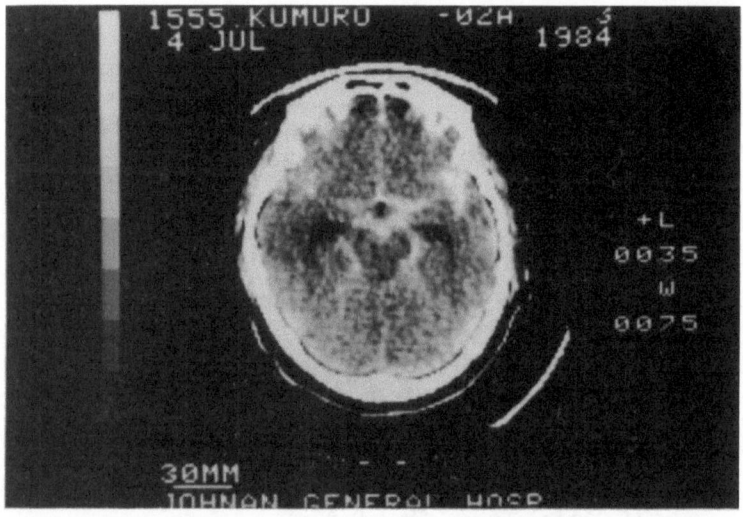

Fig. VIII-13. A 51-year-old male. CT on day 0. Note high density in various cisterns

posterior communicating artery junction (more often distally). About five sixths of these aneurysms protrude posterolaterally. Therefore, the pterional approach proposed by Yaşargil (Yaşargil and Fox 1975) is appropriate in most cases.

Fig. VIII-4 shows our modification of the pterional approach. **Fig. VIII-5** is the surgeon's view of craniotomy in this approach. As seen in the figures, we prefer a pedunculated flap to a free flap. The sphenoid ridge is drilled off as anteriorly as possible. The frontal lobe is retracted near the Sylvian fissure, leaving the olfactory tract intact. After suctioning the chiasmatic cistern to remove the cerebrospinal fluid and blood clots, the brain becomes slack the internal carotid also being cut **(Fig. VIII-7)**. Now the aneurysm, the posterior communicating artery, the anterior choroidal artery, and the oculomotor nerve come into view. To

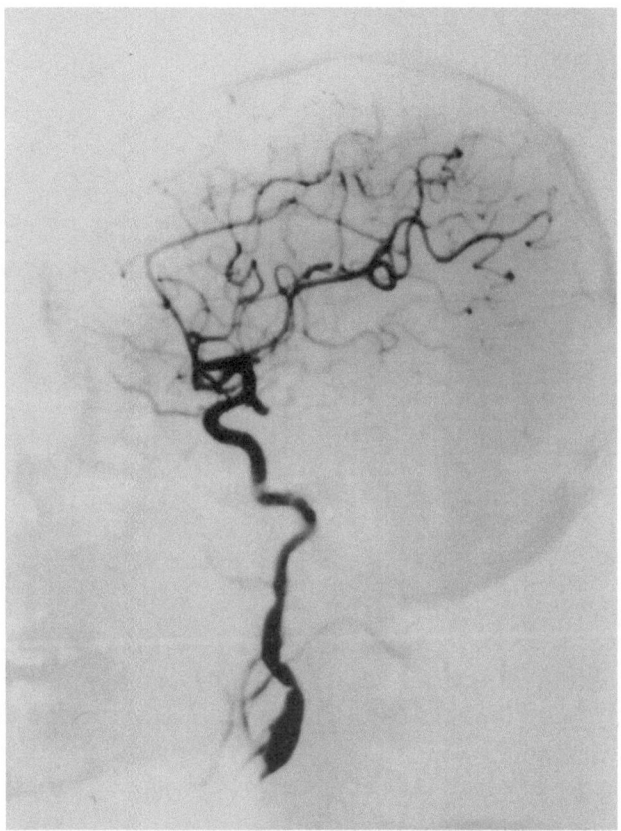

Fig. VIII-14. The same case as in Fig. 13. Carotid angiogram (Digital subtraction angiography—DSA). Note an internal carotid-posterior communicating aneurysm

(Fig. VIII-6). Use of mannitol (20%, 500 ml) or some other dehydrating agent before starting craniotomy is recommended. The arachnoid over the Sylvian fissure is cut with scissors until the internal carotid bifurcation is exposed, the arachnoid surrounding shrink the aneurysmal neck, bipolar cautery set at low levels of current is applied to the neck as mentioned above **(Fig. VIII-8)**. Now a suitable clip is applied to the neck **(Fig. VIII-9 and VIII-10)**. Care should be taken not to occlude the posterior communicating or the anterior choroidal artery. If oculomotor palsy is already present, puncture or resection of the aneurysmal dome after clipping the neck is recommended to facilitate oculomotor

geon may switch to the temporopolar approach (Sano 1980b). Namely, the bridging veins between the temporal pole and the cavernous sinus are severed, the temporal pole being re-

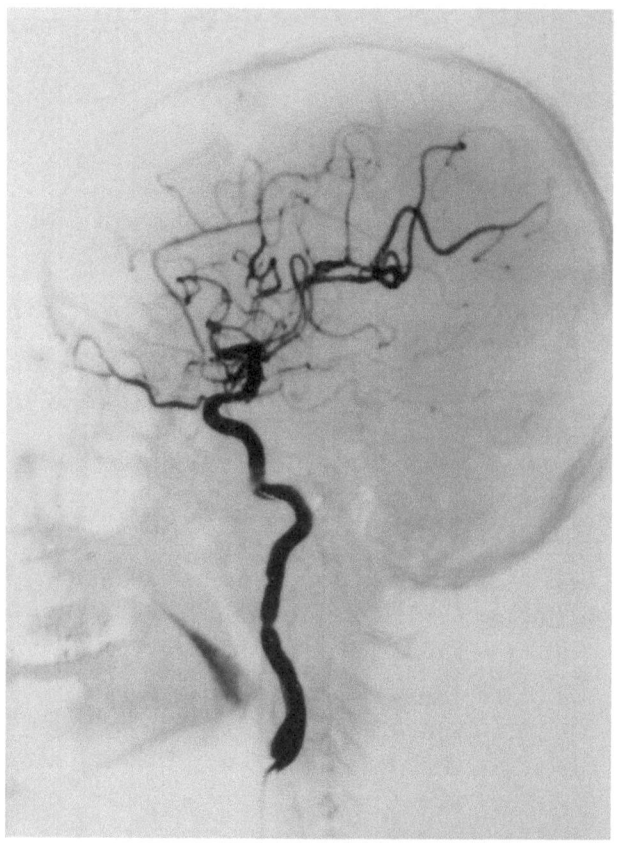

Fig. VIII-15. The postoperative angiogram. The aneurysm is obliterated and the normal structures including the posterior communicating and the anterior choroidal arteries are preserved

recovery. Use of a small otorhinolaryngological mirror to scrutinize the proper positioning of the clip is also recommended (Fig. VIII-11), as advocated by Sugita (Sugita et al. 1975). If an aneurysm is located just behind the internal carotid and is unable to be visualized by this approach, the sur-

tracted backward, and the direction of the microscope is changed to obtain a direct lateral view of the internal carotid and its branches (Fig. VIII-12). By this approach, the aneurysm and the surrounding structures will be clearly visualized.

Fig. VIII-13 is a CT (computed tomography) on day 0 of a 51-year-old man with SAH. High density expressing blood clots could be noticed in the suprasellar, ambient, interhemispheric, and Sylvian (more marked on the right)

posterior communicating artery was clipped. Furthermore as much as possible of clot was removed. **Fig. VIII-15** is a postoperative angiogram. The aneurysm was obliterated and the pos-

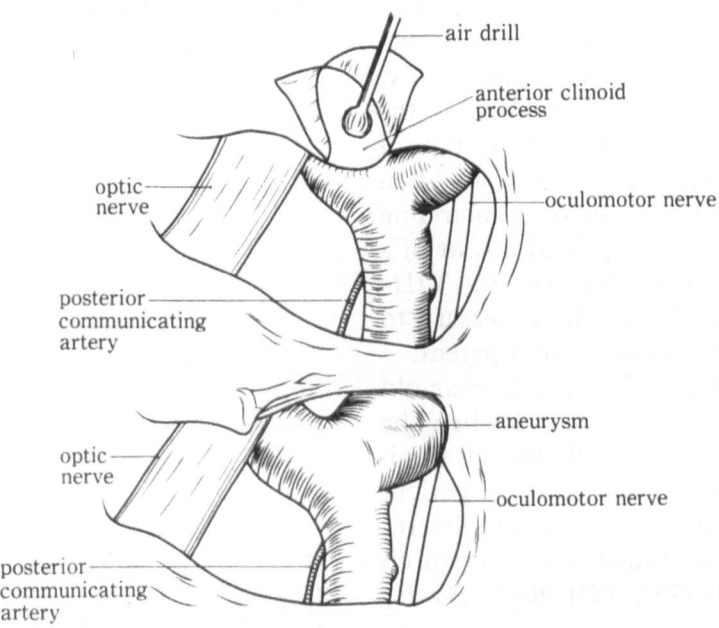

Fig. VIII-16. Clipping of a proximally located internal carotid aneurysm after drilling off the anterior clinoid process (redrawn from Sano: Cerebral aneurysms. In: Gendai Gekagaku Taikei **4**, 129–171 (1982). Tokyo: Nakayama Shoten)

cisterns. Carotid angiography (**Fig. VIII-14**) revealed an aneurysm at the internal carotid-posterior communicating artery junction on the right side. The patient was alert (Glasgow coma scale (GCS), 15) complaining of headache and neck stiffness (grade II of Hunt) and was submitted to surgery on the same day (day 0). After removal of blood clots an aneurysm protruding posterolaterally from the internal carotid just distal to the origin of the

terior communicating artery and the anterior choroidal arteries were preserved. The outcome of the patient was "good recovery" in the Glasgow outcome scale (GOS) (Jennett and Bond 1975), as of 3 months after surgery.

If an aneurysm is located so proximally that the neck is partly covered by the anterior clinoid process, the process should be drilled off to expose the neck in toto (**Fig. VIII-16**).

C. Paraophthalmic Aneurysm

Aneurysms arising from the internal carotid between the branchings of the ophthalmic and posterior communicating arteries are called "paraophthalmic aneurysms" by Pia (1979). They usually protrude posteriorly (**Fig. VIII-2 A**) and are located at an abrupt curvature of the internal carotid artery. The same operative approach as the previous one can be used for this type of aneurysm. This aneurysm may often become so large that use of an encircling clip, as shown in **Fig. VIII-17**, or of a fenestrated clip, as shown in **Fig. VIII-18** and **VIII-21**, may be necessary to preserve the internal carotid patent.

Fig. VIII-19 is a CT of a 42-year-old woman with SAH, on day 6 when she was admitted. Practically no clot (high density) could be seen and the patient showed 15 in GCS. Angiography revealed a paraophthalmic aneurysm on the right side (**Fig. VIII-20-A, B**). By

the use of a fenestrated clip, the aneurysm was clipped, preserving the internal carotid artery (**Fig. VIII-21-A, B**). The patient's outcome was good recovery in GOS, as of 3 months or 6 months postoperatively.

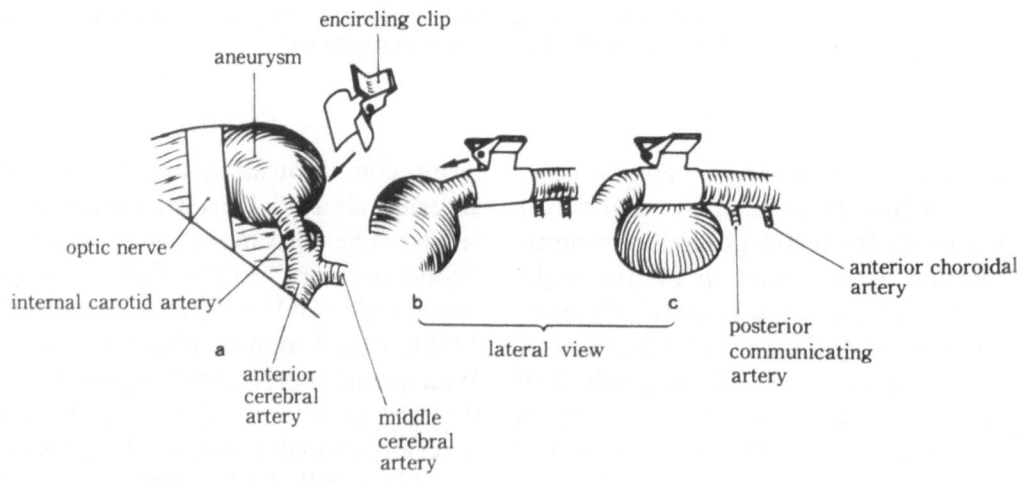

Fig. VIII-17. Applying an encircling clip to a paraophthalmic aneurysm (from Sano: Cerebral aneurysms. In: Gendai Gekagaku Taikei **4**, 129–171 (1982). Tokyo: Nakayama Shoten)

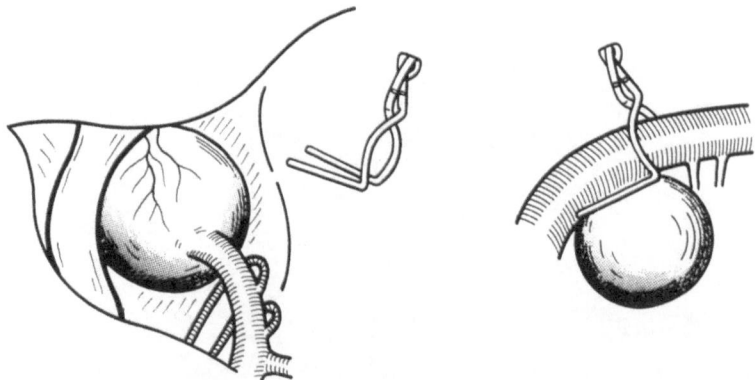

Fig. VIII-18. Applying a fenestrated clip to a paraophthalmic aneurysm

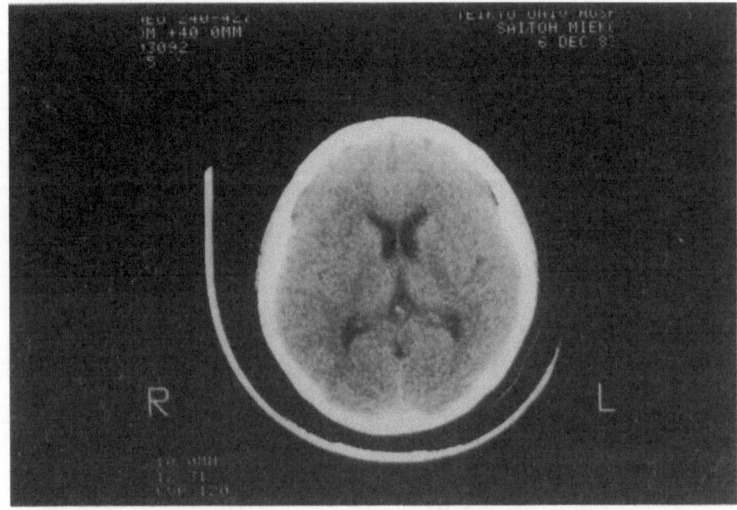

Fig. VIII-19. CT on day 6. A 42-year-old woman with SAH

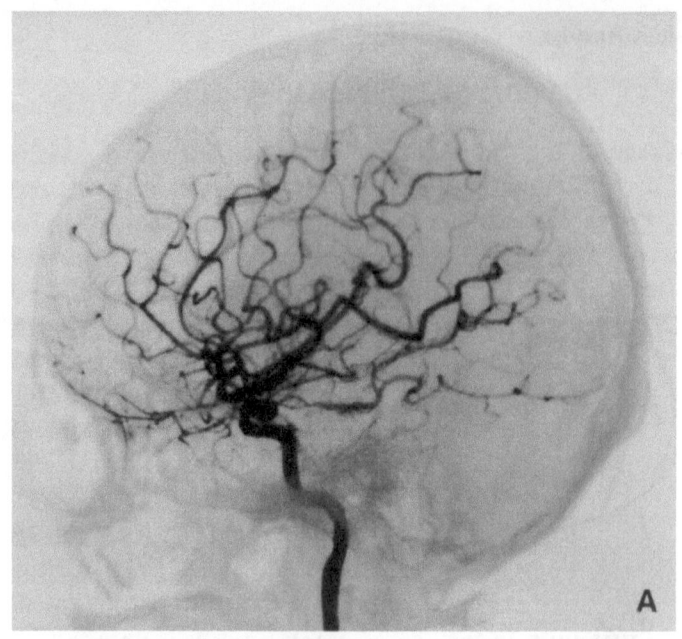

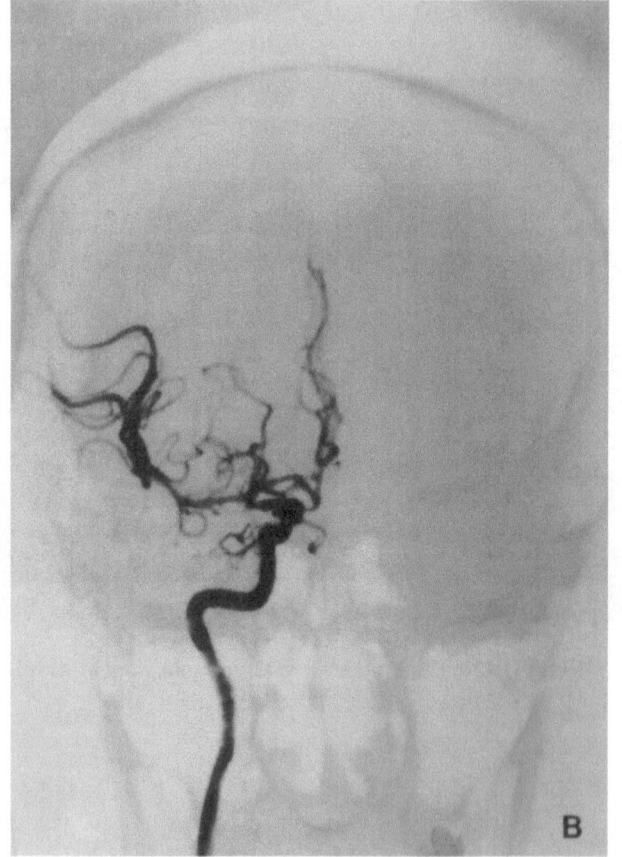

Fig. VIII-20-A, B. The same case as in Fig. 19. A paraophthalmic aneurysm on the right side

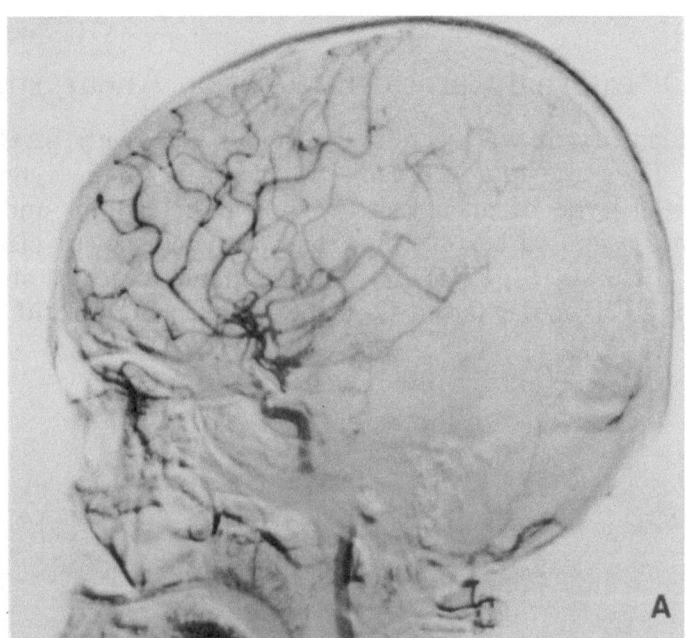

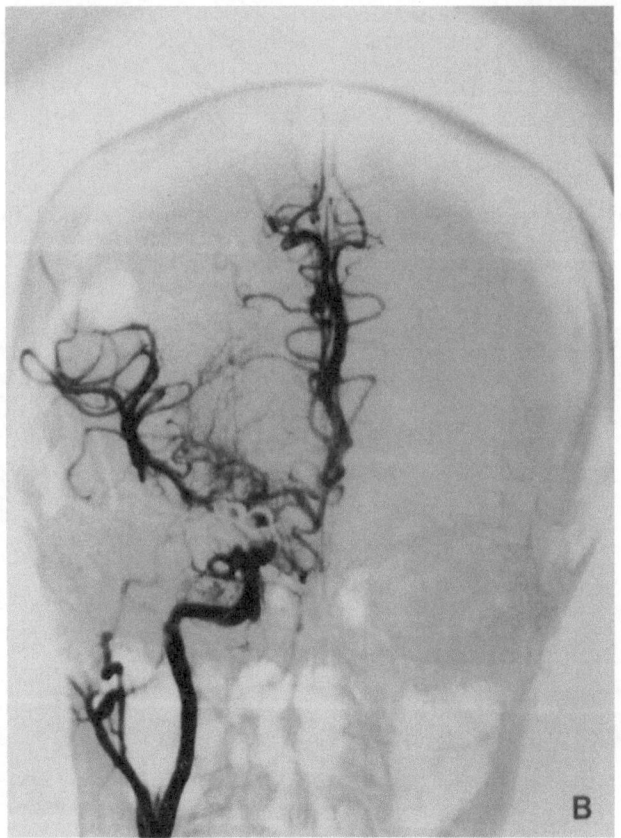

Fig. VIII-21-A, B. The aneurysm is obliterated by the use of a fenestrated clip. Note that the internal carotid is preserved

D. Internal Carotid Bifurcation Aneurysm

This type of aneurysm was seen in 3.1% ($^{46}/_{1480}$). The same approach can be used for this type of aneurysm. The aneurysm is exposed by opening the Sylvian fissure **(Fig. VIII-22)**. **Fig. VIII-23** is a CT on day 0 of a 42-year-old man with SAH. Note clots (high density) in the Sylvian and interhemispheric cisterns and in the right half of the chiasmatic cistern.

Angiography revealed an aneurysm at the internal carotid bifurcation

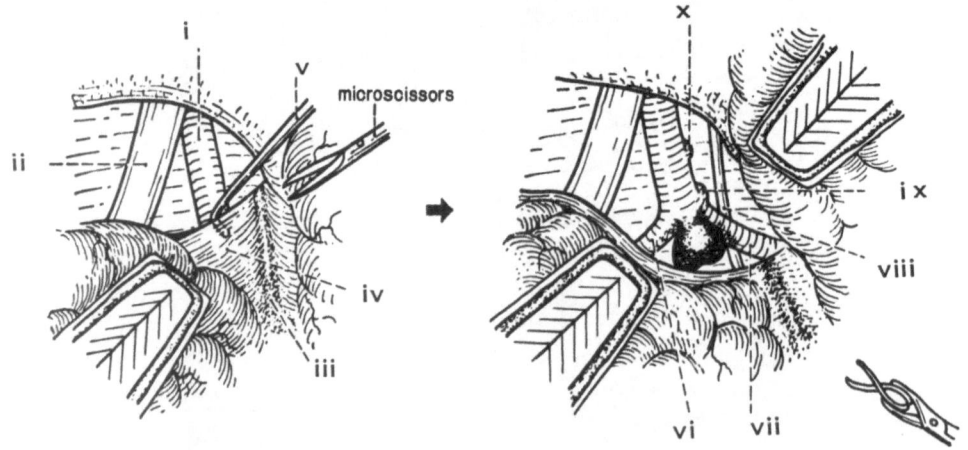

Fig. VIII-22. Approach to an aneurysm at the internal carotid bifurcation

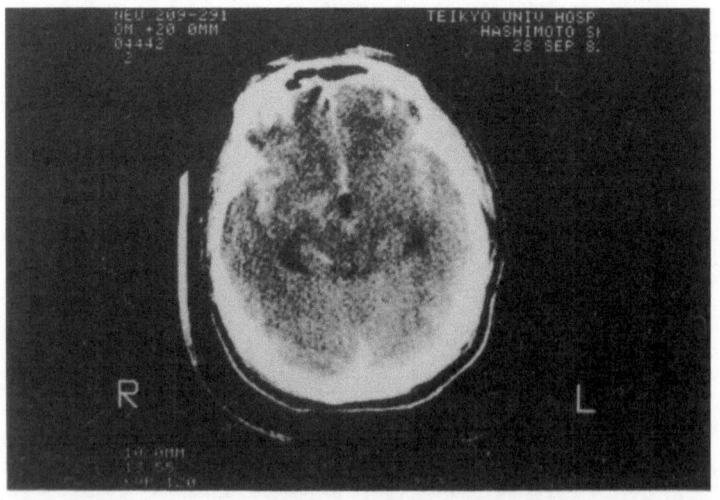

Fig. VIII-23. CT on day 0. A 42-year-old man with SAH. GCS, 14. Grade III

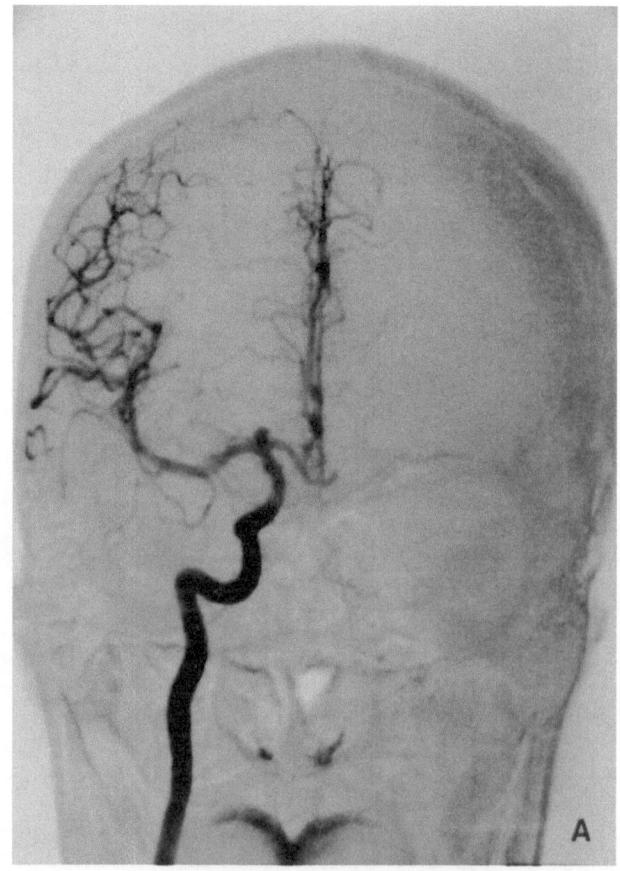

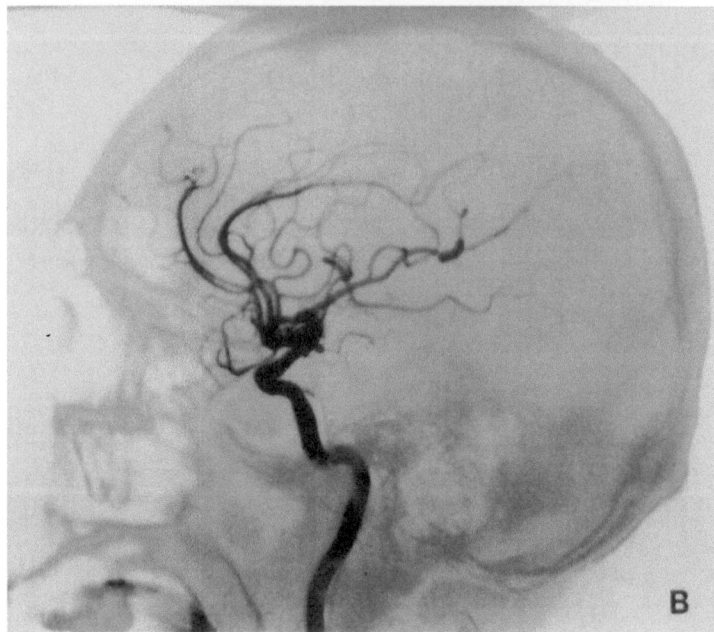

Fig. VIII-24. Angiograms revealing an aneurysm at the internal carotid bifurcation (A) and at the internal carotid-anterior choroidal junction (B)

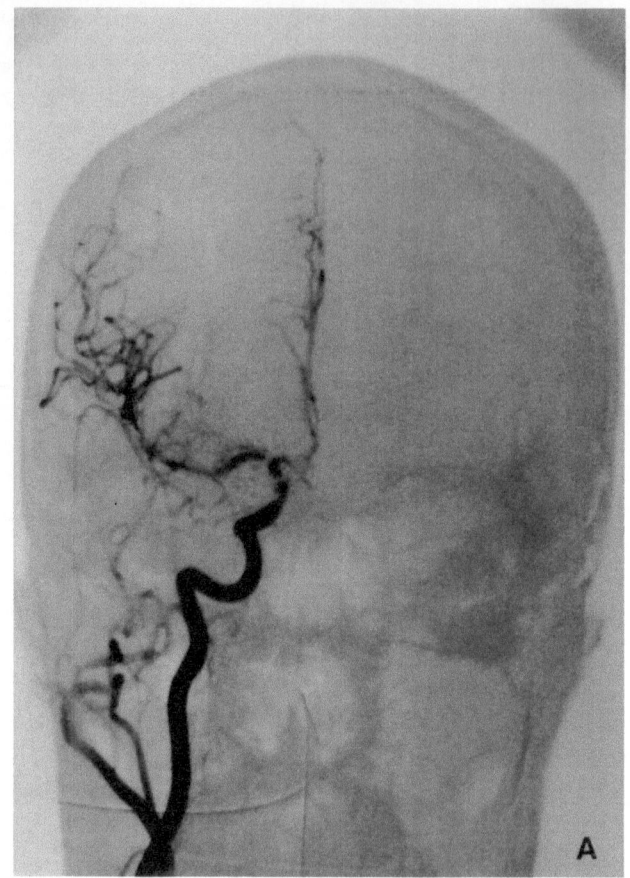

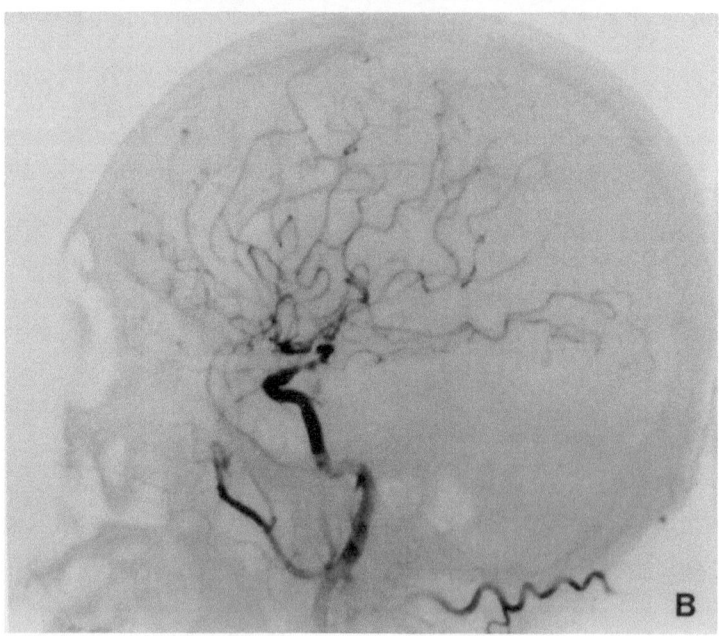

Fig. VIII-25-A, B. Postoperative angiograms showing both aneurysms obliterated

(Fig. VIII-24-A) and at the internal carotid-anterior choroidal junction (Fig. VIII-24-B). The patient underwent surgery on day 1, in grade III with GCS, 14. Both aneurysms wer successfully clipped (Fig. VIII-25-A, B). The patient later developed normal pressure hydrocephalus which was treated by ventriculo-peritoneal shunting. He was in a good physical condition with slight disorientation, 6 months postoperatively.

E. Middle Cerebral Aneurysm

This type of aneurysm is most frequent at the trifurcation or bifurcation of major branches from the middle cerebral trunk (20% or $^{296/1480}$). Small percentages of aneurysms were found in the M_1 portion (1.4% or $^{21/1480}$) and in the M_2 portion (0.3% or $^{4/1480}$).

The same operative approach as for the previous ones can be used. To open the Sylvian fissure by cutting the arachnoid, some prefer to start with the proximal portion (Fig. VIII-26-A) and some with the distal portion (Fig. VIII-26-B). We prefer the former in order to secure beforehand the preaneurysmal segment. This aneurysm is usually located at the trifurcation or bifurcation of the middle cerebral artery, and so the most important thing is not to obliterate any of the main branches.

Fig. VIII-27 is a CT on day 0 of a 66-year-old woman with SAH. Clots were seen in the Sylvian fissure and in the adjacent temporal-parietal lobes on the right side, in the interhemispheric fissure and in the right half of the ambient cistern. The patient's consciousness level was 6 in GCS, consequently, she was of grade V according to the proposal of Sano and Tamura

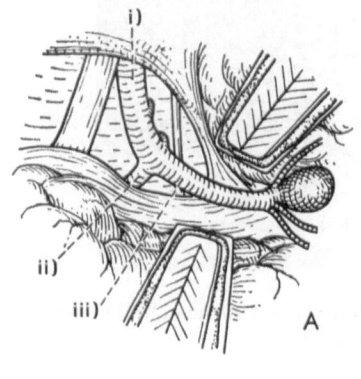

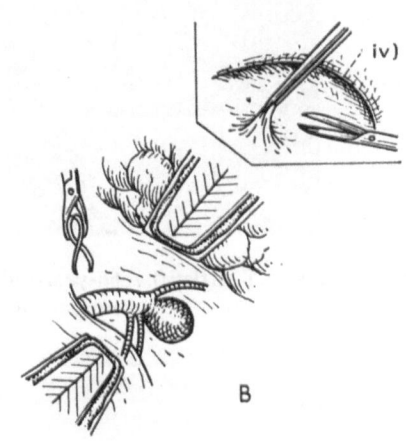

Fig. VIII-26. Approaches to middle cerebral aneurysms. A) Opening the Sylvian fissure from the proximal portion. B) Opening the Sylvian fissure from the distal portion

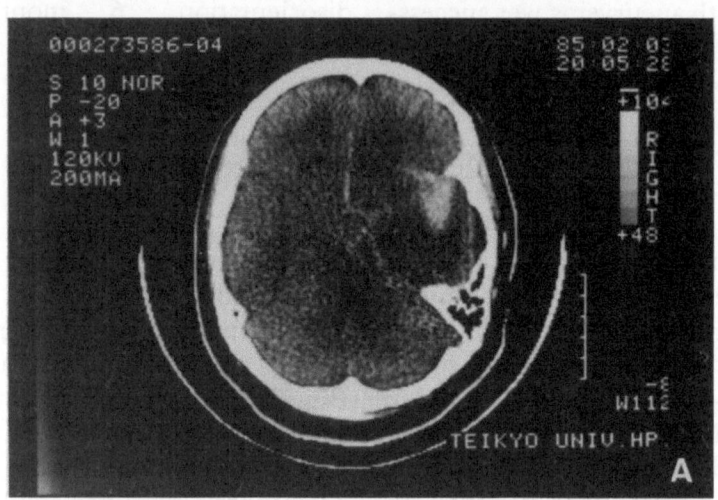

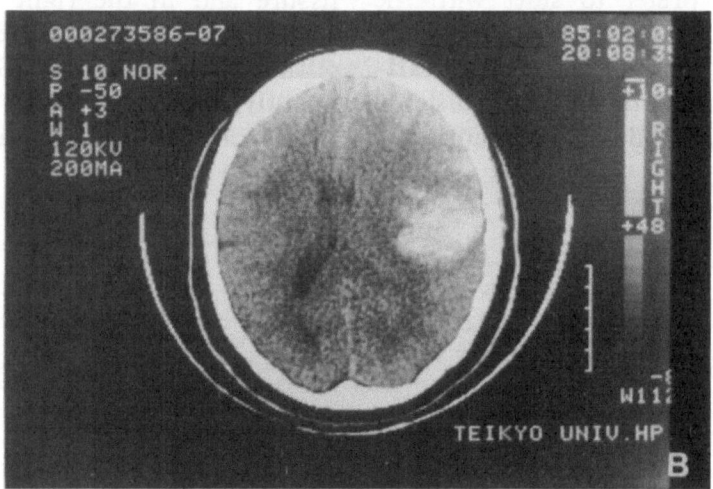

Fig. VIII-27-A, B. CT on day 0. A 66-year-old woman with SAH and intracerebral hematoma. Left hemiparesis. GCS, 6. Grade V

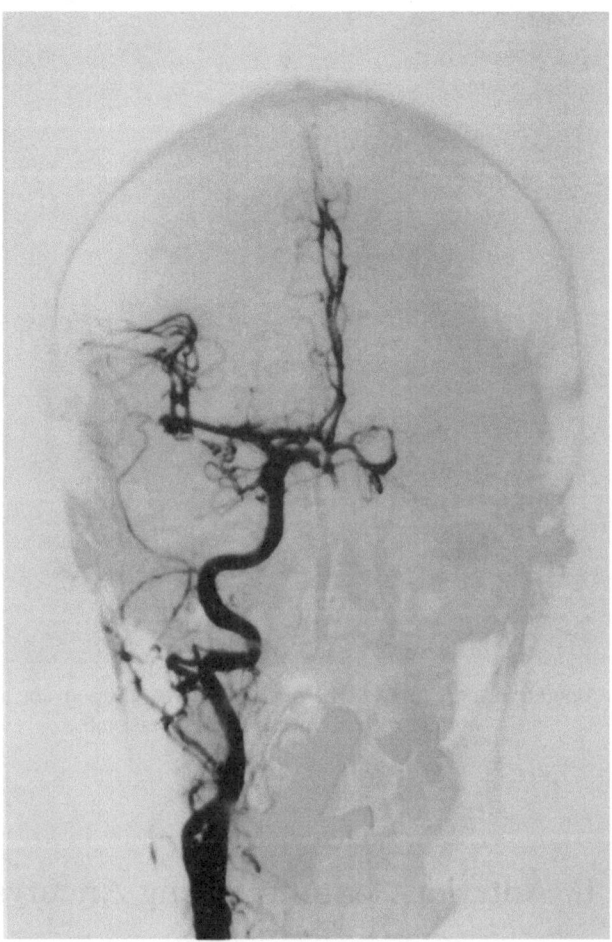

Fig. VIII-28. The same case as in Fig. 27. An aneurysm at the bifurcation of the right middle cerebral artery

(1985). She had left hemiparesis. Angiography revealed an aneurysm at the bifurcation of the right middle cerebral artery (**Fig. VIII-28**). On day 1, she was operated on, with GCS, 6 (grade V), the aneurysm being clipped and intracerebral hematoma being removed. Her consciousness level gradually improved until day 5 when cerebral vasospasm developed which was treated by induced hypertension and hypervolemia. On day 9 she had gastrointestinal bleeding. From day 13 on, her general condition ameliorated, hemiparesis improved. "Good recovery" of GOS, as of 6 months postoperatively. **Fig. VIII-29** is her postoperative angiogram, the aneurysm being clipped with preservation of normal branches of the right middle cerebral artery.

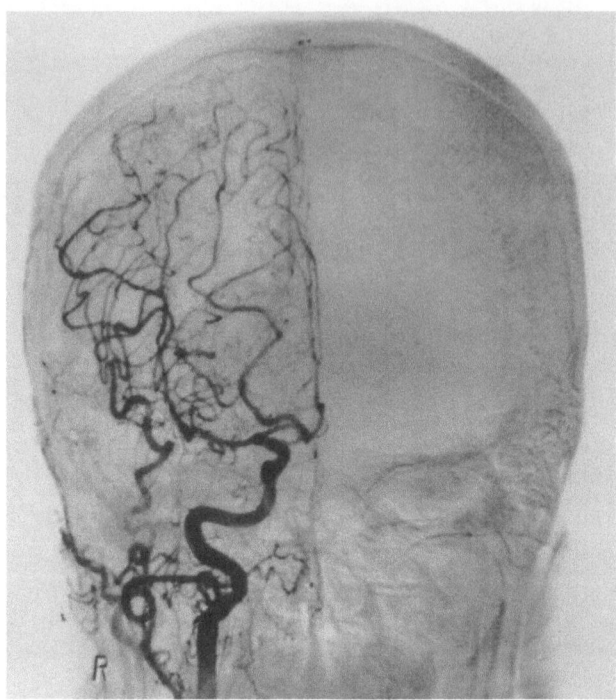

Fig. VIII-29. Postoperative angiogram. The aneurysm was clipped and branches of the right middle cerebral artery were preserved

F. Anterior Communicating Aneurysm

This is the most frequent among all intracranial aneurysms (25.3% or $^{374}/_{1480}$ in our series). The operative approach may be more medially shifted than the previous on (Fig. VIII-30-A). If the surgeon prefers the interhemispheric approach, craniotomy should be close to the midline (Fig. VIII-30-B). In the interhemispheric approach, the side of approach is always on the nondominant side. In the pterional approach, however, some prefer the nondominant side, some the side of the anterior cerebral artery (A_1) which supplies blood to the aneurysm, and others the side of the more posteriorly located A_1–A_2 junction, whose position can be determined on the angiogram.

In the interhemispheric approach (Fig. VIII-31), usually no part of the brain will be damaged. However, in this approach, one may encounter premature rupture of aneurysm before securing the preaneurysmal segment of the anterior cerebral artery. The advantage of the pterional approach, on the contrary, is that one can secure the preaneurysmal segment before dissecting the aneurysm (Fig. VIII-32) and the distance to the aneurysm is shorter

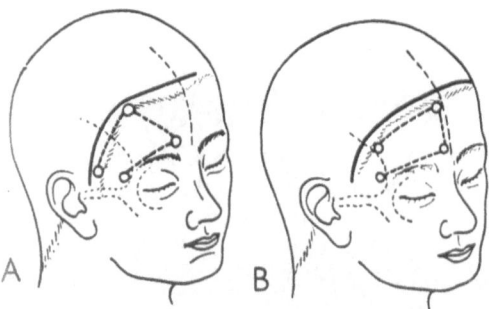

Fig. VIII-30. Skin incision and bone flap for the approach to anterior communicating aneurysms. A) Pterional approach. B) Interhemispheric approach

than in the interhemispheric approach. The disadvantage, however, is that one often has to suction the rectal gyrus in order to dissect out the neck of aneurysm **(Fig. VIII-33)**. In any case,

care should be taken not to injure the perforating branches of the anterior cerebral or the anterior communicating artery. Preferably, the anterior communicating artery is preserved patent.

Fig. VIII-34 is a CT on day 0 of a 55-year-old man with SAH. High density (clots) could be noticed in the suprasellar, chiasmatic, interhemispheric, Sylvian and ambient cisterns. Angiography revealed an anterior communicating aneurysm **(Fig. VIII-35)**. He was submitted to surgery on the same day in the condition of grade II, and GCS, 15. The aneurysm was clipped and all normal branches were preserved **(Fig. VIII-36)**. Postoperative course was uneventful and he was discharged with "good recovery" in GOS.

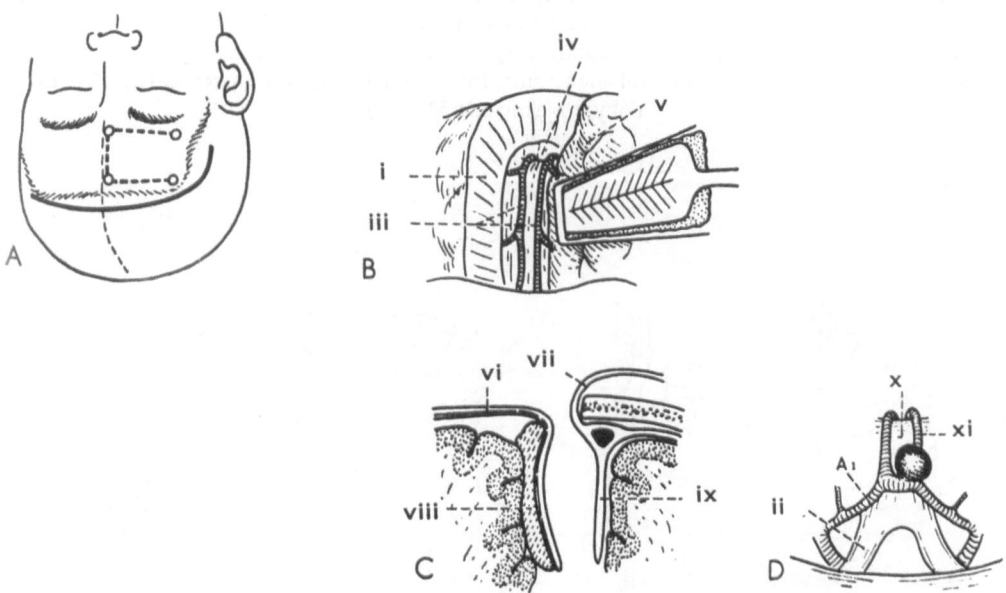

Fig. VIII-31. Interhemispheric approach. A) Skin incision and bone flap. B) Opening the interhemispheric fissure: *i* falx, *iii* anterior cerebral arteries, *iv* genu corporis callosi, *v* frontal lobe. C) Cutting the arachnoid along the falx: *vi* retractor, *vii* dura, *viii* cottonoid, *ix* falx. D) Visualizing A$_1$ portions of the anterior cerebral arteries and aneurysm: *ii* optic nerve, *xi* distal portion of the anterior cerebral artery, *x* genu corporis callosi (redrawn from Sano: Cerebral aneurysms. In: Gendai Gekagaku Taikei, **4**, 129–171 (1982). Tokyo: Nakayama Shoten)

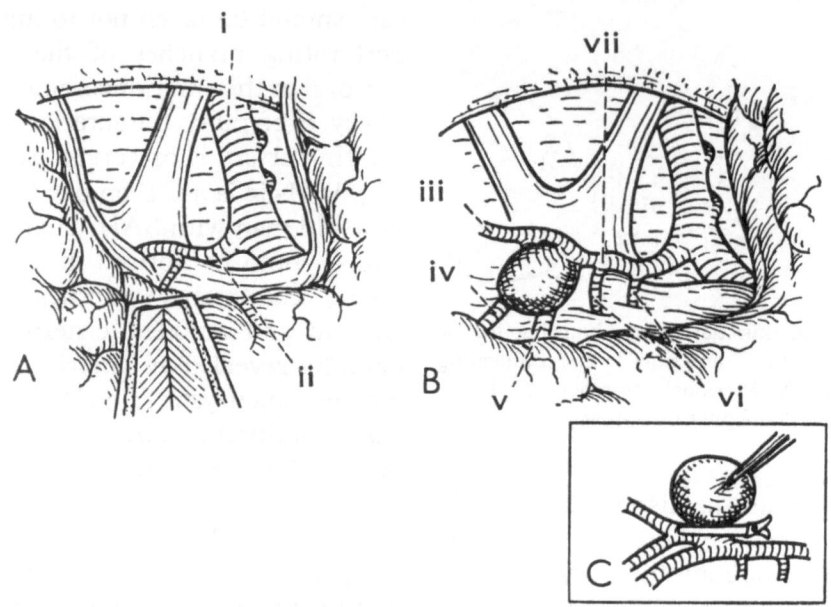

Fig. VIII-32. Pterional approach. A) Exposing the internal carotid and its bifurcation: *i* internal carotid, *ii* anterior cerebral artery (A₁). B) Tracing the anterior cerebral artery (A₁) to the anterior communicating artery and aneurysm: *iii* left A₁, *iv* left A₂, *v* right A₂, *vi* Heubner's arteries, *vii* right A₁. C) After clipping the aneurysmal neck, patency of bilateral A₁ and A₂ portions of the anterior cerebral arteries and the anterior communicating artery and its perforating branches should be checked (redrawn from Sano: Cerebral aneurysms. In: Gendai Gekagaku Taikei, **4,** 129–171 (1982). Tokyo: Nakayama Shoten)

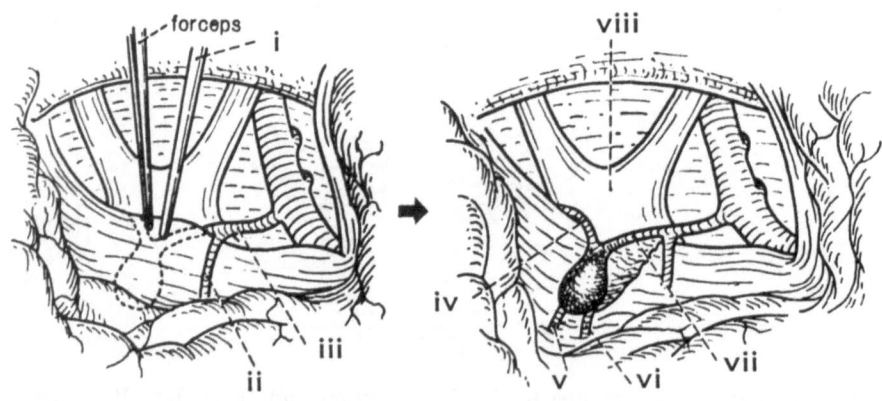

Fig. VIII-33. When an aneurysm is buried in the brain, a part of the rectal gyrus is electrocauterized and suctioned to expose the aneurysm: *i* suction tube, *ii* Heubner's artery, *iii* anterior cerebral artery (A₁), *iv* left A₁, *v* left A₂, *vi* right A₂, *vii* right A₁ (redrawn from Sano: Cerebral aneurysms. In: Gendai Gekagaku Taikei, **4,** 129–171 (1982). Tokyo: Nakayama Shoten)

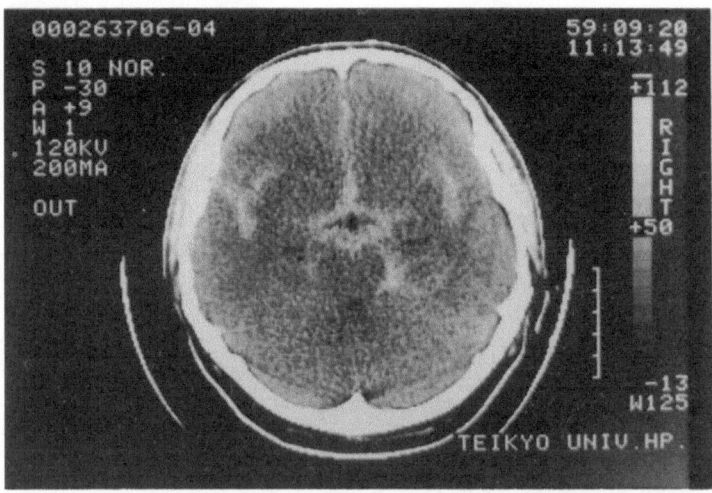

Fig. VIII-34. CT on day 0. A 55-year-old man. High density (clots) in the chiasmatic, inter-hemispheric, Sylvian, and ambient cisterns

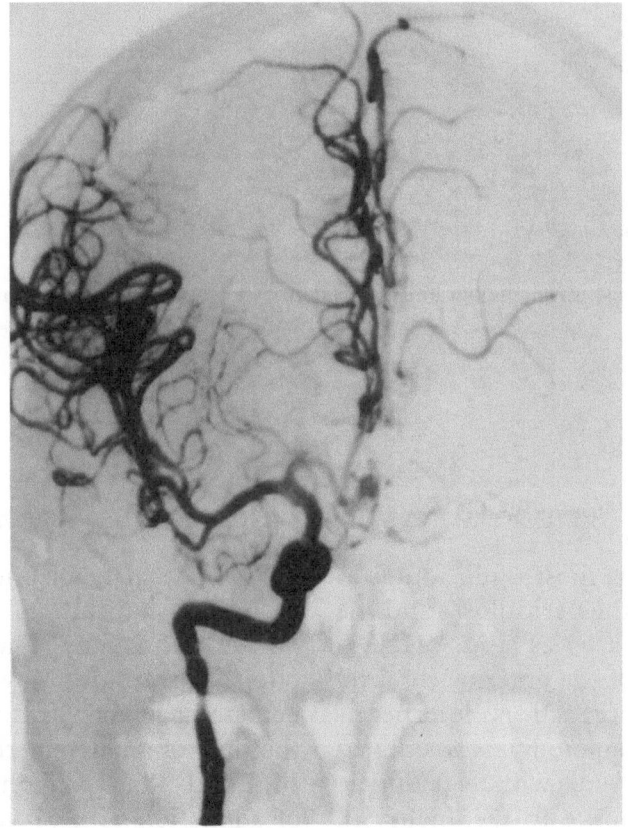

Fig. VIII-35. Angiography revealed an anterior communicating aneurysm. The same case as in Fig. 34

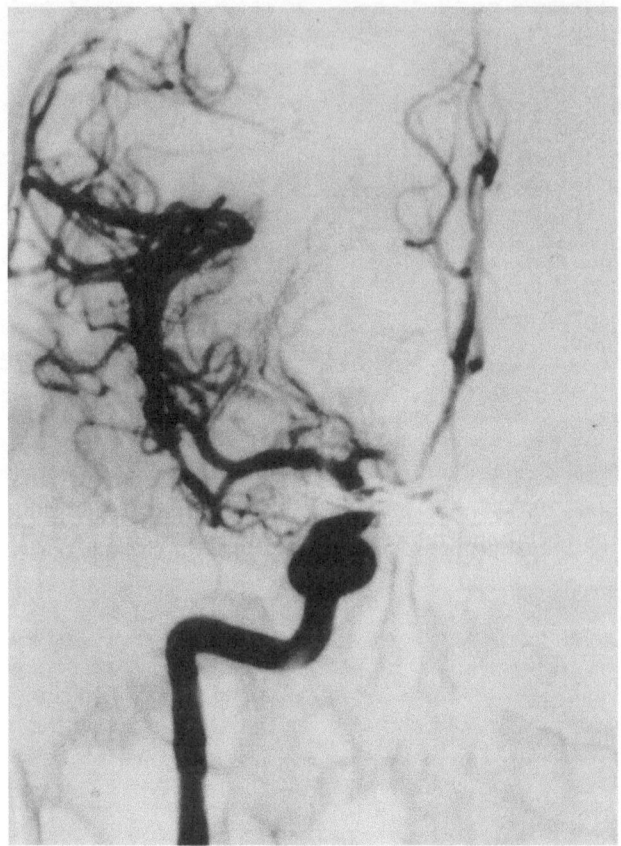

Fig. VIII-36. Postoperative angiogram. The aneurysm was clipped and normal arteries were preserved

G. Distal (Peripheral) Anterior Cerebral Artery Aneurysm

The aneurysm is most often found at the junction of the pericallosal and the callosomarginal arteries (Fig. VIII-37) or at the genu of the anterior cerebral artery. The approach is interhemispheric, and craniotomy is fashioned over the sagittal sinus with its peduncle on the side opposite to the aneurysm (Fig. VIII-37). It is better to retract the contralateral hemisphere rather than the ipsilateral hemisphere, to prevent premature rupture of the aneurysm during retraction. The aneurysm is often surrounded by intracerebral hematoma which should be removed at the time of aneurysm clipping.

Fig. VIII-38 is a CT on day 8, i.e., day of admission of a 61-year-old woman with SAH. High density (clots) was noted in the interhemispheric fissure

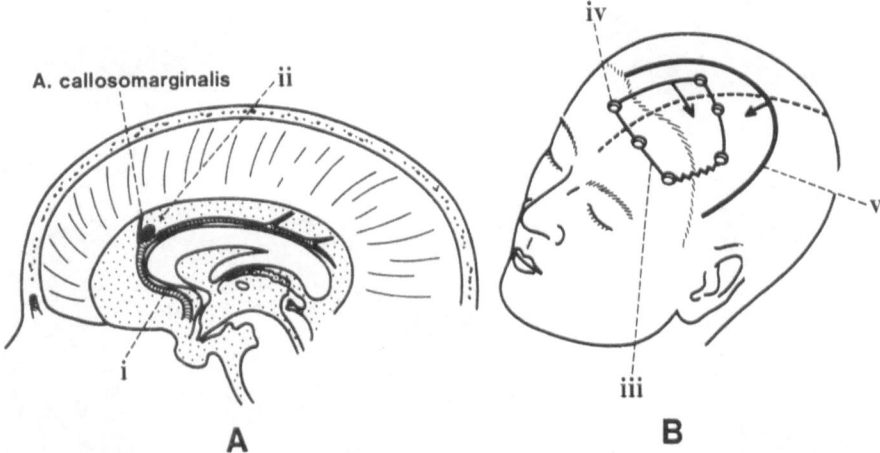

A   **B**

Fig. VIII-37. A) Aneurysm arising from the anterior cerebral-callosomarginal junction. B) Approach to the aneurysm. *i* Anterior cerebral artery, *ii* aneurysm, *iii* bone flap edge, *iv* burr hole, *v* skin incision (redrawn from Sano: Cerebral aneurysms. In: Gendai Gekagaku Taikei, **4**, 129–171 (1982). Tokyo: Nakayama Shoten)

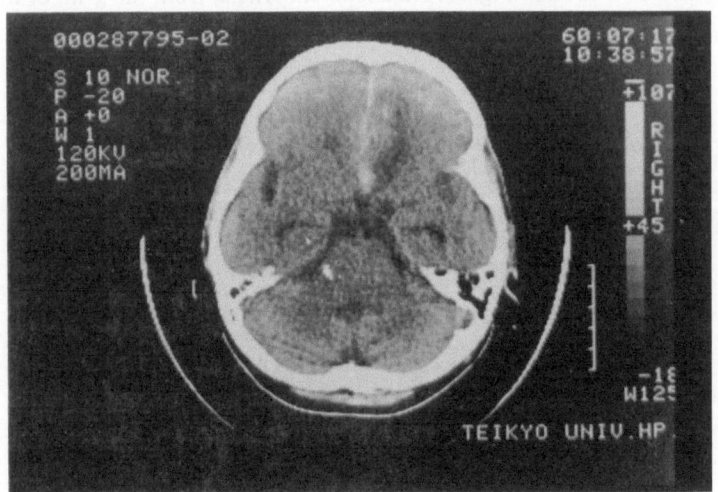

Fig. VIII-38. CT on day 8 (on admission). A 61-year-old woman. Note high density (clots) in the interhemispheric fissure and in the frontal lobes

and also in the frontal lobes. Angiography on the same day revealed an aneurysm arising from the genu of the left anterior cerebral artery (**Fig. VIII-39**). The patient's GCS was 15 and she was of grade II. She was submitted to surgery on day 10 and the aneurysm was clipped as seen in **Fig. VIII-40**. She was discharged 5 weeks later with "good recovery" of GOS.

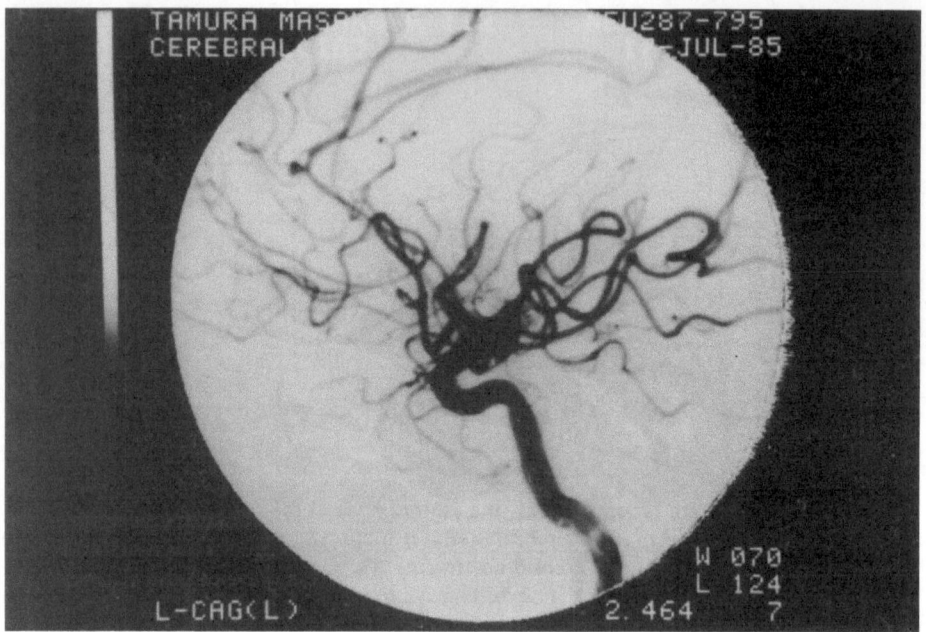

Fig. VIII-39. The same case as in Fig. 38. An aneurysm arising from the genu of the left anterior
cerebral artery was noted

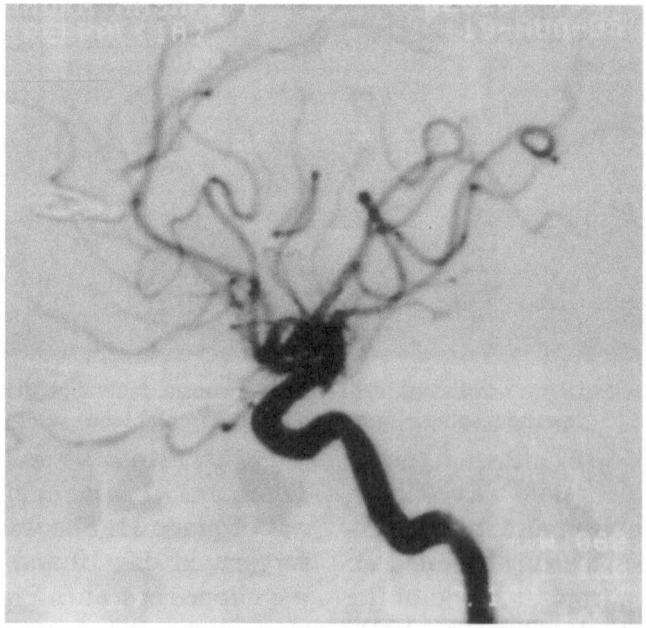

Fig. VIII-40. The same case as in Fig. 39. The aneurysm was clipped

H. Ophthalmic Aneurysm

This aneurysm arises at the junction of the internal carotid and the ophthalmic arteries. Before dissecting the aneurysm, it is advisable to explore the internal carotid artery in the neck, because, to manage premature rupture, temporary occlusion of the internal carotid in the neck may be necessary. The approach is as shown in **Fig. VIII-30-A**. The roof of the optic canal is drilled off to expose the aneurysmal neck (**Fig. VIII-41**). It may be necessary to retract the optic nerve medially or laterally to get a full exposure of the aneurysm.

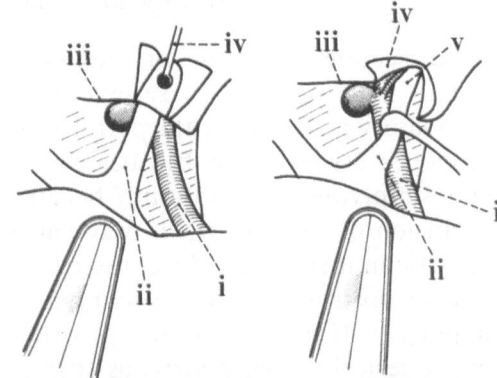

Fig. VIII-41. Ophthalmic aneurysm. The roof of the optic canal is drilled off to expose the neck of an ophthalmic aneurysm. *i* Internal carotid, *ii* optic nerve, *iii* ophthalmic aneurysm, *iv* drill or dura, *v* optic nerve (redrawn from Sano: Cerebral aneurysms. In: Gendai Gekagaku Taikei, **4**, 129–171 (1982). Tokyo: Nakayama Shoten)

I. Vertebro-Basilar Aneurysms

These aneurysms occupy about 8 to 10 percent of all intracranial aneurysms (7.8% = $^{116}/_{1480}$ in our series). The commonest sites are the distal bifurcation of the basilar artery, and the junction of the posterior inferior cerebellar artery (PICA) and the vertebral artery (**Fig. VIII-1**).

Aneurysms arising from the upper two thirds of the basilar artery, including the basilar tip, the junction of the superior cerebellar or the anterior inferior cerebellar artery with the basilar artery can be reached by the subtemporal approach (Drake 1965), or by the pterional transsylvian approach (Yaşargil *et al.* 1976), or by the temporopolar approach (Sano 1980b). For aneurysms arising from the lower one third of the basilar artery or from the vertebral artery, especially from the PICA-vertebral junction, the posterior fossa approach is appropriate. Aneurysms arising from the vertebro-basilar junction, however, are very difficult to reach by either of the posterior fossa, subtemporal or the pterional approach. They can be reached by the temporopolar approach. The transclival approach (Sano 1966) may also be one solution for aneurysms of this location in some cases.

1. Aneurysm Arising from the Distal Bifurcation of the Basilar Artery or Basilar Tip Aneurysm

In dissecting this type of aneurysm $(2.8\%$ or $^{42}/_{1480}$ in our series), it is mandatory not to injure the perforating branches arising from the basilar tip and supplying the midbrain, because occlusion of some of them may cause akinetic mutism. Therefore, the direction of protrusion of the aneurysm is important. If the aneurysm is protruding superiorly or anteriorly, as seen in **Fig. VIII-42**, i and ii, clipping of the aneurysmal neck is possible without interfering with perforating branches. If, however, the aneurysm is protruding posteriorly, as seen in **Fig. VIII-42**, iii, clipping of the aneurysmal neck may cause occlusion of the perforating branches. In the latter case, it is advisable to do wrapping of the aneurysm instead of its clipping.

The subtemporal approach popularized by Drake utilizes the skin incision

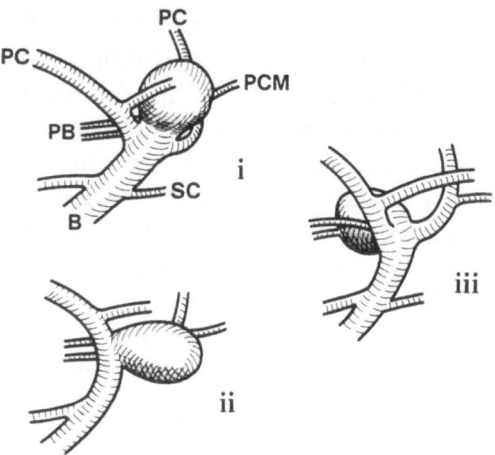

Fig. VIII-42. Various directions of protrusion of a basilar tip aneurysm. *i* Superior protrusion, *ii* anterior protrusion, *iii* posterior protrusion. *B* Basilar artery, *PB* perforating branches, *PC* posterior cerebral artery, *PCM* posterior communicating artery, *SC* superior cerebellar artery (redrawn from Sano: Cerebral aneurysms. In: Gendai Gekagaku Taikei, **4**, 129–171 (1982). Tokyo: Nakayama Shoten)

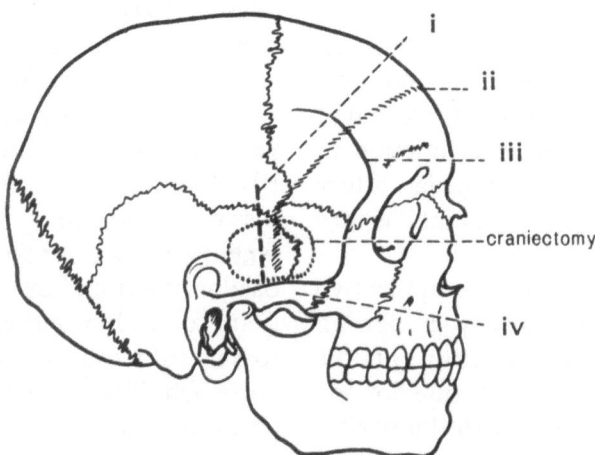

Fig. VIII-43. Skin incision and craniectomy in Drake's subtemporal approach. *i* skin incision, *ii* hair line, *iii* linea temporalis ossis frontalis, *iv* zygomatic arch

and craniotomy used in trigeminal root section via the temporal fossa as shown in **Fig. VIII-43**. The temporal lobe (usually on the nondominant side) is extensively elevated. A traction suture is put to the tentorial edge (plica petro-

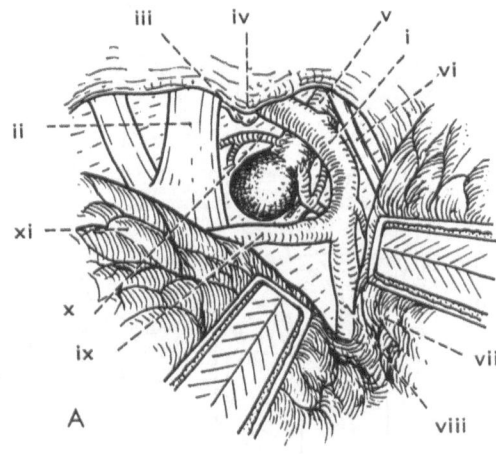

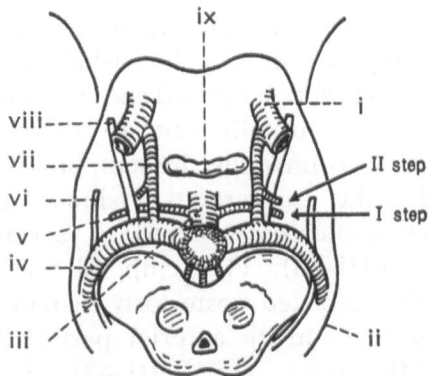

Fig. VIII-44. Subtemporal approach to a basilar tip aneurysm. I step: from the lateral side, II step: from the antero-lateral side. *i* Internal carotid, *ii* tentorium, *iii* basilar artery, *iv* posterior cerebral artery, *v* superior cerebellar artery, *vi* trochlear nerve, *vii* posterior communicating artery, *viii* oculomotor nerve, *ix* dorsum sellae (redrawn from Sano: Cerebral aneurysms. In: Gendai Gekagaku Taikei, **4**, 129–171 (1982). Tokyo: Nakayama Shoten)

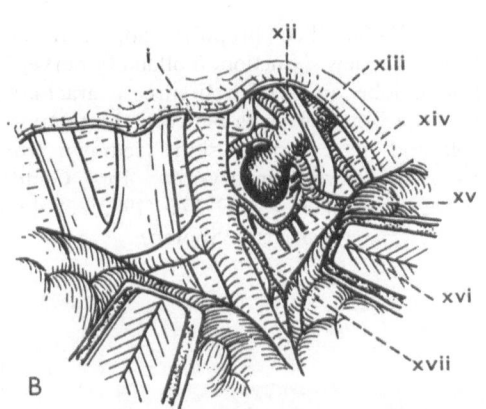

Fig. VIII-45. Surgeon's view of a basilar tip aneurysm as explored by the pterional transsylvian approach of Yaşargil. *i* Internal carotid, *ii* optic nerve, *iii* posterior clinoid, *iv* anterior clinoid, *v* basilar artery, *vi* oculomotor, *vii* temporal lobe, *viii* middle cerebral artery, *ix* anterior cerebral artery, *x* posterior cerebral artery, *xi* frontal lobe, *xii* oculomotor nerve, *xiii* basilar artery, *xiv* superior cerebellar artery, *xv* temporal lobe, *xvi* posterior communicating artery, *xvii* anterior choroidal artery

clinoidea anterior). The aneurysm is first approached from the lateral side and then from the anterolateral side (**Fig. VIII-44**). The pterional transsylvian approach utilizes the skin incision and craniotomy as shown in **Fig. VIII-4**. The Sylvian fissure is opened. The internal carotid is exposed and the posterior communicating artery is followed to the posterior cerebral artery and aneurysm (**Fig. VIII-45**). According to variations in length of the internal carotid, basilar tip aneurysm can be seen between the optic nerve and the internal carotid (**Fig. VIII-45-A**) or between the latter and the oculomotor nerve (**Fig. VIII-45-B**).

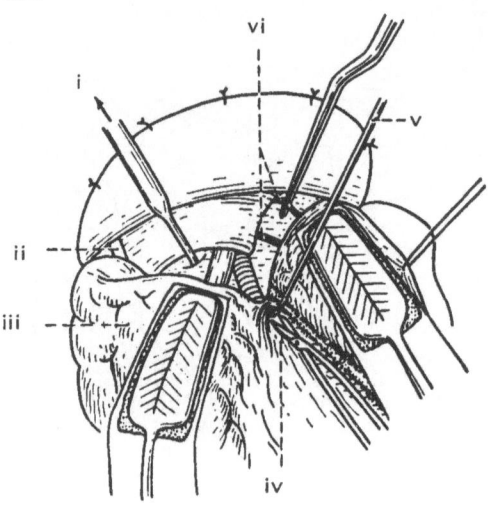

Fig. VIII-46. Temporopolar approach (1). Surgeon's view. *i* Suction, *ii* olfactory nerve, *iii* frontal lobe, *iv* Sylvian fissure, *v* arachnoid hook, *vi* bridging veins between the temporal pole and the cavernous sinus (redrawn from Sano: Cerebral aneurysms. In: Gendai Gekagaku Taikei, **4**, 129–171 (1982). Tokyo: Nakayama Shoten)

Both the subtemporal and the transsylvian approaches need very experienced hands because of the limited, narrow operative field. We devised the temporopolar approach to overcome this difficulty (Sano 1980 b). The skin incision and craniotomy are just as shown in **Fig. 4**. The operation is performed usually on the nondominant side. After opening the dura, the lower half of the Sylvian fissure is opened and the bridging veins, usually one to three in number, connecting the temporal pole and the cavernous sinus are electrocauterized and severed (**Fig. VIII-46**). The temporal pole is gently retracted posteriorly, leaving a wide space in the anterior part of the middle fossa (**Fig. VIII-47**). Care should be taken not to injure the vein of Labbé. The arachnoid covering the carotid and the posterior communicat-

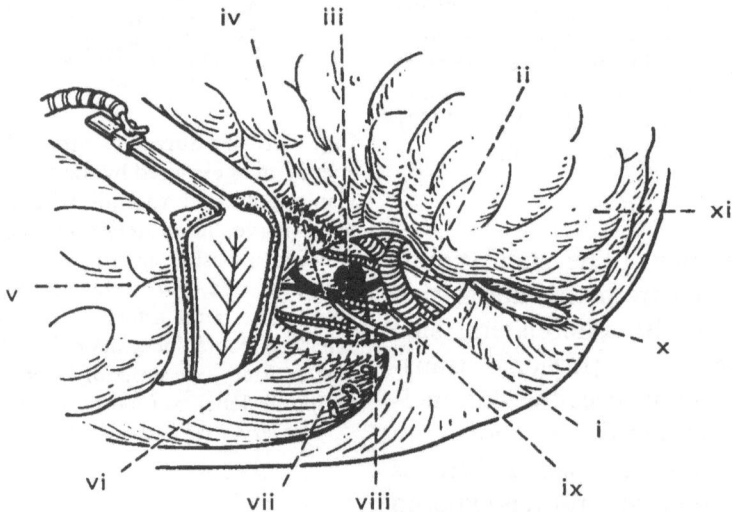

Fig. VIII-47. Temporopolar approach (2). Lateral view. *i* Internal carotid, *ii* optic nerve, *iii* aneurysm, *iv* posterior cerebral artery, *v* temporal lobe, *vi* tentorial edge, *vii* basilar artery, *viii* oculomotor nerve, *ix* posterior communicating artery, *x* olfactory nerve, *xi* frontal lobe (redrawn from Sano: Cerebral aneurysms. In: Gendai Gekagaku Taikei, **4**, 129–171 (1982). Tokyo: Nakayama Shoten)

ing arteries is meticulously cut to expose these arteries and the oculomotor nerve. The posterior communicating artery is followed backward to its junction with the posterior cerebral artery (**Fig. VIII-48**). The arachnoid

out retracting the arteries and the nerves.

Figs. VIII-49 and **VIII-50** are angiograms of a 56-year-old woman performed on admission, some 6 weeks after SAH. A basilar tip aneurysm

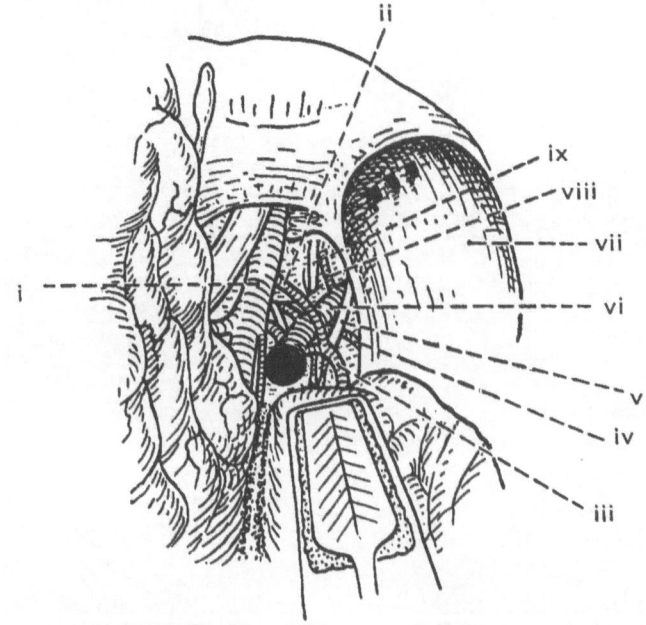

Fig. VIII-48. Temporopolar approach (3). Surgeon's view. A basilar tip aneurysm is exposed. *i* Internal carotid, *ii* posterior communicating artery, *iii* posterior cerebral artery, *iv* superior cerebellar artery, *v* oculomotor nerve, *vi* basilar artery, *vii* floor of middle fossa, *viii* contralateral anterior inferior cerebellar artery, *ix* contralateral acoustic nerve (redrawn from Sano: Cerebral aneurysms. In: Gendai Gekagaku Taikei, **4**, 129–171 (1982). Tokyo: Nakayama Shoten)

covering the area must be cut very cautiously. The bifurcation of the basilar artery, the posterior cerebral artery, the superior cerebellar artery, and often the anterior inferior cerebellar artery are thus exposed (**Fig. VIII-48**). By changing the position of the microscope and the position of the patient's head, the surgeon will have a very good view of these structures with-

could be noted. In this case, the posterior cerebral artery on the left side was supplied by the carotid system (the so-called fetal type). The aneurysm was superiorly pointed. Perforating branches were clearly noted. The aneurysm was clipped by means of the temporopolar approach.

Figs. VIII-51 and **VIII-52** are postoperative angiograms. The aneurysm

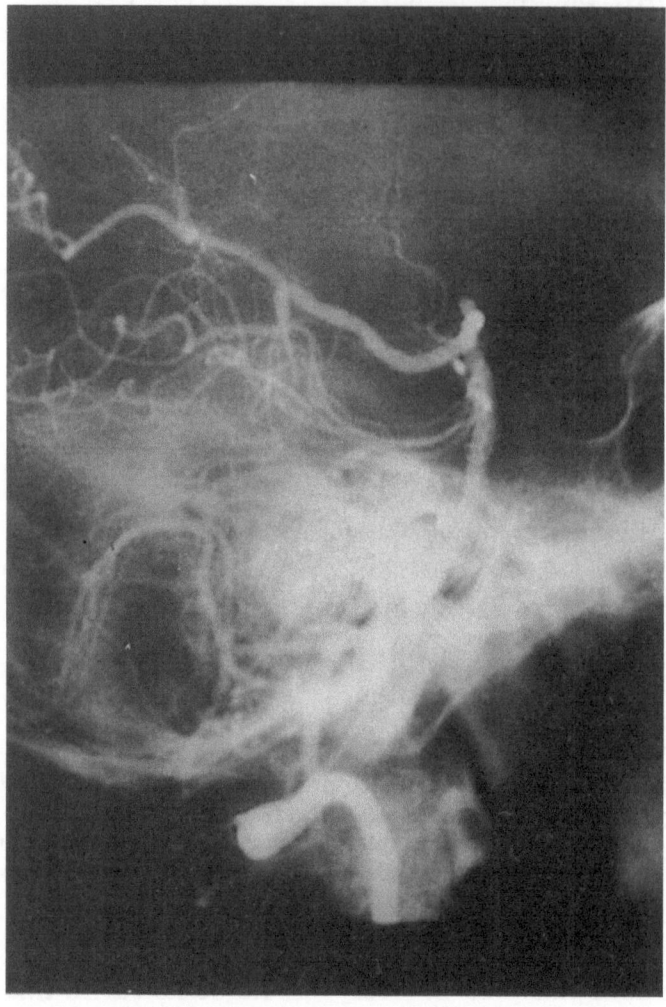

Fig. VIII-49. Aneurysm of the basilar tip. Vertebral angiogram. Lateral view. Note perforating branches. A 56-year-old woman

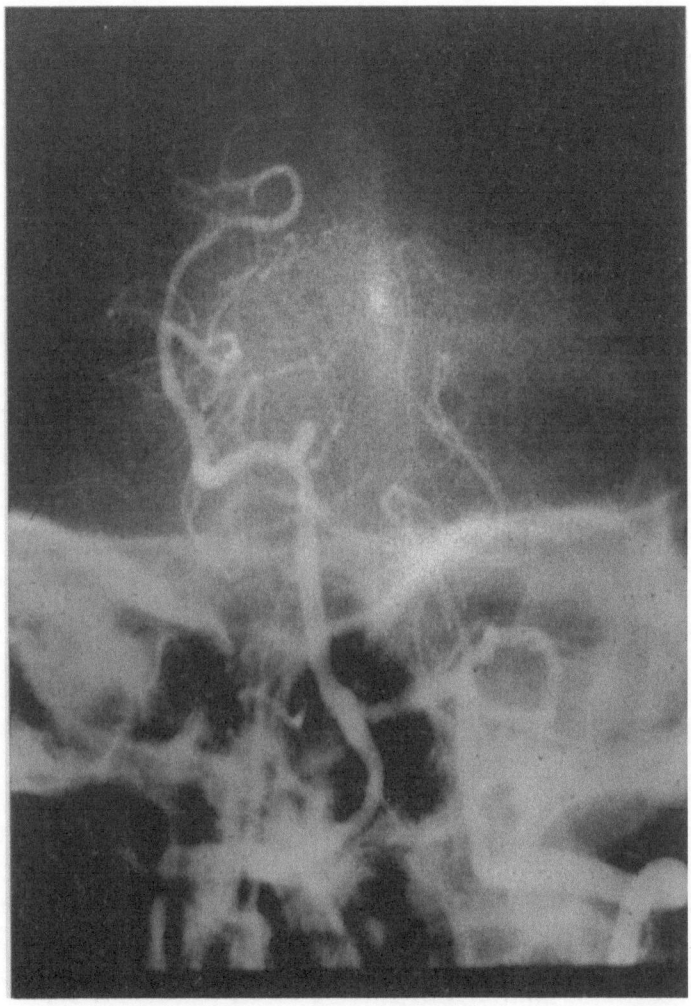

Fig. VIII-50. Aneurysm of the basilar tip. The same case as in Fig. 49. Vertebral angiogram. Anteroposterior view. Left posterior cerebral artery was supplied by the carotid system, and not demonstrated by vertebral angiography

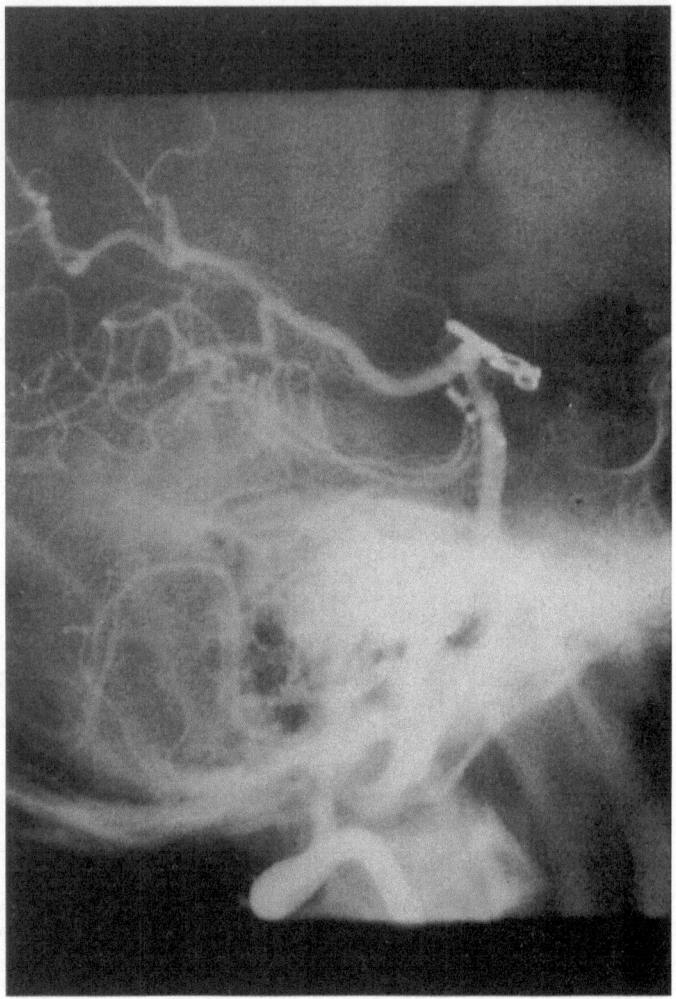

Figs. VIII-51. The same case as in Figs. 49 and 50. The aneurysm was clipped

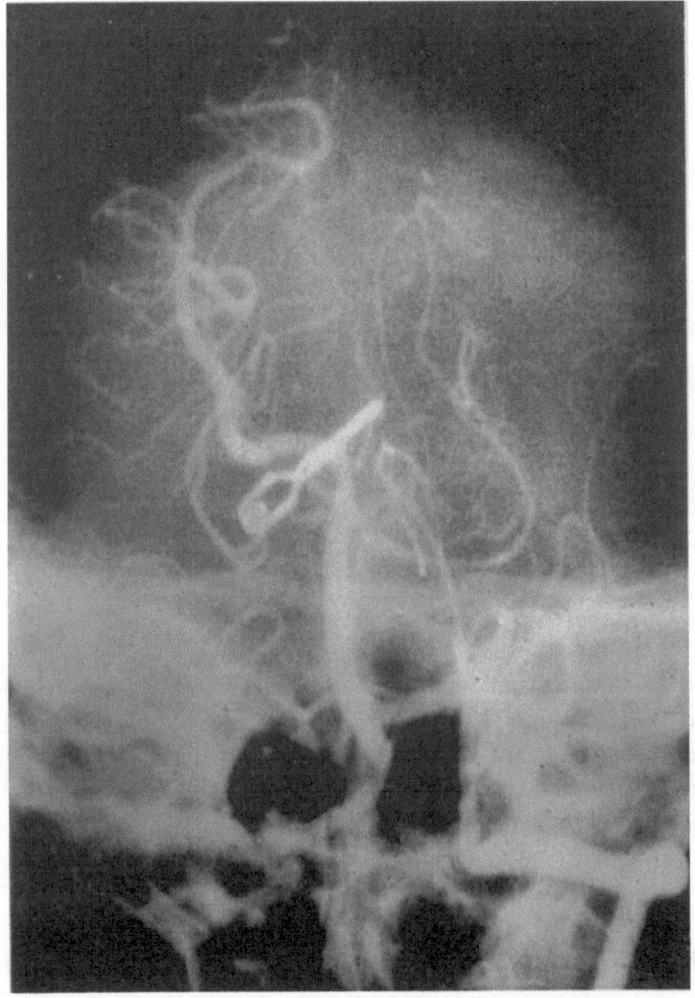

Fig. VIII-52. The same case as in Figs. 49 and 50. The aneurysm was clipped

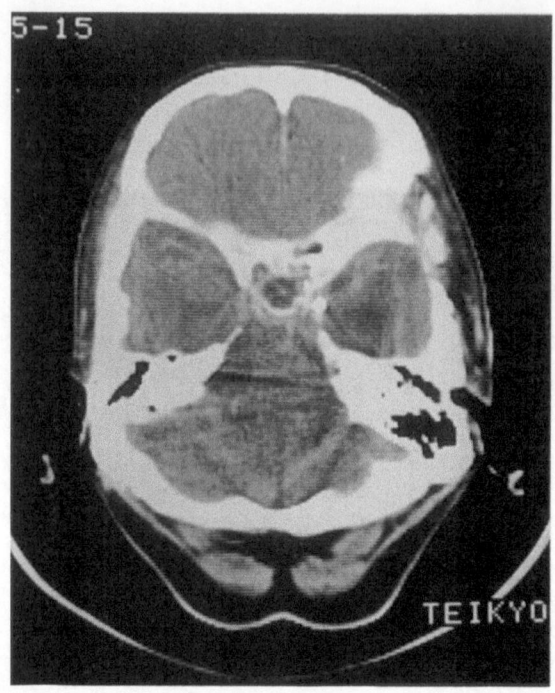

Fig. VIII-53. CT one year after the operation. The same case as in Fig. 49 and 50. No change in the temporal lobe

was clipped and all normal arteries including the perforating branches were preserved. The patient showed "good recovery" and was discharged. **Fig. VIII-53** is a CT one year after the operation. The temporal lobe showed no change which meant the temporopolar approach was not harmful to the temporal lobe.

2. Aneurysm Arising from the Superior Cerebellar-Basilar Junction

Approaches to this type of aneurysm (1.1% or $^{17}/_{1480}$ in our series) are the same as for basilar tip aneurysms, except for the side of approach which is usually ipsilateral to the side of the aneurysm.

Fig. VIII-54 shows an aneurysm arising from the junction of the left superior cerebellar and the basilar artery in a 56-year-old woman. The aneurysm was exposed by the temporopolar approach on the left side, and obliterated with two clips, preserving the superior cerebellar and other branches (**Fig. VIII-55**).

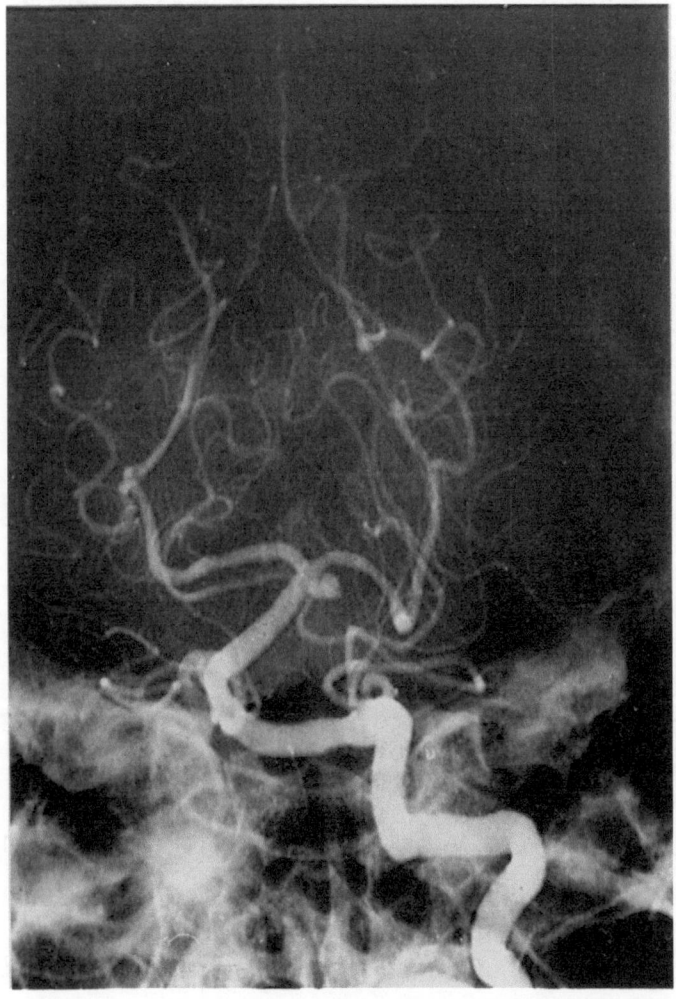

Fig. VIII-54. A 56-year-old woman. Superior cerebellar-basilar aneurysm

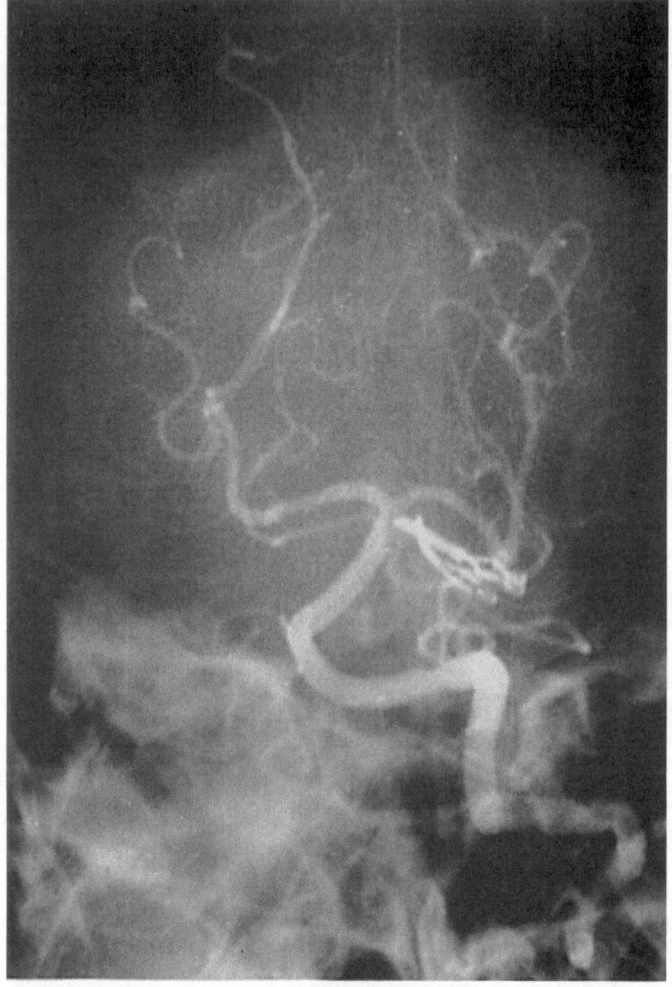

Fig. VIII-55. The same case as in Fig. 54. The aneurysm was clipped and all normal branches were
preserved

3. Aneurysms Arising from the Basilar Trunk

Aneurysms arising from the basilar trunk other than the superior cerebellar-basilar junction or the vertebro-basilar junction can be reached by the subtemporal approach or by the temporopolar approach or, if it is located in the lower basilar trunk, by the posterior fossa approach. Skin incision and bone flap for the subtemporal approach for this type of aneurysm may preferably be a question mark incision and a temporal bone flap as shown in **Fig. VIII-56**. If the aneurysm is located in the lower basilar trunk, it may be necessary to split the tentorium as illustrated in **Fig. VIII-57**.

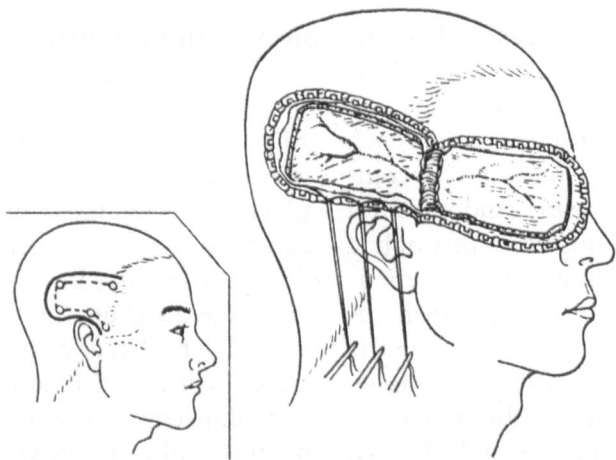

Fig. VIII-56. Question mark incision and bone flap for the subtemporal approach

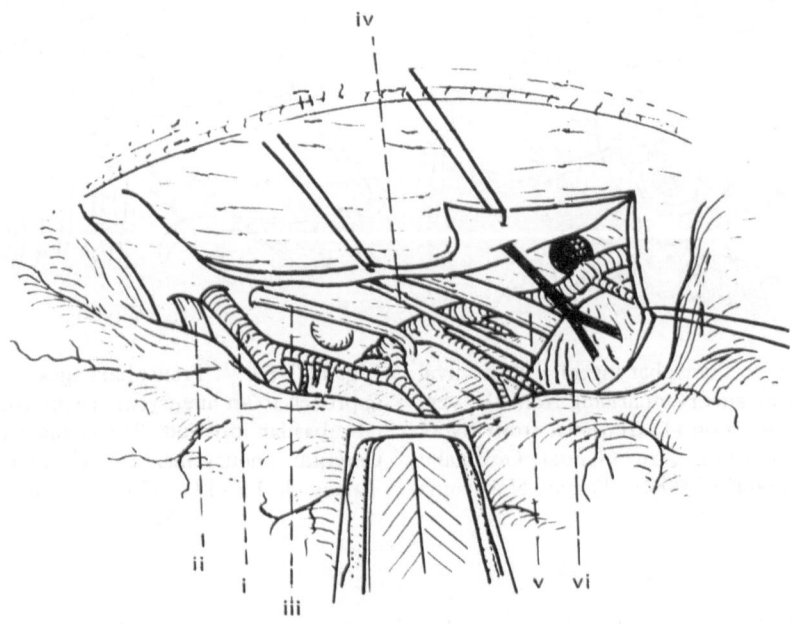

Fig. VIII-57. Surgeon's view of a basilar trunk aneurysm by means of the subtemporal approach. *i* Internal carotid, *ii* optic nerve, *iii* oculomotor nerve, *iv* trochlear nerve, *v* trigeminal nerve, *vi* cerebellum

4. Aneurysms Arising from the Vertebro-Basilar Junction

This aneurysm may be reached by the posterior fossa approach. When it is located in the midline, however, the transpharyngeal-transclival approach (Sano 1966, 1979) may be recommended in some cases. Endotracheal anesthesia is administered through a tracheostomy opening. The patient is placed in a semisitting position. A Davis retractor is used for opening the mouth and pressing down the tongue. The posterior half of the palate is incised in the midline (**Fig. VIII-58**).

made with an air drill, from the level of the posterior-inferior rim of the vomer to the level a little above the anterior rim of the foramen magnum (**Fig. VIII-59**). The epidural venous plexus is coagulated and the dura is carefully cut to expose the basilar artery. After clipping of the aneurysmal neck, the dural opening is covered with a fascia lata flap and the clival defect is sealed with dexon sutures. A gauze pack impregnated with antibiotic ointment is applied to the stuture. Then the

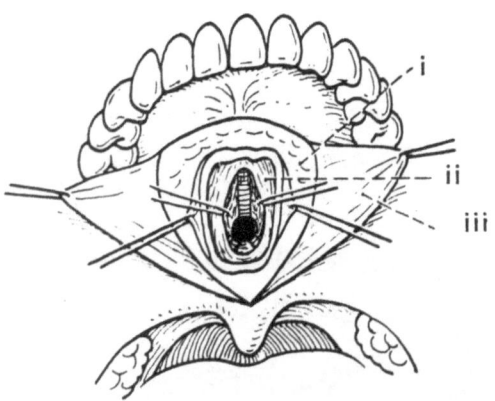

Fig. VIII-58. Transpharyngeal transclival approach to an aneurysm arising from the vertebrobasilar junction (1) (redrawn from Sano: Cerebral aneurysms. In: Gendai Gekagaku Taikei, **4**, 129–171 (1982). Tokyo: Nakayama Shoten)

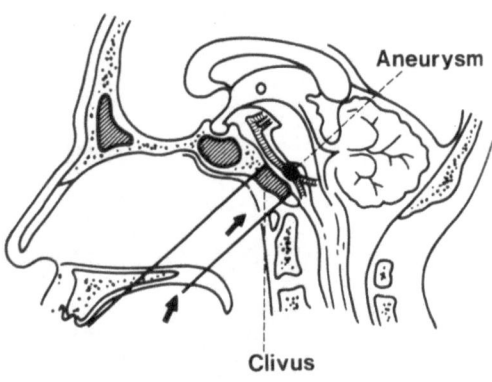

Fig. VIII-59. Transpharyngeal transclival approach to an aneurysm arising from the vertebrobasilar junction (2) (redrawn from Sano: Cerebral aneurysms. In: Gendai Gekagaku Taikei, **4**, 129–171 (1982). Tokyo: Nakayama Shoten)

The uvula is preserved. The posterior part of the bony hard palate is exposed and resected. Through this window, the nasopharyngeal mucosa in front of the clivus is incised in the midline to expose the anterior surface of the clivus where a window, about 2 cm in width, is

palate incision is closed with dexon sutures. The patient is kept in a semisitting position and continuous spinal drainage is fashioned during the first two postoperative weeks. The gauze pack is removed in these two weeks.
Fig. VIII-60 is an aneurysm of this

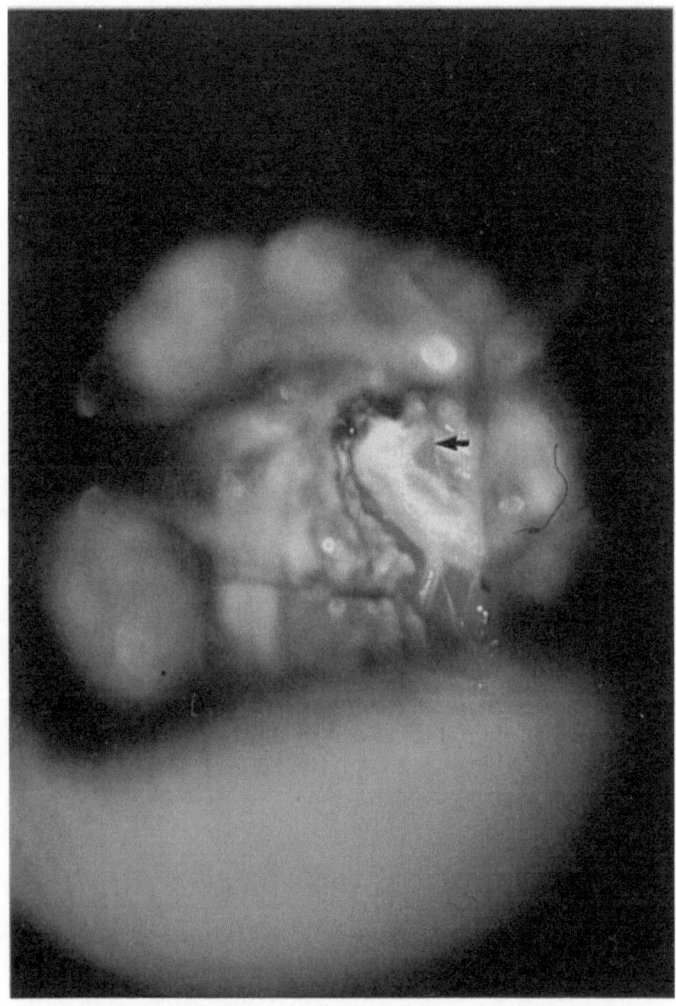

Fig. VIII-60. A 52-year-old man. Aneurysm (arrow) arising from the vertebro-basilar junction exposed by the transpharyngeal transclival approach

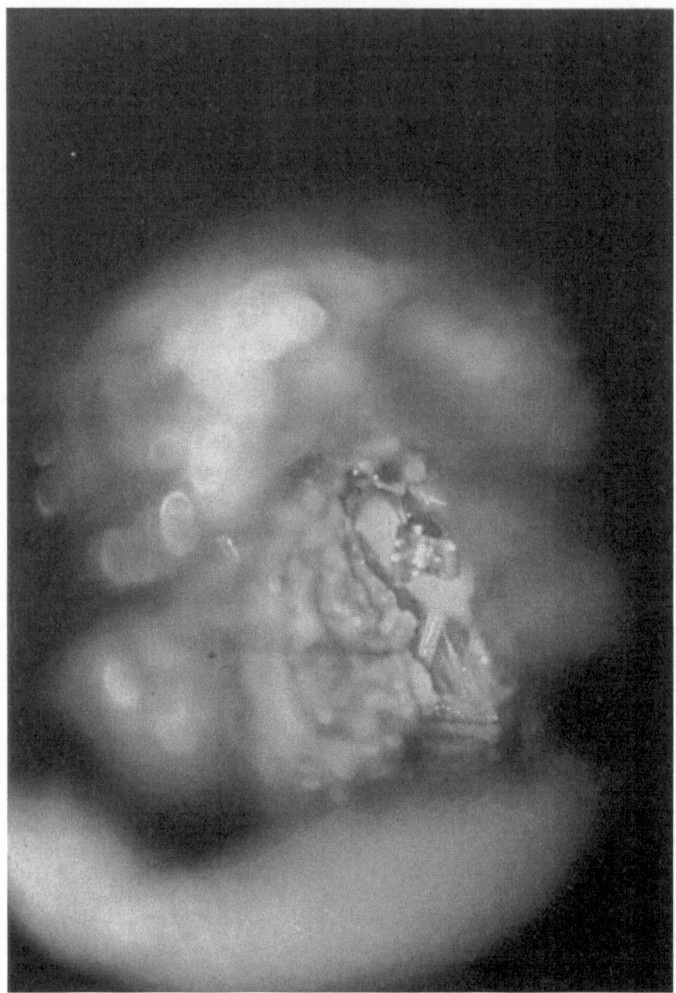

Fig. VIII-61. The same case as in Fig. 60. The aneurysm was clipped

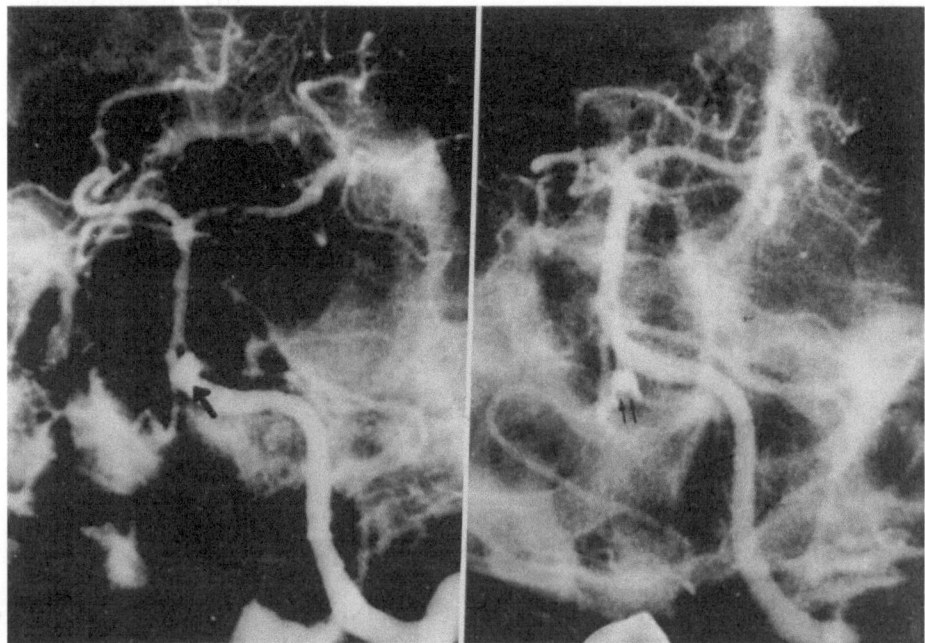

Fig. VIII-62. The same case as in Fig. 60. Left: preoperative angiogram. Single arrow indicates the aneurysm. Right: postoperative angiogram. Double arrow indicates the head of a clip obliterating the aneurysm (Figs. 60 and 61: from Sano: Basilar artery aneurysms—transoral transclival approach. In: Cerebral aneurysms (Pia HW, Langmaid C, Zierski J eds); pp 326–328, Berlin, Springer, 1979

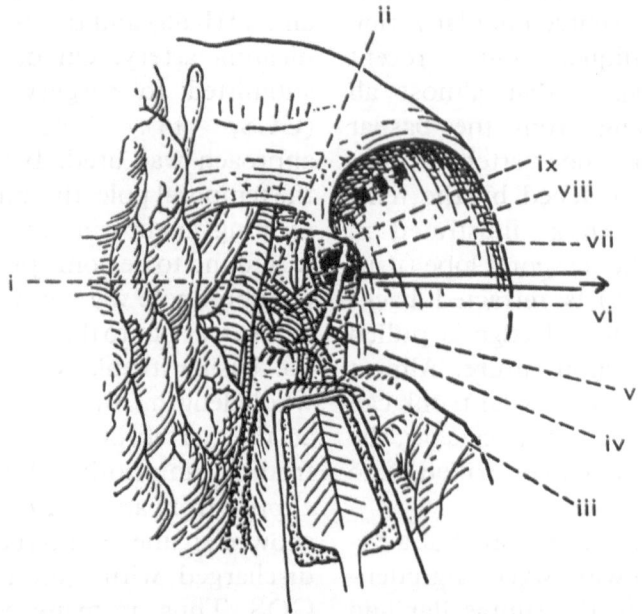

Fig. VIII-63. Exploration of an aneurysm arising from the vertebrobasilar junction by the temporopolar approach. *i* Internal carotid, *ii* posterior communicating artery, *iii* posterior cerebral artery, *iv* superior cerebellar artery, *v* oculomotor nerve, *vi* basilar trunk, *vii* floor of the middle fossa, *viii* anterior inferior cerebellar artery on the opposite side, *ix* the 7th and 8th nerves on the opposite side

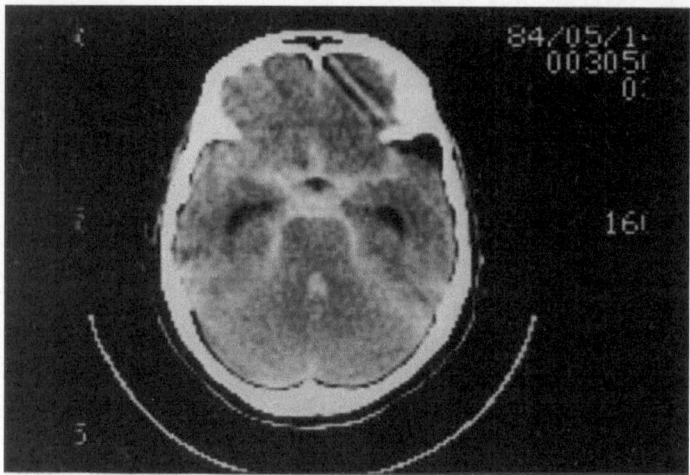

Fig. VIII-64. CT on day 0. A 51-year-old woman with SAH

location exposed by this approach which was clipped (**Fig. VIII-61**). **Fig. VIII-62** shows pre- and postoperative angiograms. The disadvantage of this approach is cerebrospinal fluid leakage and subsequent meningeal infection which prevented popularization of the technique. Our recent experiences suggest that almost all aneurysms arising from the basilar trunk including the vertebrobasilar junction can be reached by the temporopolar approach as illustrated in **Fig. VIII-63**. The temporal lobe (usually nondominant) is retracted backward and the tentorial edge is pulled laterally by a traction suture. Almost the total length of the basilar trunk can be brought into view. If necessary, the posterior clinoid should be drilled off to have a better exposure.

Fig. VIII-64 is a CT on day 0 of a 51-year-old woman with SAH. High density was noted in the suprasellar and ambient cisterns and in the 4th ventricle. She was transferred to us on day 5. Her consciousness level was 14 in

GCS. She also showed EKG findings of myocardial infarction so that she was treated conservatively for four weeks. Angiography done in the meantime showed aneurysms arising from the vertebrobasilar junction (**Figs. VIII-65 and VIII-66**) and the anterior communicating artery. On day 43, she was submitted to surgery in grade III (GCS, 14). The temporopolar approach was used. Before retracting the temporal pole, the anterior communicating aneurysm was clipped. This was found to be nonruptured. Then the temporal pole was retracted, a traction suture was put to the tentorial edge, and the basilar trunk was exposed in its whole length. An aneurysm was found at the vertebrobasilar junction, protruding anteriorly. This was easily clipped from above (**Fig. VIII-67**). She showed an uneventful recovery and was discharged with "good recovery" in GOS. Thus, in many cases of basilar trunk aneurysms, the temporopolar approach is useful.

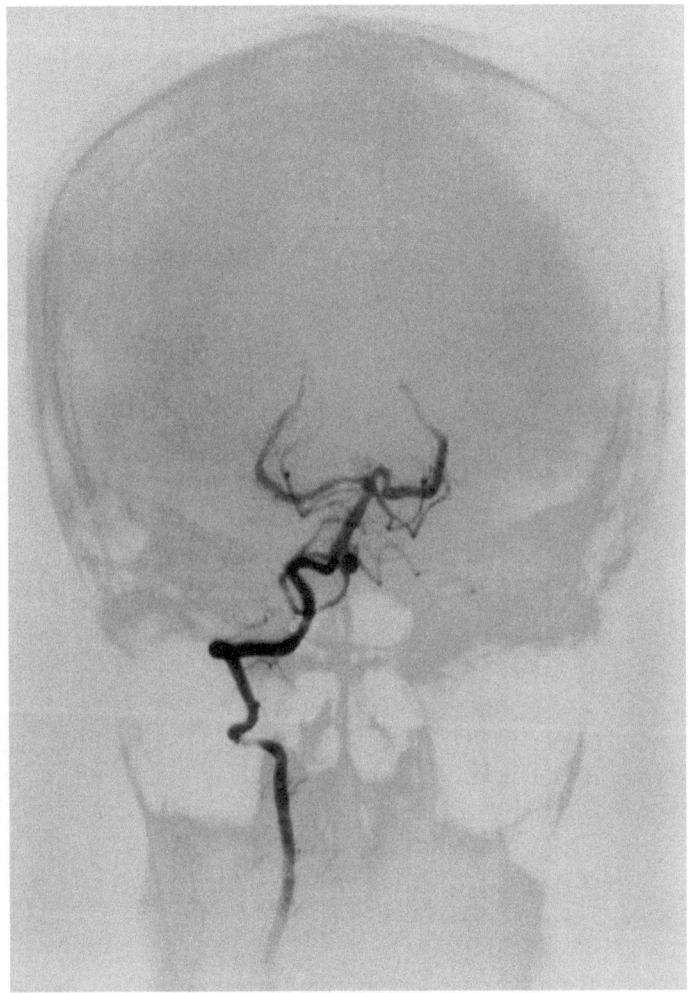

Fig. VIII-65. Aneurysm at the vertebrobasilar junction. Antero-posterior view. A 51-year-old woman. The same case as in Fig. 64

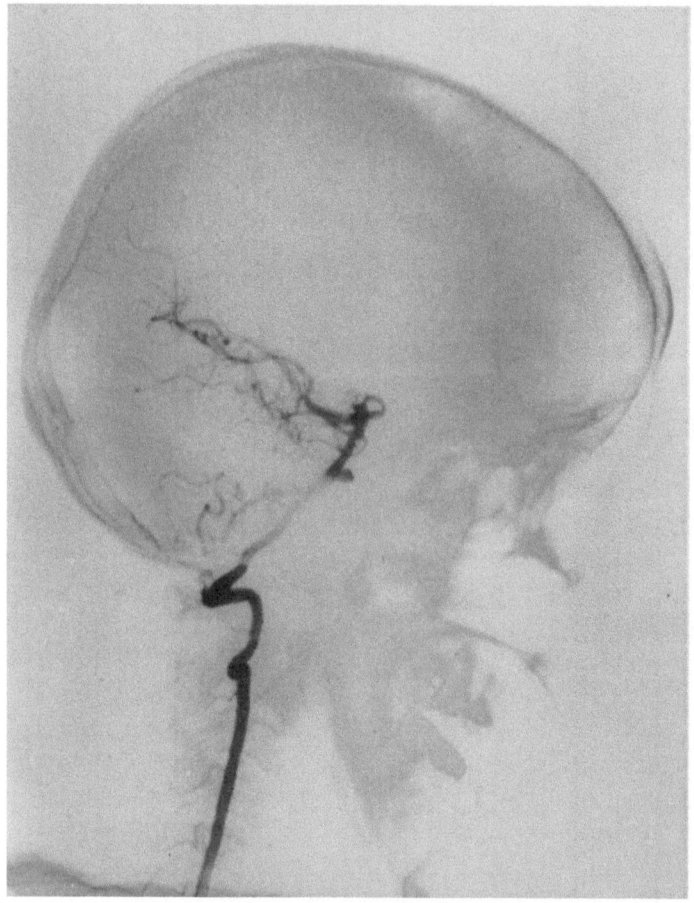

Fig. VIII-66. Aneurysm at the vertebrobasilar junction. Lateral view. The same case as in Fig. 65

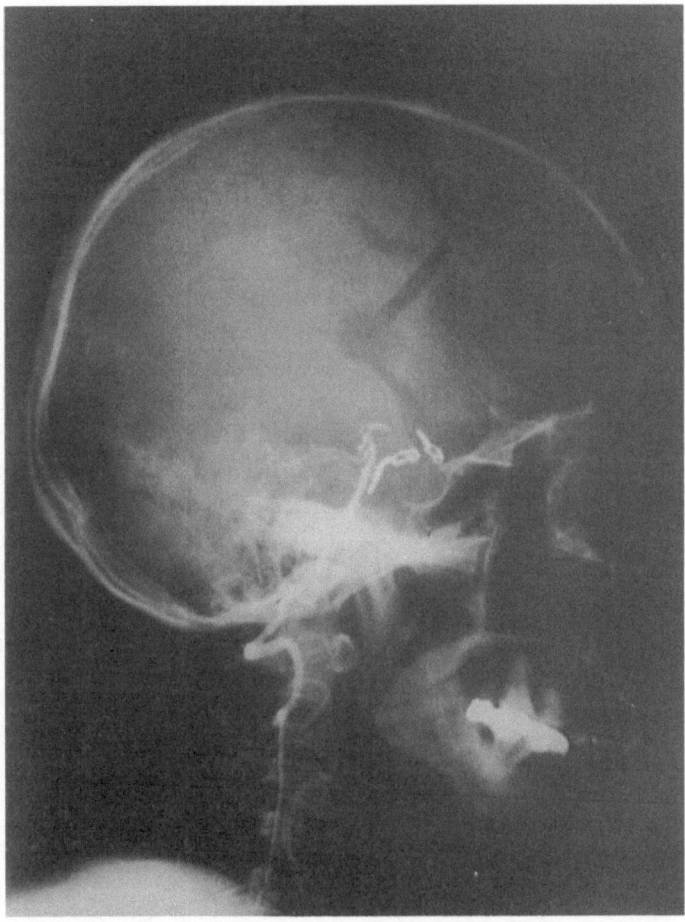

Fig. VIII-67. The aneurysm was clipped (the lower clip). The upper clip obliterated the anterior communicating aneurysm. The same case as in Figs. 65 and 66

5. *Aneurysms Arising from the PICA-Vertebral Junction*

The aneurysms arising from the posterior inferior cerebellar artery (PICA)-vertebral artery junction occupied 2.4% ($^{35}/_{1480}$) of intracranial aneurysms in our series. For surgery of this type of aneurysm, the lateral recumbent position or the supine lateral position

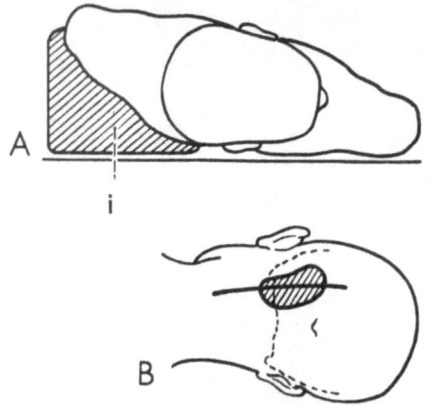

Fig. VIII-68. A) Supine lateral position. *i* Pillow under the shoulder. B) Skin incision and craniectomy

(Fig. VIII-68-A) is preferable. Posterior fossa exploration is done by a straight incision medial to the mastoid process and suboccipital craniectomy (Fig. VIII-68-B). As seen in Fig. VIII-69, this aneurysm is usually covered by the lower cranial nerves. Therefore, care should be taken not to injure these nerves and not to occlude the posterior inferior cerebellar artery (PICA) which supplies blood to the brain stem and the cerebellum.

Fig. VIII-70 is a CT on day 0 of a 50-year-old woman with SAH. High density (clot) was noticed in the right lateral ventricle, the 3rd ventricle and the 4th ventricle which were enlarged. There was disturbed consciousness (GCS, 13, grade III). Because of acute hydrocephalus, ventricular drainage was fashioned. Angiography on day 4 revealed an aneurysm arising from the PICA-vertebral junction on the left

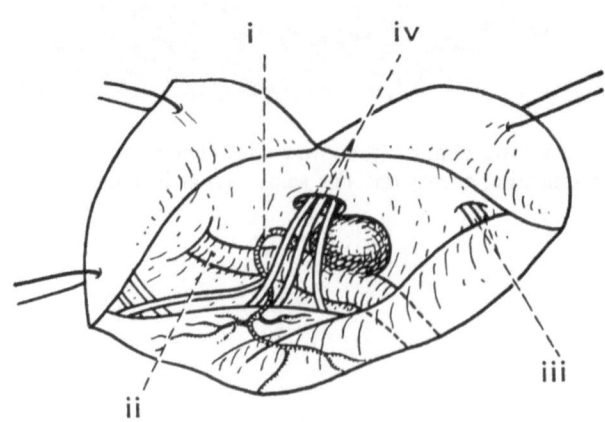

Fig. VIII-69. Surgeon's view of a PICA-vertebral aneurysm exposed by the posterior fossa approach. *i* PICA, *ii* vertebral artery, *iii* 7th and the 8th nerves, *iv* the 9th, 10th and 11th nerves

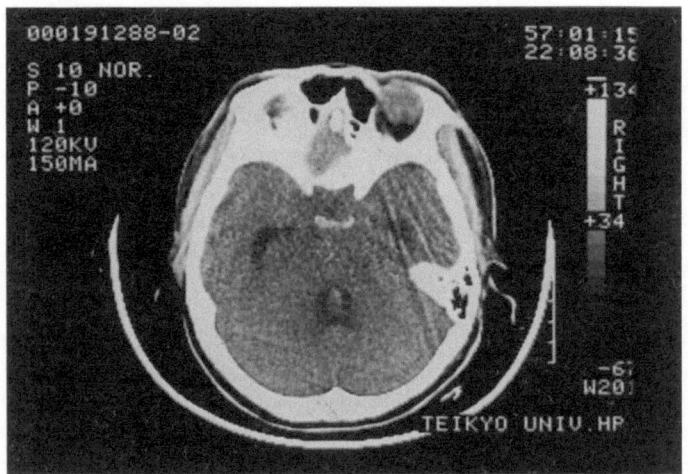

Fig. VIII-70. A 50-year-old woman. CT on day 0. Enlarged 4th ventricle with clot

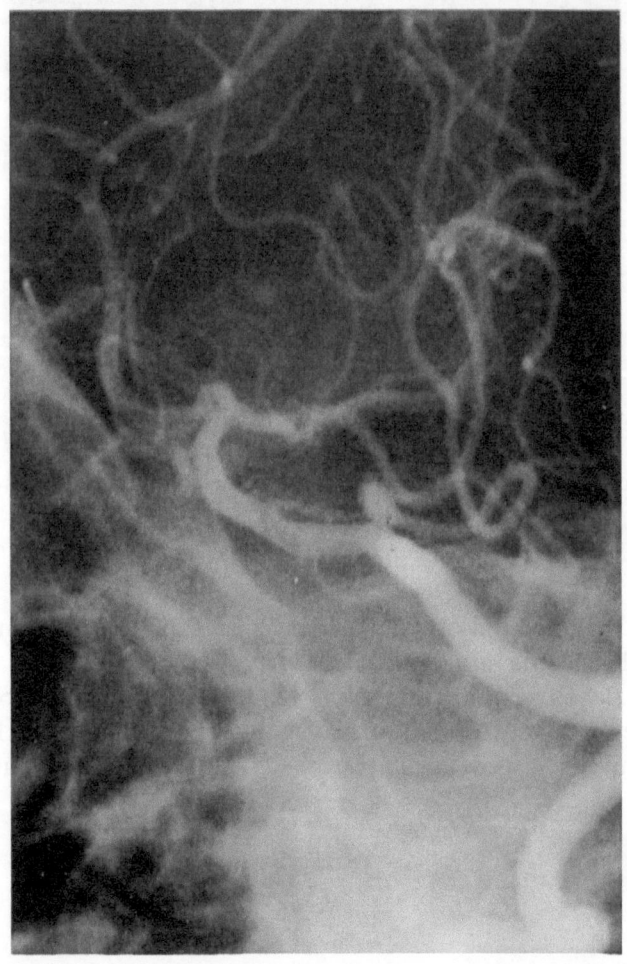

Fig. VIII-71. Angiography. An aneurysm arising from the PICA-vertebral junction on the left side. The same case as in Fig. 70

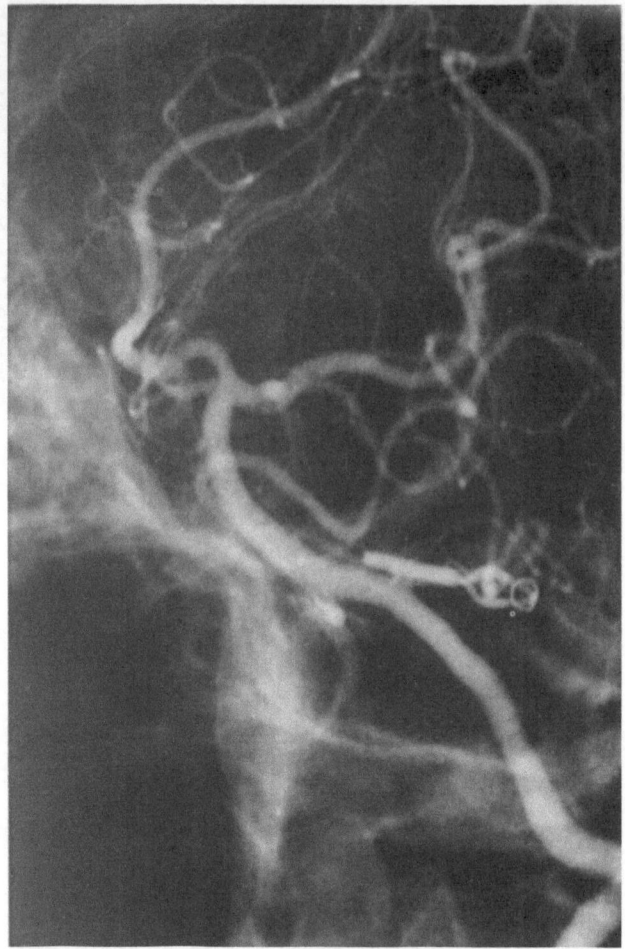

Fig. VIII-72. The aneurysm was clipped. Normal arteries including the PICA were preserved. The same case as in Fig. 71

side (**Fig. VIII-71**). The patient's consciousness level became ameliorated. She, however, developed meningitis. Therefore surgery was postponed until the meningitis subsided. On day 40 she was submitted to surgery and the aneurysm was clipped preserving the PICA (**Fig. VIII-72**). She showed uneventful recovery and her final status was "good recovery" in GOS.

IX. RÉSUMÉ

Rupture of intracranial saccular aneurysms is the most frequent cause of subarachnoid hemorrhage (SAH). If conservatively treated, 60 to 70% of the cases with ruptured aneurysms eventually die. Important factors (in the acute stage of SAH) influencing the mortality and morbidity of patients with aneurysmal SAH are listed in **Table IX-1**. When hemorrhage occurs in the subarachnoid space, intracranial pressure (ICP) inevitably rises dependent upon the amount and distribution of the subarachnoid blood. The increased ICP heralds a series of symptoms which may be termed direct effects of SAH or acute ischemic neurological deficits (AINDs). AINDs include transient global ischemia, impaired autoregulation, decrease in cerebral blood flow (CBF), systemic hypertensive response or Cushing response, vascular engorgement of the brain, brain swelling, microcirculatory disturbance, excessive release of excitatory neurotransmitters to cause neuronal hypermetabolism in decreased CBF, resulting in selective vulnerability of the brain, etc. AINDs affect the neurological conditions of the patient which can be expressed as grades. If the status of a patient with SAH is in a good grade, such as grade I

Table IX-1. SAH due to aneurysm rupture (acute stage)

Amount and distribution of blood
  Increased ICP
  AIND (acute ischemic neurological deficit)
    → patient's conditions (grades)
    transient global ischemia, impaired autoregulation,
    decrease in CBF, systemic hypertensive (Cushing) response,
    vascular engorgement, brain swelling, microcirculatory
    disturbance, selective vulnerability of the brain etc
  Delayed cerebral vasospasm
    → DIND (delayed ischemic neurological deficit)
Sites of hemorrhage ⎫
Acute hydrocephalus ⎬→ grades
Intracerebral hematoma ⎭
Rebleeding
Patient's other pathological conditions
Age

of Hunt, it means the patient has minimal AIND. Grade III, IV, or V implies that AINDs are of a considerable or severe degree. When the subarachnoid clot is of a considerable amount, delayed cerebral vasospasm or arterial spasm will occur around a week or so after SAH, accompanied by neurological symptoms or delayed ischemic neurological deficits (DINDs).

This vasospasm and DIND, the so-called direct effect of SAH or AIND, and thirdly, rebleeding of aneurysms are the three greatest causes of mortality and morbidity in SAH patients.

According to Kassell and Torner (1984), among the causes of disability and death in 1,272 patients with SAH, vasospasm occupies 33.5%, direct effect 25.5%, and rebleeding 17.3%.

In addition, the sites of aneurysms and their rupture (*e.g.*, the brain stem), acute hydrocephalus, and intracerebral hematomas, all of which should affect the grades, will hold sway over the outcome of the patient. Other pathological conditions, such as diabetes mellitus, atherosclerosis, or pulmonary edema, gastrointestinal bleeding, shock, etc, and the age of the patient may have an influence upon the mortality and morbidity.

Since the most influential factors on the outcome of SAH patients are rebleeding, cerebral vasospasm, and direct effects of subarachnoid blood, the most essential therapies of SAH patients are 1. prevention of rebleeding of aneurysms, and 2. prevention or treatment of cerebral vasospasm, and 3. if possible, to lessen the direct effects of SAH.

Theoretically, rebleeding can be prevented by clipping of aneurysms in the very early stage. Concurrent intracerebral hematomas can also be removed by early surgery. Acute hydrocephalus, if present, can be treated by ventricular drainage or shunting.

As was stated in the previous chapters, cerebral vasospasm is caused by free radical reactions initiated by clot lysis in the subarachnoid space. Therefore, if all or most of subarachnoid blood clots can be removed by early surgery with irrigation and drainage of the subarachnoid space, cerebral vasospasm should be prevented by early surgery. This, however, is at variance with the reality. There may also be a possibility of decreasing the direct effects of SAH, at least to some extent, by washing out subarachnoid clots and by irrigating and draining the subarachnoid space to remove or dilute noxious substances and at the same time to lower ICP, and by the use of so-called cerebral protectors such as barbiturates or more preferably nizofenone (Y-9179).

As seen from the statistics in the chapter VI, the results of surgery are the same at any timing of operation in grade I and II patients, especially in grade I patients. This means that in those patients, AIND is minimal and surgical intervention in the early stage of SAH is not different from that in the chronic stage of SAH. In case of grade III, IV, or V patients who have already incurred a considerable degree of AIND, surgery in the early stage may exert a harmful influence on the brain. If the patient is relatively young, the brain may soon recover from the infliction. In elderly patients, however, the brain may not be able to fully tolerate surgical interventions and as a result, AINDs may aggravate. This may be the reason why the result of early surgery for elderly patients of grade III or higher grade is rather poor.

Cerebral vasospasm or DIND is the leading cause of mortality and morbidity in SAH patients. It appears in 40–60% of patients with SAH. Cerebral vasospasm has several characteristics.

1. It usually appears after some time lag from the onset of subarachnoid hemorrhage (SAH). In our angiographical studies, it appeared first on day 4 (the day of SAH being day 0) and, on the average, on day 7.6 ± 2.5 in the natural course. Even postoperative vasospasm appeared with nearly the same interval from the onset of SAH, namely, on the average, on day 6.7 ± 1.6.

2. There is a close correlation between the development of vasospasm and subarachnoid clots, as pointed out by many investigators. One may say "Without subarachnoid clots, no vasospasm".

3. In arteries of the other parts of the body which have bled, there is no sustained arterial contraction comparable to chronic cerebral vasospasm.

Any theory concerning the pathogenesis of vasospasm should be able to explain these characteristics. And the following theory, may fulfill this requirement.

All aerobic animals, including the human being, live using oxygen molecules as the main oxidant of energy substrate. They are provided with double or triple defense mechanisms against active oxygen species (hydrogen peroxide, H_2O_2; hydroxyl radical, $\cdot OH$; superoxide anion, O_2^-; singlet oxygen, $'O_2$) which are generated in the reduction process of the oxygen molecules and also have mechanisms to protect against the oxidation-peroxidation of body constituents by these active oxygen species. Existing animals are those which have thus survived the several million year process of evolution. These defense mechanisms, however, are very poor or almost undetectable in the cerebrospinal fluid (CSF) of the higher mammals including man.

For instance, the glutathione peroxidase level in the human serum is 0.126 ± 0.014 unit/ml, whereas that in human CSF is 0.003 ± 0.001 unit/ml. And vitamin E is measured as 0.77 mg/dl in the human serum and is undetectable in the human CSF; catalase is 0.054 unit/min/ml in the human serum and undetectable in the CSF. These substances are known as potent free radical scavengers, or antioxidants, found in most tissues. During SAH, these defense mechanisms are at first supplied by the extravasated blood itself, but rapidly decrease and after 3 days become insufficient to prevent the free radical chain reactions from propagating. One of the adverse phenomena caused by this lack of free radical scavengers in the CSF is, we believe, chronic or delayed cerebral vasospasm here discussed. Therefore, cerebral vasospasm can be regarded as a sort of deficiency disease—due to deficient defense mechanisms in the CSF.

Then, the pathogenesis of cerebral vasospasm in SAH may be summarized as follows: Free radical reactions, especially lipid peroxidation, are initiated by clot lysis because of insufficient defense mechanisms in the CSF against free radicals. These defense mechanisms are especially insufficient after day 3. Each constituent of these free radical reactions (such as lipid hydroperoxides) has vasocontractile capacity. Furthermore, the endothelium and the media of the artery incur free radical injury (toxic effects of free radicals). Because of endothelial damage and of inhibitory effects of the lipid hydroperoxides on prostacyclin (PGI_2) synthesis, the arterial wall PGI_2 drops to a very low value. Vasoconstrictive substances act on the arterial wall not only from the side of the subarachnoid space, but from the lumen side as well, because there is constant adhesion and aggregation of platelets to the damaged endothelium and these platelets produce thromboxane A_2 and other vasoconstrictive substances which are unopposed by PGI_2. Thus, the artery falls into the state of chronic vasospasm with characteristic histological changes.

The cause of cerebral vasospasm is, in this way, multifactorial. Therefore, prevention or treatment of the vasospasm will be multifarious, as described in Chapter VII.

Our principle for the surgical treatment of ruptured aneurysms in the acute stage is currently as follows:

1. During the first 3 days (48 hours) of SAH, namely on day 0 through 2, microsurgery of aneurysms with removal of subarachnoid blood clots is indicated for any case except for grade IV and V cases and probably for grade III cases of the age of 60 or 65 years and over. After clipping of aneurysms, treatment with induced hypertension and intravascular volume expansion,

and additional use of medicaments such as cerebral protectors, free radical scavengers, thromboxane A_2 inhibitors or antagonists, calcium channel blockers, etc are recommended.

2. After day 3, surgery for cases in grades III, IV, and V should be postponed until they show improvement of neurological conditions by various conservative treatments. Microsurgery is indicated for cases in grades I and II at any time.

3. Any case showing neurological deterioration (to higher grades) should be submitted to computerized tomography (CT) and angiographical examinations to detect vasospasm, which is the most probable cause of the deterioration. If vasospasm is present, surgery should be postponed until the vasospasm begins to subside and disturbance of consciousness begins to improve; however, this postponement cannot be for too long, because rebleeding of aneurysms may occur when vasospasm begins to subside. The optimal timing of surgery for cases with preoperative vasospasm may be the second week after the onset of vasospasm.

4. If CT demonstrates an intracerebral hematoma caused by ruptured aneurysm, immediate surgery (clipping of the aneurysm and removal of the hematoma) is indicated for that case, because the prognosis of the conservatively treated cases is usually poor and, besides, surgery resulted in a useful social life for about a half of these cases.

One of the main reasons why many neurosurgeons hesitate to perform early surgery (which is the best means to prevent rebleeding of aneurysms) even on grade I or II patients, is fear of postoperative vasospasm, which is actually only remotely related to surgery itself. Therefore, if future investigations yield effective drugs for prevention or cure of chronic cerebral vasospasm (DIND) as well as for the treatment of AIND, the treatment of ruptured intracranial aneurysms will be changed so that very early surgery to prevent rebleeding of aneurysms combined with removal of as much subarachnoid clot as possible and fashioning of cisternal drainage will be performed—accompanied by ample administration (systemic or topical) of such drugs. Then the results of treatment of ruptured intracranial aneurysms will be greatly ameliorated. Probably, Winston Churchill's famous words which he delivered after the victory in Egypt in November, 1942, may rightly be applied to the present status of treatments of DIND and AIND; "This is not the end. It is not even the beginning of the end. But it is, perhaps, the end of the beginning".

REFERENCES

Abdel-Halim MS, Sjöquist B, Anggard E (1978) Inhibition of prostaglandin synthesis in rat brain. Acta Pharmacol Toxicol 43: 266–272
— Holst H von, Meyerson B, Sachs C, Anggard E (1980a) Prostaglandin profiles in tissue and blood vessels from human brain. J Neurochem 34: 1331–1333
— Lunden I, Cseh G, Anggard E (1980b) Prostaglandin profiles in nervous tissue and blood vessels of brain of various animals. Prostaglandins 19: 249–258
Adams HP, Kassell NF, Torner JC, Nibbelink DK, Sahs AL (1981a) Early management of aneurysmal subarachnoid hemorrhage. A report of the cooperative aneurysm study. J Neurosurg 54: 141–145
Adams HP Jr, Nibbelink DW, Torner JC, Sahs AL (1981b) Antifibrinolytic therapy in patients with aneurysmal subarachnoid hemorrhage—A report of the cooperative aneurysm study. Arch Neurol 38: 25–29
— Kassell NF, Torner JC, Sahs AL (1983) CT and clinical correlations in recent aneurysmal subarachnoid hemorrhage: A preliminary report of the cooperative aneurysm study. Neurology (Cleveland.) 33: 981–988
Adams JE, Prawirohardjo S (1959) Fate of red blood cells injected into cerebrospinal fluid pathways. Neurology 9: 561–564
Adesuyi SA, Cockrell CS, Gamache DA, Ellis EF (1985) Lipoxygenase metabolism of arachidonic acid in brain. J Neurochem 45: 770–776
Aharony D, Smith JB, Smith EF, Lefer AM (1981) Effect of arachidonic acid hydroperoxides on vascular and nonvascular smooth muscle. Prostaglandins and Medicine 7: 527–535
Aishita H, Morimura T, Obata T, Miura Y, Miyamoto T, Tsuboshima M, Mizushima Y (1983) ONO-3144, a new anti-inflammatory drug and its possible mechanism of action. Arch int Pharmacodyn 261: 316–327
Alexander S, Kerr FWL (1964) Blood pressure responses in acute compression of the spinal cord. J Neurosurg 21: 485–491
Alksne JF, Greenhoot JH (1974) Experimental catecholamine-induced chronic cerebral vasospasm. Myonecrosis in vessel wall. J Neurosurg 41: 440–445
— Smith RW (1977a) Experimental models of spasm. Clin Neurosurg 24: 216–227
— — (1977b) Iron-acrylic compound for stereotaxic aneurysm thrombosis. J Neurosurg 47: 137–141
— Fingerhut A, Rand RW (1966) Magnetically controlled metallic thrombosis of intracranial aneurysms. Surgery 60: 212–218
Allcock JM, Drake CG (1965) Ruptured intracranial aneurysms. The role of arterial spasm. J Neurosurg 22: 21–29
Allen GS, Henderson LM, Chou SN et al (1974a) Cerebral arterial spasm. Part 1. In vitro contractile activity of vasoactive agents on canine basilar and middle cerebral arteries. J Neurosurg 40: 433–441
— — — (1974b) Cerebral arterial spasm. Part 2. In vitro contractile activity of serotonin in human serum and CSF on the canine basilar artery and its blockage by methysergide and phenoxybenzamine. J Neurosurg 40: 442–450
— Gross CJ, French LA et al (1976) Cerebral arterial spasm. Part 5. In vitro contractile activity of vasoactive agents including human CSF on human basilar and anterior cerebral arteries. J Neurosurg 44: 594–600
— Ahn HS, Preziosi TJ, Battye R, Boone SC, Chou SN, Kelly DL, Weir BK, Crabbe RA, Lavik PJ, Rosenbloom SB, Dorsey FC, Ingram CR, Mellits DE, Bertsch LA, Boisvert PJ, Hundley MB, Johnson RK, Strom

JA, Transou CR (1983) Cerebral arterial spasm—A controlled trial of nimodipine in patients with subarachnoid hemorrhage. N Engl J Med 308: 619–624

Almaani WS, Richardson AE (1978) Multiple intracranial aneurysms: Identifying the ruptured lesion. Surg Neurol 9: 303–305

Alvord EC, Thorn RB (1976) Natural history of subarachnoid hemorrhage: early prognosis. Clin Neurosurg 24: 167–175

Ames A III, Wright RL, Kowada M, Thurston JM, Majno G (1968) Cerebral ischemia II: The no-reflow phenomenon. Am J Pathol 52: 437–453

— (1975) Incidence and significance of vascular occlusion in focal and diffuse ischemia. In: Langfitt TW, McHenry LC Jr, Reivich M, Wollman H (eds) Cerebral circulation and metabolism. Springer, Berlin Heidelberg New York, pp 551–554

Ansell GB, Spanner S (1977) Functional metabolism of brain phospholipids. International Review of Neurobiology 20: 1–29

Aritake K, Wakai S, Asano T, Takakura K (1983) Peroxidation of arachidonic acid and brain edema. Brain and Nerve 35: 965–973

Asano M, Hidaka H (1979) Contractile response of isolated rabbit aortic strips to unsaturated fatty acid peroxides. J Pharmacol Exp Ther 208: 347–353

Asano T, Sano K (1977) Pathogenetic role of no-reflow phenomenon in experimental subarachnoid haemorrhage in dogs. J Neurosurg 46: 454–466

— Tamura A, Nagamune A, Taku Y, Kim S, Sano K (1976) The effect of acutely raised ICP on cerebral microcirculation. Neurol Med Chir (Tokyo) 16: 388–395

— — Sano K, Mii K (1978) The role of humoral agents released by platelet aggregation in the pathogenesis of transient ischemic attacks. Neurol Med Chir 18: 59–66

— Tanishima T, Sasaki T, Sano K (1980) Possible participation of free radical reactions initiated by clot lysis in the pathogenesis of vasospasm after subarachnoid hemorrhage. In: Wilkins RH (ed) Cerebral arterial spasm. Williams & Wilkins, Baltimore, pp 190–201

— Sasaki T, Sano K (1981) Lipid peroxidation and cerebral vasospasm. Brain and Nerve 33: 33–46

— — Ochiai C, Takakura K (1982) *In vitro* evaluation of the inhibitory action of PGI$_2$ to vasoconstrictions induced by various prostaglandins, serotonin and hemoglobin using the canine basilar artery. Neurol Med Chir 22: 507–512

— (1983) Metabolic changes of membrane lipids following cerebral ischemia with special reference to the possible pathways of arachidonate metabolism. Brain and Nerve 35: 41–50

— Johshita H, Koide T, Takakura K (1984a) Amelioration of ischemic cerebral oedema by a free radical scavenger, AVS, 1,2-Bis(nicotinamide)-propane. An experimental study using a regional ischaemia model in cats. Neurol Res 6: 163–168

— Sasaki T, Koide T, Takakura K, Sano K (1984b) Experimental evaluation of the beneficial effect of an antioxidant on cerebral vasospasm. Neurol Res 6: 49–53

— Matsui T, Basugi N, Tamura A, Takakura K, Sano K (1984c) The effect of indomethacin on cortical specific gravity during regional ischemia and recirculation. In: Go KG, Baethmann A (eds) Recent progress in the study and therapy of brain edema. Plenum Publishing Co, New York, pp 617–626

— Koide T, Takakura K (1984d) Modulation of the eicosanoid synthetic activity of the rat brain microvessel by 15-hydroperoxyarachidonic acid. In: Bes A, Braquet P, Paoletti R, Siesjö BK (eds) Cerebral Ischemia. Excerpta Medica, Amsterdam, pp 363–368

— Shigeno T, Hanamura T, Koide T, Matsushita H, Watanabe E, Mima JT, Johshita H, Usui M, Takakura K (1985a) Alteration of brain capillary function in cerebral ischemia: Role of capillary Na$^+$, K$^+$-ATPase in ischemic edema formation. J Cereb Blood Flow Metab 5 [Suppl] 1: S 63–S 64

— Gotoh O, Koide T, Takakura K (1985b) Ischemic brain edema following occlusion of the middle cerebral artery in the rat. II. Alteration of the eicosanoid synthesis profile of brain microvessels. Stroke 16: 110–113

— Koide T, Matsushita H, Takakura K (1985c) Enhancement of the Na$^+$, K$^+$-ATPase activity of the brain microvessel by arachidonic acid and its hydroperoxide, 15-HPAA. In: Inaba Y, Klatzo I, Spatz M (eds)

Brain edema. Springer, Berlin Heidelberg New York Tokyo, pp 336–343

Asano T, Johshita H, Gotoh O, Usui M, Koide T, Shigeno T, Takakura K (1985d) The pathomechanism underlying ischemic brain edema: The role of Na^+, K^+-ATPase of the brain microvessel. Neurol Surg (Tokyo) 13: 1147–1159

Ask-Upmark I, Ingvar D (1950) A follow-up examination of 138 cases of subarachnoid hemorrhage. Acta Med Scand 138: 15–31

Astrup J, Heuser D, Lassen NA, Nilsson B, Norberg G, Siesjö BK (1978) Evidence against H^+ and K^+ as main factors for the control of cerebral blood flow: a microelectrode study. In: Cerebral vascular smooth muscle and its control, Ciba Foundation Symposium 56. Elsevier, Amsterdam, pp 313–331

— Sørensen PM, Sørensen HR (1981) Oxygen and glucose consumption related to Na^+-K^+ transport in canine brain. Stroke 12: 726–1730

Au AM, Chan PH, Fishman RA (1985) Stimulation of phospholipase A_2 activity by oxygen-derived free radicals in isolated brain capillaries. J Cell Biochem 27: 449–453

Auer LM (1984) Acute operation and preventive nimodipine improve outcome in patients with ruptured cerebral aneurysms. Neurosurgery 15: 57–66

Ausman JL, Diaz FG, Malik GM, Fielding AS, Son CS (1985) Current management of cerebral aneurysms: Is it based on facts or myths? Surg Neurol 24: 625–635

Aveldano MI, Bazán NG (1975) Rapid production of diacylglycerols enriched in arachidonate and stearate during early brain ischemia. J Neurochem 25: 919–920

Averet N, Rigoulet M, Cohadon F (1984) Modifications of synaptosomal Na^+-K^+-ATPase activity during vasogenic brain edema in the rabbit. J Neurochem 42: 275–277

Bagley C Jr (1928) Blood in cerebrospinal fluid. Resultant functional and organic alterations in central nervous system: experimental data. Arch Surg 17: 18–81

Barber AA, Bernheim F (1967) Lipid peroxidation: its mesurement, occurrence, and significance in animal tissues. Adv Gerontol Res 2: 355–403

Barrows LJ, Hunter FT, Banker BQ (1955) The nature and clinical significance of pigments in the cerebrospinal fluid. Brain 78: 59–80

Bayliss WM (1902) On the local reactions of the arterial wall to changes of internal pressure. J Physiol (London) 28: 220–231

Bazán NG Jr (1970) Effects of ischemia and electroconvulsive shock on free fatty acid pool in the brain. Biochim Biophys Acta 218: 1–10

— (1976) Free arachidonic acid and other lipids in the nervous system during early ischemia and after electroshock. In: Porcellati G (ed) Function and metabolism of phospholipids in the central and peripheral nervous systems. Plenum, New York, pp 317–336

— Bazán HEP, Kennedy WG, Joel CD (1971) Regional distribution and rate of production of free fatty acids in rat brain. J Neurochem 18: 1387–1393

Bebin J, Currier RD (1957) Cause of death in ruptured intracranial aneurysms. Arch Intern Med 99: 771–790

Benos D (1982) Amiloride: a molecular probe of sodium transport in tissues and cells. Am J Physiol (Cell Physiol) 11: C 131–C 145

Bergvall U, Steiner L, Forster DMC (1973) Early pattern of cerebral circulatory disturbances following subarachnoid hemorrhage. Neuroradiology 5: 24–32

Bering EA, Sato O (1963) Hydrocephalus: Changes in formation and absorption of cerebrospinal fluid within the cerebral ventricles. J Neurosurg 20: 1050–1063

Berridge MJ (1984) Inositol triphosphate and diacylglycerol as second messengers. Biochem J 220: 345–360

Betz AL (1983 a) Sodium transport in capillaries isolated from rat brain. J Neurochem 41: 1150–1157

— (1983 b) Sodium transport from blood to brain: Inhibition by furosemide and amiloride. J Neurochem 41: 1158–1164

— Firth JA, Goldstein GW (1980) Polarity of the blood-brain barrier: Distribution of enzymes between the luminal and antiluminal membranes of brain capillary endothelial cells. Brain Res 192: 17–28

Bhakoo KK, Crockard A, Lascelles PC, Avery SF (1984) Prostaglandin synthesis and oedema formation during reperfusion following experimental brain ischemia in the gerbil. Stroke 15: 891–895

Birkle DL, Bazán NG (1984) Lipoxygenase- and cyclooxygenase-reaction products and incorporation into glycerolipids of radiolabeled arachidonic acid in the bovine retina. Prostaglandins 27: 203–216

— Wright KF, Ellis CK, Ellis EF (1981) Prostaglandin levels in isolated brain microvessels and in normal and norepinephrine-stimulated cat brain homogenates. Prostaglandins 21: 865–876

Birse SH, Tom MI (1960) Incidence of cerebral infarction associated with ruptured aneurysms. Neurology 10: 101–106

Bisgaard H, Kristensen J, Sondergaard J (1982) The effect of leucotriene C_4 and D_4 on cutaneous blood flow in humans. Prostaglandins 23: 797–801

Bishai I, Coceani F (1981) Transformations of prostaglandin H_2 in the cat brain. Biochim Biophys Acta 664: 1–9

Bito LZ, Davson H, Hollingsworth JR (1976) Facilitated transport of prostaglandins across the blood-cerebrospinal fluid and blood-brain barrier. J Physiol 253: 273–285

Black PM (1986) Hydrocephalus and vasospasm after subarachnoid hemorrhage from ruptured intracranial aneurysms. Neurosurgery 18: 12–16

Blackwell GJ, Flower RJ (1983) Inhibition of phospholipase. Br Med Bull 39: 260–264

Blomqvist P, Wieloch T (1985) Ischemic brain damage in rats following cardiac arrest using a long-term recovery model. J Cereb Blood Flow Metab 5: 420–431

Boehme DH, Kosecki R, Carson S, Stern F, Marks N (1977) Lipoperoxidation in human and rat brain tissue: developmental and regional studies. Brain Res 136: 11–21

Borst P, Loos JA, Christ EF et al (1962) Uncoupling activity of long-chain fatty acids. Biochim Biophys Acta 62: 509–518

Bosisio E, Galli C, Galli G, Nicosia S, Spagnuolo C, Tose L (1976) Correlation between release of free arachidonic acid and prostaglandin formation in brain cortex and cerebellum. Prostaglandins 11: 773–781

Botterell EH, Lougheed WM, Scott JW, Vandewater SL (1956) Hypothermia, and interruption of carotid, or carotid and vertebral circulation, in the surgical manage-ment of intracranial aneurysms. J Neurosurg 13: 1–42

Du Boulay G (1963) Distribution of spasm in the intracranial arteries after subarachnoid hemorrhage. Acta Radiol 1: 257–266

— Gado M (1974) The protective value of spasm after subarachnoid haemorrhage. Brain 97: 153–156

Boullin DJ (1980) Cerebral vasospasm. Wiley, New York

— Mohan J, Grahame-Smith DG (1976) Evidence for the presence of a vasoactive substance (possibly involved in the aetiology of cerebral arterial spasm) in cerebrospinal fluid from patients with subarachnoid haemorrhage. J Neurol Neurosurg Psychiat 39: 756–766

— Bunting S, Blaso WP et al (1979) Responses of human and baboon arteries to prostaglandin endoperoxides and biologically generated and synthetic prostacyclin. Their relevance to arterial spasm in man. Br J Clin Pharmacol 7: 139–147

Boullin J, Brandt L, Ljunggren B, Tagari P (1981) Vasoconstrictor activity in cerebrospinal fluid from patients subjected to early surgery for ruptured intracranial aneurysms. J Neurosurg 55: 237–245

Boveris A (1977) Mitochondrial production of superoxide radical and hydrogen peroxide. Adv Exp Biol Med 78: 67–82

Bradbury M (1979) The concept of a blood-brain barrier. John Wiley, New York

Brandt L, Ljunggren B, Andersson KE, Hindfelt B (1981 a) Individual variations in response of human cerebral arterioles to vasoactive substances, human plasma, and CSF from patients with aneurysmal SAH. J Neurosurg 55: 431–437

— — — et al (1981 b) Vasoconstrictive effects of human post-hemorrhagic cerebrospinal fluid on cat pial arterioles in situ. J Neurosurg 54: 351–356

— — Saveland H, Nillson PE, Anderson K-E, Vinge E (1985) Results of early aneurysm operation and intravenous Nimodipine. In: Auer LM (ed) Timing of aneurysm surgery. Walter de Gruyter, Berlin, pp 515–521

Branston NM, Strong AJ, Symon L (1977) Extra-cellular potassium activity, evoked potential and tissue blood flow. J Neurol Sci 32: 305–321

Braughler JM, Duncan LA, Goodman T (1985) Calcium enhances *in vitro* free radical-induced damage to brain synaptosomes, mitochondria, and cultured spinal cord neurons. J Neurochem 45: 1288–1293

Brierley JB (1973) Pathology of cerebral ischemia. In: McDowell FH, Brennan RW (eds) Cerebral vascular diseases. Grune & Stratton, New York, pp 59–75

— (1979) Ischemic necrosis along brain arterial boundary zones: some aspects of its etiology. Adv Neurol 26: 155–162

— Excell BJ (1966) The effects of profound systemic hypotension upon the brain of M. Rhesus: physiological and pathological observations. Brain 89: 269–298

— Brown AW, Excell BJ, Meldrum BS (1969) Brain damage in the rhesus monkey resulting from profound arterial hypotension. 1. Its nature, distribution and general physiological correlates. Brain Res 13: 68–100

— Adams JH, Graham DI, Simpsom JA (1971) Neocortical death after cardiac arrest. A clinical, neurophysiological, and neuropathological report of two cases. Lancet 2: 560–565

Brown AW, Brierley JB (1972) Anoxic-ischaemic cell change in rat brain light microscopic and fine-structural observations. J Neurol Sci 16: 59–84

Brown FK (1956) Cardiovascular effects of acutely raised intracranial pressure. Am J Physiol 185: 510–514

Brunson B, Robertson JT, Morgan H, Friedman BI (1973) The measurement of cerebral infarction edema with sodium 22. Stroke 4: 461–464

Buckell M (1964) Demonstration of substances capable of contracting smooth muscle in the haematoma fluid from certain cases of ruptured cerebral aneurysm. J Neurol Neurosurg Psychiat 27: 198–199

Bunting S, Gryglewski R, Moncada S, Vane JR (1976) Arterial walls generate from prostaglandin endoperoxides a substance (prostaglandin X) which relaxes strips of mesenteric and coeliac arteries and inhibits platelet aggregation. Prostaglandins 12: 897–913

Cantu RC, Ames A III (1969) Distribution of vascular lesions caused by cerebral ischemia. Relation to survival. Neurology 19: 128–132

Caronna JJ (1979) Diagnosis, prognosis, and treatment of hypoxic coma. Adv Neurol 26: 1–15

Chacko S, Conti MA, Adelstein RS (1977) Effect of phosphorylation of smooth muscle myosin on actin activation and Ca^{2+} regulation. Proc Natl Acad Sci USA 74: 129–133

Chan PH, Fishman RA (1978) Brain edema: Induction in cortical slices by polyunsaturated fatty acids. Science 201: 358–360

— — (1980) Transient formation of superoxide radicals in polyunsaturated fatty acid-induced brain swelling. J Neurochem 35: 1004–1007

— — Caronna J, Schmidley JW, Prioleau G, Lee J (1983a) Induction of brain edema following intracerebral injection of arachidonic acid. Ann Neurol 13: 625–632

— Kerlan R, Fishman RA (1983b) Reductions of ρ-aminobutyric acid and glutamate uptake and $(Na^+ + K^+)$-ATPase activity in brain slices and synaptosomes by arachidonic acid. J Neurochem 40: 309–316

— Schmidley JW, Fishman RA, Longar SM (1984a) Brain injury, edema, and vascular peremeability changes induced by oxygen-derived free radicals. Neurology 34: 315–320

— Fishman RA, Longar S, Chen S, Yu A (1985) Cellular and molecular effects of polyunsaturated fatty acids in brain ischemia and injury. Prog Brain Res 63: 227–235

Chan RC, Durity FA, Thompson GB, Nugent RA, Kendall M (1984b) The role of the prostacyclin-thromboxane system in cerebral vasospasm following induced subarachnoid hemorrhage in the rabbit. J Neurosurg 61: 1120–1128

Chance B, Sies H, Boveris A (1979) Hydroperoxide metabolism in mammalian organs. Physiol Rev 59: 527–605

Chapleau CE, White RP (1979) Effects of prostacyclin on the canine isolated basilar artery. Prostaglandins 17: 573–580

— — Robertson JT (1980) Cerebral vasospasm: Effects of prostaglandin synthetase inhibitors *in vitro*. Neurosurgery 6: 155–159

Chaplin ER, Free RG, Goldstein GW (1981) Inhibition by steroids of the uptake of potassium by capillaries isolated from rat brain. Biochem Pharmacol 30: 241–245

Chiang J, Kowada M, Ames A III, Wright RL, Majno G (1968) Cerebral ischemia III. Vascular changes. Am J Pathol 52: 455–476

Chyatte D, Rusch N, Sundt TM (1983) Prevention of chronic experimental cerebral vasospasm with ibuprofen and high-dose methylprednisolone. J Neurosurg 59: 925–932

Clower BR, Smith RR, Haining JL, Lockard DJ (1981) Constrictive endarteropathy following experimental subarachnoid hemorrhage. Stroke 12: 501–508

— Sullivan DM, Smith RR (1984) Intracranial vessels lack vasa vasorum. J Neurosurg 61: 44–48

Connolly RC (1961) The incidence of cerebral ischaemia in spontaneous subarachnoid haemorrhage. J Neurol Neurosurg Psychiat 24: 294

Conti MA, Adelstein RS (1980) Phosphorylation by cyclic adenosine 3′ : 5′-monophosphate-dependent protein kinase regulates myosin light chain kinase. Fed Proc 39: 1569–1573

Conway LW, McDonald LW (1972) Structural changes of the intradural arteries following subarachnoid hemorrhage. J Neurosurg 37: 715–723

Cooper AJL, Pulsinelli WA, Duffy TE (1980) Glutathione and ascorbate during ischemia and postischemic reperfusion in rat brain. J Neurochem 35: 1242–1245

Cornog JL Jr, Gonatas NK, Frierman JR (1967) Effects of intracerebral injection of ouabain on the fine structure of rat cerebral cortex. Am J Pathol 51: 573–590

Crockard HA, Bhakoo KK, Lascelles PT (1982) Regional prostaglandin levels in cerebral ischaemia. J Neurochem 38: 1311–1314

Crompton MR (1962) Intracerebral haematoma complicating ruptured cerebral berry aneurysm. J Neurol Neurosurg Psychiat 25: 378–386

— (1964 a) Cerebral infarction following the rupture of cerebral berry aneurysms. Brain 87: 263–279

— (1964 b) The pathogenesis of cerebral infarction following the rupture of aneurysms. Brain 87: 491–510

Cseh G, Szabo IK, Lang T, Palkovits M (1978) Distribution of prostaglandins E and F in

different regions of the rat brain. Brain Res Bull 3: 293–297

Cushing H (1901) Concerning the definite regulatory mechanism of the vaso-motor centre which controls blood pressure during cerebral compression. Bull Johns Hopk Hosp 12: 290–292

Cutler RG (1984) Antioxidants, aging, and longevity. In: Pryor WA (ed) Free radicals in biology, vol VI. Academic Press, Orlando, pp 371–428

Cutler RWP, Page L, Calicich J, Watters GV (1968) Formation and absorption of cerebrospinal fluid in man. Brain 91: 707–720

Cuypers J, Matakas F (1974) The effect of postischemic hyperemia on intracranial pressure and the no-reflow phenomenon. Acta Neuropathol (Berl) 29: 73–84

Dahle LK, Hill EG, Holman RT (1962) The thiobarbituric acid reaction and the autoxidations of polyunsaturated fatty acid methyl esters. Arch Biochem Biophys 98: 253–261

Dandy WE (1938) Intracranial aneurysm of the internal carotid artery. Cured by operation. Ann Surg 107: 654–659

— (1939) The treatment of internal carotid aneurysms within the cavernous sinus and the cranial chamber. Ann Surg 109: 689–711

— (1942) Results following ligation of the internal carotid artery. Arch Surg 45: 521–533

Davis KR, New PFJ, Ojemann RG et al (1976) Computed tomographic evaluation of hemorrhage secondary to intracranial aneurysm. Am J Roentgenol 127: 143–153

Davson H (1956) Physiology of the ocular and cerebrospinal fluids. J. & A. Churchill Ltd., London

Demediuk P, Saunders RD, Clendenon NR, Means ED, Anderson DK, Horrocks LA (1985) Changes in lipid metabolism in traumatized spinal cord. Prog Brain Res 63: 211–226

Demopoulos HB, Flamm ES, Seligman ML, Poser R, Pietronigro O, Ransohoff J (1975) Molecular pathology of lipids in CNS membranes. In: Jöbsis FF (ed) Oxygen and physiological function. Professional Information Library, Dallas, pp 491–508

— — — Mitamura JA, Ransohoff J (1979) Membrane perturbations in CNS injury:

Theoretical basis for free radical damage and a review of the experimental data. In: Popp AJ, Bourke RS, Nelson LR, Kimelberg HK (eds) Neural trauma. Raven Press, New York, pp 63–78

Demopoulos HB, Flamm ES, Pietronigro DD, Seligman ML (1980) The free radical pathology and the microcirculation in the major central nervous system disorders. Acta Physiol Scand [Suppl] 492: 91–119

Dempsey RJ, Roy MW, Meyer K, Cowen DE, Tai HH (1986) Development of cyclooxygenase and lipoxygenase metabolites of arachidonic acid after transient cerebral ischemia. J Neurosurg 64: 118–124

Denny-Brown D (1951) The treatment of recurrent cerebrovascular symptoms and the question of "vasospasm". Med Clin North Am 35: 1457–1474

Denton IC, Robertson JT, Dugdale M (1971) An assessment of early platelet activity in experimental subarachnoid hemorrhage and middle cerebral artery thrombosis in the cat. Stroke 2: 268–272

Dickinson CJ, McCubbin JW (1963) Pressor effect of increased cerebrospinal fluid pressure and vertebral artery occlusion with and without anaesthesia. Circ Res 12: 190–202

Dietrich WD, Busto R, Ginsberg MD (1984) Cerebral endothelial microvillia: Formation following global forebrain ischemia. J Neuropathol Exp Neurol 43: 72–83

Doczi T, Ambrose J, O'Laoire S (1984) Significance of contrast enhancement in cranial computerized tomography after subarachnoid hemorrhage. J Neurosurg 60: 335–342

Dott NM (1933) Intracranial aneurysms: cerebral arterio-radiography: surgical treatments. Edinburgh Med J 40: 219

Dougherty JH, Rawlinson DG, Levy DE, Plum F (1981) Hypoxic-ischemic brain injury and the vegetative state: Clinical and neuropathologic correlation. Neurology 31: 991–997

Drake CG (1965) Surgical treatment of ruptured aneurysms of the basilar artery. Experiences with 14 cases. J Neurosurg 23: 457–473

— (1981) Management of cerebral aneurysm. Stroke 12: 273–283

Ducker TB, Simmons RL, Anderson RW (1968) Increased intracranial pressure and pulmonary edema. Part 3: The effect of increased intracranial pressure on the cardiovascular hemodynamics of chimpanzees. J Neurosurg 29: 475–483

Duckrow RB, LaManna JS, Rosenthal M (1981) Disparate recovery of resting and stimulated oxidative metabolism following transient ischemia. Stroke 12: 677–685

Echlin FA (1942) Vasospasm and focal cerebral ischemia. An experimental study. Arch Neurol Psychiat 47: 77–96

— (1965) Spasm of basilar and vertebral arteries caused by experimental subarachnoid hemorrhage. J Neurosurg 23: 1–11

— (1968) Current concepts in the etiology and treatment of vasospasm. Clin Neurosurg 15: 133–160

Ecker A, Riemenschneider PA (1951) Arteriographic demonstration of spasm of the intracranial arteries with special reference to saccular arterial aneurysms. J Neurosurg 8: 660–667

Edgar AD, Strosznajder J, Horrocks LA (1982) Activation of ethanolamine phospholipase A$_2$ in brain during ischemia. J Neurochem 39: 1111–1116

Edvinsson L, MacKenzie ET (1977) Amine mechanisms in the cerebral circulation. Pharmacol Rev 28: 275–348

— Hardebo JE, Owman C (1978) Influence of the cerebrovascular sympathetic innervation on regional flow, autoregulation, and blood-brain barrier function. In: Cerebral vascular smooth muscle and its control, Ciba Foundation Symposium 56. Elsevier, Amsterdam, pp 69–95

Eisenberg HM, Suddith RL (1979) Cerebral vessels have the capacity to transport sodium and potassium. Science 206: 1083–1085

Ekström-Jodal B (1970) On the relation between blood pressure and blood flow in the canine brain with particular regard to the mechanism responsible for cerebral blood flow autoregulation. Thesis Acta Physiol Scand [Suppl] 350: 1–61

— Häggendal E, Linder LE, Nilsson NJ (1972) Cerebral blood flow autoregulation at high arterial pressures and different levels of carbon dioxide tension in dogs. Eur Neurol 6: 6–10

Ellis EF, Nies AS, Oates JA (1977) Cerebral arterial smooth muscle contraction by thromboxane A$_2$. Stroke 8: 480–482

— Wei EP, Kontos HA (1979) Vasodilation of cat cerebral arterioles by prostaglandins D_2, E_2, G_2, and I_2. Am J Physiol 237: H 381–H 385

— Wright KF, Wei EP, Kontos HA (1981) Cyclooxygenase products of arachidonic acid metabolism in cat cerebral cortex after experimental concussive injury. J Neurochem 37: 892–896

Endo S, Suzuki J (1977) Experimental cerebral vasospasm after subarachnoid hemorrhage. Stroke 8: 702–707

Enseleit WH, Domer FR, Jarrot DM, Baricos WH (1984) Cerebral phospholipid content and Na^+, K^+-ATPase activity during ischemia and postischemic reperfusion in the mongolian gerbil. J Neurochem 43: 320–327

Espinosa F, Weir B, Overton T, Castor W, Grace M, Boisvert D (1984 a) A randomized placebo-controlled double-blind trial of nimodipine after SAH in monkeys. Part 1: Clinical and radiological findings. J Neurosurg 60: 1167–1175

— — Shnitka T, Overton T, Boisvert D (1984 b) A randomized placebo-controlled double-blind trial of nimodipine after SAH in monkeys. Part 2: Pathological findings. J Neurosurg 60: 1176–1185

Evans JP (1956) Increased intracranial pressure; its physiology and management. Surg Clin N Amer 36: 233–242

Falconer MA (1954) Surgical pathology of spontaneous intracranial haemorrhage due to aneurysms and arteriovenous malformations. Proc Royal Soc Med 147: 693–700

Fantone JC, Ward PA (1982) Role of oxygen-derived free radicals and metabolites in leukocyte-dependent inflammatory reactions. Am J Pathol 107: 397–418

Farhat SM, Schneider RC (1967) Observations on the effect of systemic blood pressure on intracranial circulation in patients with cerebrovascular insufficiency. J Neurosurg 27: 441–445

Farias RN, Bloj B, Morero RD, Sineriz F, Trucco RE (1975) Regulation of allosteric membrane-bound enzymes through changes in membrane lipid composition. Biochim Biophys Acta 415: 231–251

Faustmann PM, Dermietzel R (1985) Extravasation of polymorphonuclear leukocytes from the cerebral microvasculature. Inflammatory response induced by alphabungarotoxin. Cell Tissue Res 242: 399–407

Fearnsides EG (1916) Intracranial aneurysms. Brain 39: 224–296

Fein JM, Flor WJ, Cohan SL, Parkhurst J (1974) Sequential changes of vascular ultrastructure in experimental cerebral vasospasm. Myonecrosis of subarachnoid arteries. J Neurosurg 41: 49–58

Ferguson GG, Harper AM, Fitch W, Rowan JO, Jennett B (1972) Cerebral blood flow measurements after spontaneous subarachnoid haemorrhage. Europ Neurol 8: 15–23

Fischer EG, Ames A III (1972) Studies on mechanisms of impairment of cerebral circulation following ischemia: Effect of hemodilution and perfusion pressure. Stroke 3: 538–542

— — Hedley-Whyte ET, O'Gorman S (1977) Reassessment of cerebral capillary changes in acute global ischemia and their relationship to the "no-reflow phenomenon". Stroke 8: 36–39

— — Lorenzo A (1979) Cerebral blood flow immediately following brief circulatory stasis. Stroke 10: 423–427

Fisher CM, Robertson GH, Ojemann RG (1977) Cerebral vasospasm with ruptured saccular aneurysm. The clinical manifestations. Neurosurgery 1: 245–248

— Kistler JP, Davis JM (1980) Relation of cerebral vasospasm to subarachnoid hemorrhage visualized by computerized tomographic scanning. Neurosurgery 6: 1–9

Fitch W, McDowall G, Keaney NP, Pickerodt VWA (1977) Systemic vascular responses to increased intracranial pressure. 2. The "Cushing" response in the presence of intracranial space-occupying lesions: systemic and cerebral haemodynamic studies. J Neurol Neurosurg Psychiat 40: 843–852

Flamm ES, Demopoulos HB, Seligman ML, Poser RG, Ransohoff J (1978) Free radicals in cerebral ischemia. Stroke 9: 445–447

Fletcher TM, Taveras JM, Pool JL (1959) Cerebral vasospasm in angiography for intracranial aneurysms. Incidence and significance in 100 consecutive angiograms. Arch Neurol 1: 38–47

Flower RJ (1974) Drugs which inhibit prostaglandin biosynthesis. Pharmacol Rev 26: 33–67

— Blackwell GJ (1976) The importance of phospholipase-A_2 in prostaglandin biosynthesis. Biochem Pharmacol 25: 285–291

Forbes HS, Cobb S (1937) Vasomotor control of cerebral vessels. Proc Ass Nerv ment Dis 18: 201–217

— Wolff HG (1928) Cerebral circulation III. The vasomotor control of cerebral vessels. Arch Neurol Psychiat (Chicago.) 19: 1057–1086

Fourcans B (1974) Role of phospholipids in transport and enzymic reactions. Adv Lipid Res 12: 147–226

Fox JL, Ko JP (1978) Cerebral vasospasm: a clinical observation. Surg Neurol 10: 269–275

Fraser RAR, Stein BM, Barrett RE, Pool JL (1970) Noradrenaergic mediation of experimental cerebrovascular spasm. Stroke 1: 356–362

Freytag E (1966) Fatal rupture of intracranial aneurysms. Survey of 250 medicolegal cases. Arch Path 81: 418–424

Fridovich I (1976) Oxygen radicals, hydrogen peroxide and oxygen toxicity. In: Pryor W (ed) Free radicals in biology. Academic Press, New York, pp 239–277

Fujita S (1985) Computed tomographic grading with Hounsfield number related to delayed vasospasm in cases of ruptured cerebral aneurysm. Neurosurgery 17: 609–612

Fujita Y, Shingu T, Yamada K, Araki O, Matsunaga M, Mori K, Kawano T (1980) Noxious free radicals derived from oxyhemoglobin as a cause of prolonged vasospasm. Neurol Med Chir (Tokyo) 20: 137–144

— — Kurihara M, Miyake H, Kono T, Mori K (1985) Na-K-activated adenosine triphosphatase activity and lipoperoxide metabolites in microvessels and parenchymas of the ischemic brain. In: Inaba Y, Klatzo I, Spatz M (eds) Brain edema. Springer, Berlin Heidelberg New York Tokyo, pp 344–353

Fujiwara S, Kuriyama H (1984) Hemolysate-induced contraction in smooth muscle cells of the guinea pig basilar artery. Stroke 15: 503–510

— Kassell NF, Sasaki T, Nakagomi T, Lehman RH (1986) Hemoglobin selectively inhibits the endothelium-dependent vasodilation of rabbit basilar artery. J Neurosurg 64: 445–452

Fukasawa H (1969) Hemodynamical studies of cerebral arteries by means of mathematical analysis of arterial casts. Tohoku J Exptl Med 99: 255–268

Furchgott RF (1983) Role of endothelium in responses of vascular smooth muscle. Circ Res 53: 557–573

— Zawadzki JV (1980) The obligatory role of endothelial cells in the relaxation of arterial smooth muscle by acetylcholine. Nature 288: 373–376

— Cherry PO, Zawadzki JV, Jothianandan D (1984) Endothelial cells as mediators of vasodilation of arteries. J Cardiovasc Pharmacol 6: S 336–S 343

Furlow TW, Hallenbeck JM (1978) Indomethacin prevents impaired perfusion of the dog's brain after global ischaemia. Stroke 9: 591–594

Gale PH, Egan RW (1984) Prostaglandin endoperoxide synthase-catalyzed oxidation reactions. In: Pryor WA (ed) Free radicals in biology, vol. VI. Academic Press, Orlando, pp 1–38

Galli C, Spagnuolo C, Petroni A (1980) Factors affecting brain prostaglandin formation. In: Samuelsson B, Ramwell PW, Paoletti R (eds) Advances in prostaglandin and thromboxane research, vol. 8. Raven Press, New York 8: 1235–1239

Garcia JH, Cox JV, Hudgins WR (1971) Ultrastructure of the microvasculature in experimental cerebral infarction. Acta Neuropath 18: 273–285

Garland CJ, Keatinge WR (1982) Constrictive actions of ACH-5HT and histamine on bovine coronary artery inner and outer muscle. J Physiol 327: 363–376

Gaudet RJ, Levine L (1979) Transient cerebral ischemia and brain prostaglandins. Biochem Biophys Res Commun 86: 893–901

— — (1980) Effect of unilateral common carotid artery occlusion on levels of prostaglandins D_2, F_{2a}, and 6-keto-prostaglandin F_{1a} in gerbil brain. Stroke 11: 648–652

— Alam I, Levine L (1980) Accumulation of cyclooxygenase products of arachidonic acid metabolism in gerbil brain during reperfusion after bilateral common carotid artery occlusion. J Neurochem 35: 653–658

Gazendam J, Go KG, Zanten AK, van (1979) The effect of intracerebral ouabain administration on the composition of edema fluid isolated from cats with cold-induced brain edema. Brain Res 175: 279–290

Gecse A, Ottlecz A, Merei Z, Telegdy G, Joo F, Dux E, Karushina I (1982) Prostacyclin and prostaglandin synthesis in isolated brain capillaries. Prostaglandins 23: 287–297

Gelmers HJ, Beks JWF, Journee HL (1979) Regional cerebral blood flow in patients with subarachnoid hemorrhage. Acta Neurochir (Wien) 47: 245–251

Gerritsen ME, Parks TP, Printz MP (1980) Prostaglandin endoperoxide metabolism by bovine cerebral microvessels. Biochim Biophys Acta 619: 196–206

Ginsberg MC, Myers RE (1972) The topography of impaired microvascular perfusion in the primate brain following total circulatory arrest. Neurology 22: 998–1011

Ginsberg MD, Mela L, Wrobel-Kuhl K, Reivich M (1977) Mitochondrial metabolism following bilateral cerebral ischemia in the gerbil. Ann Neurol 1: 519–527

— Graham DI, Busto R (1985) Regional glucose utilization and blood flow following graded forebrain ischemia in the rat: correlation with neuropathology. Ann Neurol 18: 470–481

Goehlert UG, Ng Ying, Kin NMK, Wolfe LS (1981) Biosynthesis of prostacyclin in rat cerebral microvessels and the choroid plexus. J Neurochem 36: 1192–1201

Goetzl EJ (1981) Oxygenation products of arachidonic acid as mediators of hypersensitivity and inflammation. Med Clin North Am 65: 809–828

Goldstein GW (1979) Relation of potassium transport to oxidative metabolism in isolated brain capillaries. J Physiol (Lond.) 286: 185–195

— Betz AL (1983) Recent advances in understanding brain capillary function. Ann Neurol 14: 389–395

Goodman SJ, Becker DP, Seelig J (1972) The effects of mass-induced intracranial pressures on arterial hypertension and survival in awake cats. J Neurosurg 37: 514–527

Gotoh O, Asano T, Takakura K (1983) Ischemic brain edema and the glutathione system. In: Ishii S, Nagai H, Brock M (eds) Intracranial pressure V. Springer, Berlin Heidelberg New York Tokyo, pp 399–404

— Koide T, Asano T, Takakura K, Tamura A, Sano K (1984) A model to study ischemic brain edema in rats and the influence of drugs. In: Go KG, Baethmann A (eds) Recent progress in the study and therapy of brain edema. Plenum Press, New York, pp 499–508

— Asano T, Koide T, Takakura K (1985) Ischemic brain edema following occlusion of the middle cerebral artery in the rat. I. The time courses of brain water, sodium, and potassium contents and blood-brain barrier permeability to ^{125}I-albumin. Stroke 16: 101–109

Greenfield JC, Tindall GT (1965) Effect of acute increase in intracranial pressure on blood flow in the internal carotid artery of man. J Clin Invest 44: 1343–1351

Grotenhuis JA, Bettag W, Fiebach BJO, Dabir K (1984) Intracarotid slow bolus injection of nimodipine during angiography for treatment of cerebral vasospasm after SAH. J Neurosurg 61: 231–240

Grotta J, Ackerman R, Correia J, Fallick G, Chang J (1982) Whole blood viscosity parameters and cerebral blood flow. Stroke 13: 296–301

Grubb RL, Raichle ME, Eichling J, Gado M (1977) Effects of subarachnoid hemorrhage on cerebral blood volume, blood flow, and oxygen utilization in humans. J Neurosurg 46: 446–453

Gryglewski RJ, Palmer RMJ, Moncada S (1986) Superoxide anion is involved in the breakdown of endothelium-derived vascular relaxing factor. Nature 320: 454–456

Gurdjian ES, Thomas LM (1969) Cerebral vasospasm. S G O 129: 931–948

Gurusinghe NT, Richardson AE (1984) The value of computerized tomography in aneurysmal subarachnoid hemorrhage. J Neurosurg 60: 763–770

Häggendal E, Löfgren J, Nilsson NJ, Zwetnow NN (1970 a) Effects of varied cerebrospinal fluid pressure on cerebral blood flow in dogs. Acta physiol Scand 79: 262–271

— — — — (1970 b) Prolonged cerebral hyperemia after periods of increased cerebrospinal fluid pressure in dogs. Acta Physiol Scand 79: 272–279

Hagen AA, Gerber JN, Sweeley CC, White RP, Robertson JT (1977) Levels and disappearance of prostaglandin $F_{2\alpha}$ in cerebral spinal fluid: A clinical and experimental study. Stroke 8: 672–675

— White RP, Robertson JT (1979) Synthesis of prostaglandins and thromboxane B_2 by cerebral arteries. Stroke 10: 306–309

Hallenbeck JM, Furlow TW (1979) Prostaglandin I_2 and indomethacin prevent impairment of post-ischemic brain reperfusion in the dog. Stroke 10: 629–637

— Dutka AJ, Tanishima T, Kochanek PM, Kumaroo KK, Thompson CB, Obrenovitch TP, Contreras TJ (1986) Polymorphonuclear leukocyte accumulation in brain regions with low blood flow during the early post-ischemic period. Stroke 17: 246–253

Halliwell B, Gutteridge JMC (1984 a) Oxygen toxicity, oxygen radicals, transition metals and disease. Biochem J 219: 1–14

— — (1984 b) Lipid peroxidation, oxygen radicals, cell damage, and antioxidant therapy. Lancet 23: 1396–1397

Hamberg M, Hedqvist P, Strandberg K, Svensson J, Samuelsson B (1975) Prostaglandin endoperoxides. IV. Effects on smooth muscle. Life Sci 16: 451–462

Hammarström S (1983) Leukotrienes. Ann Rev Biochem 52: 355–377

Hammes EM (1944) Reaction of the meninges to blood. Arch Neurol Psychiat 52: 505–514

Hanamura T, Asano T, Shigeno T, Mima T, Takakura K (1986) Prostaglandin profiles in relation to local circulatory changes following focal cerebral ischemia in cats. Paper presented in 6th International Conference on Prostaglandins and Related Compounds. Florence, Italy. Paper in preparation

Handa J, Yoneda S, Matsuda M, Handa H (1974) Effects of prostaglandins A_1, E_1, E_2, and $F_{2\alpha}$ on the basilar artery of cats. Surg Neurol 2: 251–255

Hansotia PL (1985) Persistent vegetative state. Review and report of electrodiagnostic studies in eight cases. Arch Neurol 42: 1048–1052

Harman D (1962) Role of free radicals in mutation, cancer, aging, and the maintenance of life. Radiat Res 16: 753–763

Harper AM (1966) Autoregulation of cerebral blood flow: Influence of the arterial blood pressure on the blood flow through the cerebral cortex. J Neurol Neurosurg Psychiat 29: 398–403

— Deshmukh VD, Rowan JO, Jenett WB (1972) Influence of sympathetic nervous activity on cerebral blood flow. Arch Neurol 27: 1–6

Harreveld A van, Ochs S (1956) Cerebral impedance changes after circulatory arrest. Am J Physiol 187: 180–192

Hashi K (1986) Intravenous high dose hydrocortisone therapy for the treatment of delayed cerebral ischemia after SAH.—A double-blind study. International workshop in intracranial aneurysms. April 3, Tokyo

— Matsuoka Y, Tanaka K, Ohkawa N (1980) Treatment of cerebral vasospasm with large doses of hydrocortisone. In: Wilkins RH (eds) Cerebral arterial spasm. Williams & Wilkins, Baltimore

Harris RJ, Bayhan M, Branston NM, Watson A, Symon L (1982) Modulation of the pathophysiology of primate focal cerebral ischaemia by indomethacin. Stroke 13: 17–24

Hayashi S, Nehls DG, Kieck CF, Vielma J, DeGirolami U, Crowell RM (1984) Beneficial effects of induced hypertension on experimental stroke in awake monkeys. J Neurosurg 60: 151–157

Heilbrun MP, Olesen J, Lassen NA (1972) Regional cerebral blood flow studies in subarachnoid hemorrhage. J Neurosurg 37: 36–44

Heistad DD, Marcus ML, Ehrhardt JC, Abboud FM (1976) Effect of stimulation of carotid chemoreceptors on total and regional blood flow. Circ Res 38: 20–25

— — Sandberg S, Abboud FM (1977) Effect of sympathetic nerve stimulation on cerebral blood flow and on large cerebral arteries of dogs. Circ Res 41: 342–350

— — Abboud FM (1978) Experimental attempts to upmask effects of neural stimuli on cerebral blood flow. In: Cerebral vascular smooth muscle and its control. Ciba Foundation Symposium 56. Elsevier, Excerpta Medica, North Holland, Amsterdam, pp 97–118

Heros RC, Zervas NT, Varsos V (1983) Cerebral vasospasm after subarachnoid hemorrhage: An update. Ann Neurol 14: 599–608

Hertz L (1981) Features of astrocytic function apparently involved in the response of central nervous tissue to ischemia-hypoxia. J Cereb Blood Flow Metab 1: 143–153

Herz DA, Baez S, Shulman K (1975) Pial microcirculation in subarachnoid hemorrhage. Stroke 6: 417–424

Heuser D (1978) The significance of cortical extracellular H^+, K^+, and Ca^{++} activities for regulation of local cerebral blood flow under conditions of enhanced neuronal activity. In: Cerebral vascular smooth muscle and its control. Ciba Foundation Symposium 56. Elsevier. Excerpta Medica, North Holland, Amsterdam, pp 339–348

Higgs GA, Vane JR (1983) Inhibition of cyclooxygenase and lipoxygenase. Br Med Bull 39: 265–270

Hilal SK, Maudsley AA, Simon HE, Perman WH, Bonn J, Mawad ME, Silver AJ, Ganti SR (1983) *In vivo* NMR imaging of tissue sodium in the intact cat before and after acute cerebral stroke. AJNR 4: 245–249

Hillered L, Ernstar L (1983) Respiratory activity of isolated rat brain mitochondria following *in vitro* exposure to oxygen radicals. J Cereb Blood Flow Metab 3: 207–214

— Smith M-L, Siesjö BK (1985) Lactic acidosis and recovery of mitochondrial function following forebrain ischemia in the rat. J Cereb Blood Flow Metab 5: 259–266

Hirashima Y, Moto A, Endo S, Takaku A, Ishikawa A (1985) The activities of phospholipase A, PI-specific phospholipase C, lipase, lysophospholipase and acylCoA: lysophospholipid acyltransferase in ischemic brain microsomal fraction. Brain and Nerve 37: 385–391

Hirata F, Axelrod J (1980) Phospholipid methylation and biological signal transmission. Science 209: 1082–1090

Hirata Y, Matsukado Y, Fukumura A (1982) Subarachnoid enhancement secondary to subarachnoid hemorrhage with special reference to the clinical significance and pathogenesis. Neurosurgery 11: 367–371

Hoff JT, Reis DJ (1970) Localization of regions mediating the Cushing response in CNS of cat. Arch Neurol 23: 228–240

Holst H von, Granstrom E, Hammarstrom S, Samuelsson B, Steiner L (1982) Effect of leukotrienes C_4, D_4, prostacyclin and thromboxane A_2 on isolated human cerebral arteries. Acta Neurochir (Wien) 62: 177–185

Hoshi T, Shimizu T, Kito K, Yamasaki N, Takahashi K, Takahashi M, Okada T, Kasuya H, Kitamura K (1984) Immunological study of late cerebral vasospasm in subarachnoid hemorrhage. Detection of immunoglobulins, C 3, and fibrinogen in cerebral arterial walls by immunofluorescence method. Neurol Med Chir (Tokyo) 24: 647–654

Hossmann K-A (1982) Treatment of experimental cerebral ischemia. J Cereb Blood Flow Metab 2: 275–297

— (1985) The pathophysiology of ischemic brain edema. In: Inaba Y, Klatzo I, Spatz M (eds) Brain edema. Springer, Berlin Heidelberg New York Tokyo, pp 367–384

— Kleihues P (1973) Reversibility of ischemic brain damage. Arch Neurol 29: 375–384

— Takagi S (1976) Osmolality of brain in cerebral ischemia. Exp Neurol 51: 124–131

— Sakaki S, Komoto K (1976) Cerebral uptake of glucose and oxygen in the cat brain after prolonged ischemia. Stroke 7: 301–305

— — Zimmermann V (1977) Cation activities in reversible ischemia of the cat brain. Stroke 8: 77–81

Hothersall JS, Greenbaum AL, McLean P (1982) The functional significance of the pentose phosphate pathway in synaptosomes: protection against peroxidative damage by catecholamines and oxidants. J Neurochem 39: 1325–1332

Hugenholtz H, Elgie RG (1982) Considerations in early surgery on good-risk patients with ruptured intracranial aneurysms. J Neurosurg 56: 180–185

Hughes JT, Schianchi PM (1978) Cerebral artery spasm. A histological study at necropsy of the blood vessels in cases of subarachnoid hemorrhage. J Neurosurg 48: 515–525

Hunt WE (1979) Grading of risk in intracranial aneurysms. In: Sano K, Ishii S (eds) Recent progresses in neurological surgery. Excerpta Medica, Amsterdam, pp 169–175

— (1983) Clinical assessment of SAH. J Neurosurg 59: 550

— Hess RM (1968) Surgical risk as related to time of intervention in the repair of intracranial aneurysms. J Neurosurg 28: 14–20

— Kosnik EJ (1974) Timing and perioperative care in intracranial aneurysm surgery. Clin Neurosurg 21: 79–89

— Miller CA (1977) The results of early operation for aneurysm. Clin Neurosurg 24: 208–215

Hunter FE, Gebicki JM, Hoffsten PE, Weinstein J, Scott A (1963a) Swelling and lysis of rat liver mitochondria induced by ferrous ions. J Biol Chem 238: 828–835

Hunter FE Jr, Scott A, Hoffsten PE, Guerra F, Weinstein J, Schneider A, Schutz B, Fink J, Ford L, Smith E (1963b) Studies on the mechanism of ascorbate-induced swelling and lysis of isolated liver mitochondria. J Biol Chem 239: 604–613

— — — Gebicki JM, Weinstein J, Schneider A (1963c) Studies on the mechanism of swelling, lysis, and disintegrations of isolated liver mitochondria exposed to mixtures of oxidized and reduced glutathione. J Biol Chem 239: 614–621

Ianotti F, Crockard A, Ladds G, Symon L (1981) Are prostaglandins involved in experimental ischemic edema in gerbils? Stroke 12: 301–306

Ikeda M, Yoshida S, Busto R, Santiso M, Ginsberg MD (in press) Polyphosphoinositides as a probable source of brain free fatty acids accumulated at the onset of ischemia. J Neurochem

Ikuta F, Yoshida Y, Ohama E, Oyanagi K, Takeda S, Yamazaki K, Watabe K (1983) Revised pathophysiology on BBB damage: The edema as an ingeniously provided condition for cell motility and lesion repair. Acta Neuropathol (Berl) [Suppl] VIII: 103–110

Ingvar DH, Brun, Arne, Johansson L, Samuelsson SM (1978) Survival after severe cerebral anoxia with destruction of the cerebral cortex: the apallic syndrome. Ann NY Acad Sci 315: 184–214

Ishii R (1979) Regional cerebral blood flow in patients with ruptured intracranial aneurysms. J Neurosurg 50: 587–594

Ishii S, Tsuji H, Ozawa K, Kondo Y, Evans JP (1967) Brain edema. Some clinical and experimental correlation. In: Klatzo I Seitelberger F (eds) Brain edema. Springer New York, pp 32–66

Ito M (1980) Significance of oxyhemoglobin, fibrin degradation products and breakdown products of white ghost in the pathogenesis of cerebral vasospasm. Treatment of vasospasm with gabenate mesilate and diphenhydramine. Neurol Med Chir (Tokyo) 20: 225–236

Ito U, Spatz M, Walker JT Jr, Klatzo I (1975) Experimental cerebral ischemia in Mongolian gerbils. I. Light microscopic observations. Acta Neuropathol (Berl) 32: 209–223

— Go KG, Walker JT Jr, Spatz M, Klatzo I (1976) Experimental cerebral ischemia in Mongolian gerbils. III. Behaviour of the blood-brain barrier. Acta Neuropathol (Berl) 34: 1–6

Ito Z, Matsuoka S, Moriyama T, Hen R, Nakajima K (1975) Factors related to level of consciousness in the acute stage of ruptured intracranial aneurysms. Brain and Nerve (Tokyo) 27: 895–901

Jackson IJ (1949) Aseptic hemogenic meningitis. An experimental study of aseptic meningeal reactions due to blood and its breakdown products. Arch Neurol Psychiat 65: 572–589

Jakubowski J, Bell BA, Symon L, Zawirski MB, Francis DM (1982) A primate model of subarachnoid hemorrhage: change in regional cerebral blood flow, autoregulation, carbon dioxide reactivity, and central conduction time. Stroke 13: 601–611

James IM (1968) Changes in cerebral blood flow and in systemic arterial pressure following spontaneous subarachnoid hemorrhage. Clin Sci 35: 11–22

Jane JA, Winn HR, Richardson AE (1977) The natural history of intracranial aneurysms; rebleeding rates during the acute and long term period and implication for surgical management. Clin Neurosurg 24: 176–184

Jefferson G (1938) On the saccular aneurysms of the internal carotid artery in the cavernous sinus. Br J Surg 26: 267–302

Jennett B, Bond M (1975) Assessment of outcome after severe brain damage. Lancet (i) 480–484

— Teasdale G (1977) Aspects of coma after severe head injury. Lancet (i) 878–881

— — (1981) Management of head injuries. FA Davis, Philadelphia, pp 77–93

Johansson BB (1981) Indomethacin and cerebrovascular permeability to albumin in acute hypertension and cerebral embolism in the rat. Exp Brain Res 42: 331–336

Johnston IH, Rowan JO, Harper AM, Jennett WB (1972) Raised intracranial pressure and cerebral blood flow. I. Cisterna magna infusion in primates. J Neurol Neurosurg Psychiat 35: 285–296

— — — — (1973) Raised intracranial pressure and cerebral blood flow. 2. Supratentorial and infratentorial mass lesions in primates. J Neurol Neurosurg Psychiat 36: 161–170

— — (1974) Raised intracranial pressure and cerebral blood flow. 3. Venous outflow tract pressures and vascular resistances in experimental intracranial hypertension. J Neurol Neurosurg Psychiat 37: 392–402

Johshita H, Asano T, Takakura K Experimental evaluation of the role of cyclooxygenase pathway in the pathogenesis of ischemic brain edema. Paper in submission

Joris I, Majno G (1981) Endothelial changes induced by arterial spasm. Am J Pathol 102: 346–358

Kagström E, Greitz T, Hanson J, et al (1966) Changes in cerebral blood flow after subarachnoid haemorrhage. Excerpta Med ICS 110: 629–633

— Smith ML, Siesjö BK (1983a) Local cerebral blood flow in the recovery period following complete cerebral ischemia in the rat. J Cereb Blood Flow Metab 3: 170–182

— — — Recirculation in the rat brain following incomplete ischemia. J Cereb Blood Flow Metab 3: 183–192

Kajikawa H, Ohta T, Yoshikawa Y, Funatsu N, Yamamoto M, Someda K (1979) Cerebral vasospasm and hemoglobins—clinical and experimental studies. Neurol Med Chir (Tokyo) 19: 61–71

Kakiuchi S, Yamazaki R (1970) Calcium dependent phosphodiesterase activity and its activating factor (PAF) from brain. Biochem Biophys Res Commun 41: 1104–1110

Kalimo H, Olsson Y, Paljarvi L, Soderfeldt B (1982) Structural changes in brain tissue under hypoxid-ischemic conditions. J Cereb Blood Flow Metab 2 [Suppl]: S 19–S 22

Kamitani T, Little MH, Ellis EF (1985) Effect of leukotrienes, 12-HETE, histamine, bradykinin, and 5-hydroxytryptamine on in vivo rabbit cerebral arteriolar diameter. J Cereb Blood Flow Metab 5: 554–559

Kapp J, Mahaley MS Jr, Odom GL (1968) Cerebral spasm. Part 3. Partial purification and characterization of a spasmogenic substance in feline platelets. J Neurosurg 29: 350–356

Kapp JP, Robertson JT, White RP (1976) Spasmogenic qualities of prostaglandin $F_{2\alpha}$ in the cat. J Neurosurg 44: 173–175

Kassell NF (1986) Intracranial aneurysms and subarachnoid hemorrhage: World-wide experience. Paper spoken in the 11th Annual Meeting of Japan Stroke Society

— Drake CG (1982) Timing of aneurysm surgery. Neurosurgery 10: 514–519

— Torner JC (1983) Aneurysmal rebleeding: a preliminary report from the cooperative aneurysm study. Neurosurgery 13: 479–481

— — (1984) The international cooperative study on timing of aneurysm surgery-an update. Stroke 15: 566–570

— Peerless SJ, Drake CG (1980) Cerebral vasospasm: Acute proliferative vasculopathy? I. Hypothesis. In: Wilkins RH (ed) Cerebral arterial spasm. Williams & Wilkins, Baltimore, pp 85–87

— — Durward QJ, Beck DW, Drake CG, Adams HP (1982) Treatment of ischemic deficits from vasospasm with intravascular volume expansion and induced arterial hypertension. Neurosurgery 11: 337–343

— Torner JC, Adams HP (1984) Antifibrinolytic therapy in the acute period following aneurysmal subarachnoid hemorrhage. J Neurosurg 61: 225–230

— Sasaki T, Colohan ART, Nazar G (1985) Cerebral vasospasm following aneurysmal subarachnoid hemorrhage. Stroke 16: 562–572

Katzman R, Pappius HM (1973) Brain electrolytes and fluid metabolism. Williams & Wilkins, Baltimore

— Clasen R, Klatzo I, Meyer JS, Pappius HM, Waltz AG (1977) Report of joint committee for stroke resources. IV. Brain edema in stroke. Study group on brain edema in stroke. Stroke 8: 512–540

Kaye AH, Tagari PC, Teddy PJ, Adams CBT, Blaso WP, Boullin DJ (1984) CSF smooth-muscle constrictor activity associated with cerebral vasospasm and mortality in SAH patients. J Neurosurg 60: 927–934

Kee DB, Wood JH (1984) Rheology of the cerebral circulation. Neurosurgery 15: 125–131

Kellogg EW III, Fridovich I (1977) Liposome oxidation and erythrocyte lysis by enzymatically generated superoxide and hydrogen peroxide. J Biol Chem 252: 6721–6728

Kelly PJ, Gorten RJ, Grossmann RG, et al (1977) Cerebral perfusion, vascular spasm, and outcome in patients with ruptured intracranial aneurysms. J Neurosurg 47: 44–49

Kempinsky WH (1958) Experimental study of distant effects of acute focal brain injury. Arch Neurol Psychiat 79: 376–389

Kety SS, Shenkin HA, Schmidt CF (1948) The effects of increased intracranial pressure on cerebral circulatory functions in man. J Clin Invest 27: 493–499

Kim H, Mizukami M, Kawase T, Takemae T, Araki G (1979) Time course of vasospasm—its clinical significance. Neurol Med Chir (Tokyo) 19: 95–102

Kim S, Sano K (1977) The role of platelets and erythrocytes in disturbances of cerebral microcirculation. Neurol Med Chir (Tokyo) Part II.2: 135–144

Kimelberg HK, Bourke RS (1984) Mechanisms of astrocytic swelling. In: Bes A, Braquet P, Paoletti R, Siesjö BK (eds) Cerebral ischemia. Excerpta Medica, Amsterdam, pp 131–146

Kirino T (1982) Delayed neuronal death in the gerbil hippocampus following ischemia. Brain Res 239: 57–69

— Sano K (1984) Selective vulnerability in the gerbil hippocampus following transient ischemia. Acta Neuropathol (Berl) 62: 201–208

Kistler JP, Crowell RM, Davis KR, Heros R, Ojemann RG, Zervas T, Fisher CM (1983) The relation of cerebral vasospasm to the extent and location of subarachnoid blood visualized by CT scan: A prospective study. Neurology (Cleveland) 33: 424–436

Kiwak KJ, Moskowitz MA, Levine L (1985) Leukotriene production in gerbil brain after ischemic insult, subarachnoid hemorrhage, and concussive injury. J Neurosurg 62: 865–869

Kjällquist A, Siesjö BK, Zwetnow N (1969) Effects of increased intracranial pressure on cerebral blood flow and on cerebrospinal fluid HCO_3^-, pH, lactate, and pyruvate in dogs. Acta Physiol Scand 75: 345–352

Klatzo I (1967) Neuropathological aspects of brain edema: Presidential address. J Neuropathol Exp Neurol 26: 1–14

Knuckey NW, Fox RA, Surveyor I, Stokes BAR (1985) Early cerebral blood flow and computerized tomography in predicting ischemia after cerebral aneurysm rupture. J Neurosurg 62: 850–855

Kodama N, Mizoi K, Sakurai Y, Suzuki J (1980) Incidence and onset of vasospasm. In: Wilkins RH (ed) Cerebral arterial spasm. William & Wilkins, Baltimore, pp 361–365

Kofke WA, Nemoto EM, Hossmann KA, Taylor F, Kessler PD, Stezoski SW (1979) Brain blood flow and metabolism after global ischemia and post-insult thopental therapy in monkeys. Stroke 10: 554–560

Kogure K, Morooka H, Busto R, Scheinberg P (1979) Involvement of lipid peroxidation in postischemic brain damage. Neurology 29: 546

— Busto R, Scheinberg P (1981) The role of hydrostatic pressure in ischemic brain edema. Ann Neurol 9: 273–282

— Arai H, Abe K, Nakano M (1985) Free radical damage of the brain following ischemia. Prog Brain Res 63: 237–259

Koide T, Noda Y, Hata S, Sugioka K, Kobayashi S, Nakano M (1981) Contraction of the canine basilar artery following linoleic, arachidonic, 13-hydroperoxy-linoleic or 15-hydroperoxyarachidonic acid. Proc Soc Exp Biol Med 168: 399–402

— Neichi T, Takato M, Matsushita H, Sugioka K, Nakano M, Hata S (1982) Possible mechanisms of 15-hydroperoxyarachidonic acid-induced contraction of the canine basilar artery *in vitro*. J Pharm Exp Ther 221: 481–488

— Gotoh O, Asano T, Takakura K (1985) Alterations of the eicosanoid synthetic capacity of rat brain microvessels following ischemia: Relevance to ischemic brain edema. J Neurochem 44: 85–93

— Asano T, Matsushita H, Takakura K (1986) Enhancement of ATPase activity by a lipid peroxide of arachidonic acid in rat brain microvessels. J Neurochem 46: 235–242

Kontos HA, Wei EP, Navari RM, Levasseur JE, Rosenblum WI, Patterson JL Jr (1978) Responses of cerebral arteries and arterioles to acute hypotension and hypertension. Am J Physiol 234: H 371–H 383

— — Povlishock JT, *et al.* (1980) Cerebral arteriolar damage by arachidonic acid and prostaglandin G 2. Science 209: 1242–1245

— — Ellis EF, Dietrich WD, Povlishock JT (1981) Prostaglandins in physiological and in certain pathological responses of the cerebral circulation. Fed Proc 40: 2326–2330

Kosnik EJ, Hunt WE (1976) Postoperative hypertension in the management of patients with intracranial arterial aneurysms. J Neurosurg 45: 148–154

Kovachich GB, Mishra OP (1980) Lipid peroxidation in rat brain cortical slices as measured by the thiobarbituric acid test. J Neurochem 35: 1449–1452

— — (1981) Partial inactivation of Na$^+$, K$^+$-ATPase in cortical brain slices incubated in normal Krebs-Ringer phosphate medium at 1 and 10 atm oxygen pressures. J Neurochem 36: 333–335

Kramer W, Tuynman JA (1967) Acute intracranial hypertension—An experimental investigation. Brain Res 6: 686–705

— (1970) Acute lethal intracranial hypertension. Clinical and experimental observations. Psychiat Neurol Neurochir 73: 243–255

Kudo T, Suzuki S, Iwabuchi T (1981) Importance of monitoring the circulating blood volume in patients with cerebral vasospasm after subarachnoid hemorrhage. Neurosurgery 9: 514–520

Kuehl FA, Egan RW (1980) Prostaglandins, arachidonic acid, and inflammation. Science 210: 978–984

Kushinsky W, Wahl M (1978) Local chemical and neurogenic regulation of cerebral vascular resistance. Physiol Rev 58: 656–689

Kuwabara K, Sakaki S, Nakano K, Yano M, Hatakeyama T, Matsuoka K (1982) A study on the changes of superoxide dismutase activities and the amount of lipid peroxide in patients with ruptured intracranial aneurysms with special reference to the pathogenesis of vasospasm. In: Iwabuchi T (ed) Proceedings of the 11th Conference of Surgical Treatment of Stroke. Neuron-Sha (Tokyo), pp 159–164

Kwak R, Niizuma H, Ohi T, Suzuki J (1979) Angiographic study of cerebral vasospasm following rupture of intracranial aneurysms. Part I. Time of the appearance. Surg Neurol 11: 257–262

La Torre I, Patrono C, Fortuna A, Grossi-Belloni D (1974) Role of prostaglandin F$_2$ in human cerebral vasospasm. J Neurosurg 41: 293–299

Lands WEM (1979) The biosynthesis and metabolism of prostaglandins. Ann Rev Physiol 41: 633–652

— Kulmacz RJ, Marshall PJ (1984) Lipid peroxide actions in the regulation of prostaglandin biosynthesis. In: Pryor WA (ed) Free radicals in biology, vol VI. Academic Press, Orlando, pp 39–63

Langfitt TW, Weinstein JD, Kassell NF, Simeone FA (1964) Transmission of increased intracranial pressure. I. Within the craniospinal axis. J Neurosurg 21: 989–997

— Kassell NF, Weinstein JD (1965) Cerebral blood flow with intracranial hypertension. Neurology 15: 761–773

— Weinstein JD, Sklar FH, Zaren HA, Kassell NF (1968) Contribution of intracranial blood volume to three forms of experimental brain swelling. Johns Hopkins Med J 122: 261–270

— (1969) Increased intracranial pressure. Clin Neurosurg 16: 436–471

Lapetina EG, Billah MM, Cuatrecasas P (1981) The phosphatidylinositol cycle and the regulation of arachidonic acid production. Nature 292: 367–369

Larsen GL, Henson PM (1983) Mediators of inflammation. Ann Rev Immunol 1: 335–359

Lassen NA (1959) Cerebral blood flow and oxygen consumption in man. Physiol Rev 39: 183–238

— (1974) Control of cerebral circulation in health and disease. Circ Res 34: 749–760

Lazarewicz JW, Strosznajder J, Gromek A (1972) Effects of ischemia and exogenous fatty acids on the energy metabolism in brain mitochondria. Bull Acad Pol Sci (Biol) 8: 599–606

Leaf A (1959) Maintenance of concentration gradients and regulation of cell volume. Ann NY Acad Sci 72: 396–404

Lee SC, Levine L (1974) Cytoplasmic reduced nicotinamide adenine dinucleotide phosphate-dependent and microsomal reduced nicotinamide adenine dinucleotide-dependent prostaglandin E 9-ketoreductase activities in monkey and pigeon tissues. J Biol Chem 249: 1369–1375

Lehninger AL, Remmert LF (1959) An endogenous uncoupling and swelling agent in liver mitochondria and its enzymic formation. J Biol Chem 234: 2459–2464

Levasseur JE, Kontos HA, Ellis EF (1985) Reduction in cerebral arteriolar oxygen consumption by arachidonate. Am J Physiol 248: H 534–H 539

Levy DE, Brierley JB, Plum F (1975) Ischaemic brain damage in the gerbil in the absence of "no-reflow". J Neurol Neurosurg Psychiat 38: 1197–1205

— Duffy T (1977) Cerebral energy metabolism during transient ischemia and recovery in the gerbil. J Neurochem 28: 63–70

— Uitert RL van, Pike CL (1979) Delayed postischemic hypoperfusion: A potentially damaging consequence of stroke. Neurology 29: 1245–1252

Lewis DH, Maestro RF del (eds) (1980) Free radicals in medicine and biology. Acta Physiol Scand [Suppl] 492

Liliequist B, Lindqvist M, Valdimarsson E (1977) Computed tomography and subarachnoid hemorrhage. Neuroradiology 14: 21–26

Lim ST, Sage DJ (1977) Detection of subarachnoid blood clot and other thin, flat structures by computed tomography. Radiology 123: 79–84

Lindgren JA, Hokfelt T, Dahlen SE, Patrono C, Samuelsson B (1984) Leukotrienes in the rat central nervous system. Proc Natl Acad Sci USA 81: 6212–6216

Lindsay KW, Teasdale GM, Knill-Jones RP, Murray L (1982 a) Observer variability in grading patients with subarachnoid hemorrhage. J Neurosurg 56: 628–633

— — — (1982 b) Assessment of the consequences of subarachnoid haemorrhage. Acta Neurochir (Wien) 63: 59–64

— — — (1983) Observer variability in assessing the clinical features of subarachnoid hemorrhage. J Neurosurg 58: 57–62

Liszczak TM, Varsos VG, Black PM, Kistler JP, Zervas NT (1983) Cerebral arterial constriction after experimental subarachnoid hemorrhage is associated with blood components within the arterial wall. J Neurosurg 58: 18–26

Ljunggren B, Brandt L, Kagström E, Sundbarg G (1981) Results of early operations for ruptured aneurysms. J Neurosurg 54: 473–479

Lo WD, Betz L (1986) Oxygen free-radical reduction of brain capillary rubidium uptake. J Neurochem 46: 394–398

Lobato RD, Marin J, Salaices M, Burgos J, Rivilla F, Garcia AG (1980) Effect of experimental subarachnoid hemorrhage on the adrenergic innervation of cerebral arteries. J Neurosurg 53: 477–479

— — — Rivilla F, Burgos J (1980) Cerebrovascular reactivity to noradrenaline and serotonin following experimental subarachnoid hemorrhage. J Neurosurg 53: 480–485

Locksley HB (1966) Report on the cooperative study of intracranial aneurysms and subarachnoid hemorrhage. Section V, Part II. Natural history of subarachnoid hemorrhage, intracranial aneurysms and arteriovenous malformations based on 6368 cases in the cooperative study. J Neurosurg 25: 321–368

Löfgren J, Zwetnow NN (1972) Kinetics of arterial and venous hemorrhage in the skull cavity. In: Brock M, Dietz H (eds) Intracranial pressure. Springer, Berlin Heidelberg New York, pp 155–159

— Essen C von, Zwetnow NN (1973) The pressure-volume curve of the cerebrospinal fluid space in dogs. Acta Neurol Scand 49: 557–574

Logue V (1956) Surgery in spontaneous sub-arachnoid hemorrhage. Operative treatment of aneurysms on the anterior cerebral and anterior communicating artery. Br Med J 1: 473–479

Lorente de Nó R (1934) Studies on the structure of the cerebral cortex. II. Continuation of the study of the ammonic system. J Psychol Neurol 46: 113–177

Lowe D, Schieweck CHR, Meier-Ruge W, Bangerter D, Wolff JR (1975) The effect of "ouabain" on the ultrastructure of cerebral arterioles and surrounding tissue, studied by a cannulation of a cerebral artery. Res Exp Med 166: 97–114

Lowell HM, Bloor BM (1971) The effect of increased intracranial pressure on cerebro-vascular hemodynamics. J Neurosurg 34: 760–769

Lundberg N, Kjällquist A, Kullberg G, Pontén U, Sundbärg G (1974) Non-operative management of intracranial hypertension. In: Krayenbühl H, et al (eds) Advances and technical standards in neurosurgery, vol 1. Springer, Wien New York pp 3–59

Lunt GG, Rowe CE (1968) The production of unesterified fatty acid in brain. Biochim Biophys Acta 152: 681–693

Lysz TW, Needleman P (1982) Evidence for two distinct forms of fatty acid cyclooxygenase in brain. J Neurochem 38: 1111–1117

MacKnight ADC, Leaf A (1977) Regulation of cellular volume. Physiological Reviews 57: 510–573

MacMillan V (1982) Cerebral Na^+, K^+-ATPase activity during exposure to and recovery from acute ischemia. J Cereb Blood Flow Metab 2: 457–465

— Shankaran R (1984) Influence of lactate accumulation of Na^+, K^+-ATPase activity of ischemic and postischemic brain. Brain Res 303: 125–132

Maeda Y, Tani E, Miyamoto T (1981) Prosta-glandin metabolism in experimental cerebral vasospasm. J Neurosurg 655: 779–785

Maestro RF del (1980) An approach to free radicals in medicine and biology. Acta Physiol Scand [Suppl] 492: 153–169

Majewska MD, Strosznajder J, Lazarewicz J (1978) Effect of ischemic anoxia and barbi-turate anesthesia on free radical oxidation of mitochondrial phospholipids. Brain Res 158: 423–434

Mann JD, Butler AB, Rosenthal JE, Maffeo CJ, Jonson RN, Bass NH (1978) Regulation of intracranial pressure in rat, dog, and man. Ann Neurol 3: 156–165

Marcus AJ (1978) The role of lipids in platelet function: With particular reference to the arachidonic acid pathway. J Lipid Res 19: 793–824

Marion J, Wolfe LS (1978) Increase in vivo of unesterified fatty acids, prostaglandin F_{2a}, but not thromboxane B_2 in rat brain during drug induced convulsions. Prostaglandins 16: 99–110

Maroon JC, Nelson PB (1979) Hypovolemia in patients with subarachnoid hemorrhage: Therapeutic implications. Neurosurgery 4: 223–226

Marshall LF, Durity F, Lounsbury R, Graham JDI, Welsh F, Langfitt TW (1975) Experi-mental cerebral oligemia and ischemia pro-duced by intracranial hypertension. Part 1: Pathophysiology, electroencephalography, cerebral blood flow, blood-brain barrier, and neurological function. J Neurosurg 43: 308–317

Mathew NT, Meyer JS, Hartmann A (1974) Diagnosis and treatment of factors com-plicating subarachnoid hemorrhage. Neu-roradiology 6: 237–245

Matsuoka Y, Hossmann K-A (1982) Cortical impedance and extracellular volume changes following middle cerebral artery occlusion in cats. J Cereb Blood Flow Metab 2: 466–474

Matthews WF, Frommeyer WB (1955) The in vitro behavior of erythrocytes in human cerebrospinal fluid. J Lab Clin Med 45: 508–515

Maurer P, Moskowitz MA, Levine L, Melamed E (1980) The synthesis of prostaglandins by bovine cerebral microvessels. Prostaglandin and Medicine 4: 153–161

McCay PB, Gibson DD, Fong K-L, Hornbrook KR (1976) Effect of glutathione peroxidase activity on lipid peroxidation in biological membranes. Biochim Biophys Acta 431: 459–468

McGee-Russell SM, Brown AW, Brierley JB (1970) A combined light and electron micro-scope study of early anoxic-ischaemic cell change in rat brain. Brain Res 20: 193–200

McKnight RC, Hunter FE, Oehlert WH (1965) Mitochondrial membrane ghosts produced by lipid peroxidation induced by ferrous ion. J Biol Chem 240: 3439–3446

Mchedlishvili GI, Ormotsadze LG, Nikolaishvili LS, Baramidze DG (1967) Reaction of different parts of the cerebral vascular system in asphyxia. Exp Neurol 18: 239–252

— — (1979) Responses of the internal carotid artery to different endogenous vasoconstrictor substances. Blood Vessels 16: 126–134

Mead JF (1976) Free radical mechanisms of lipid damage and consequences for cellular membranes. In: Pryor WA (ed) Free radicals in biology, vol 1. Academic Press, New York, pp 51–68

Meier-Ruge W, Gygax P, Iwangoff P, Schieweck CH, Wolff J (1974) The significance of pericapillary astroglia for cerebral cortical blood flow and EEG activity. In: Cervos-Navarro J (ed) Pathology of cerebral microcirculation. Walter de Gruyter, Berlin, pp 235–243

Meinen K, Kremer B, Huecher H, Geisler W (1975) Intracranial pressure increase and changes in microcirculation of the pial and iridial vessels in correlation to EEG, ECG, and arterial blood pressure. Brain Res 86: 439–447

Michell RH, Lapetina EG (1972) Production of cyclic inositol phosphate in stimulated tissues. Nature (New Biol) 240: 258–259

— (1975) Inositol phospholipids and cell surface receptor function. Biochim Biophys Acta 415: 81–147

Mickey B, Vorstrup S, Voldby B, Lindewald H, Harmsen A, Lassen NA (1984) Serial measurement of regional cerebral blood flow in patients with SAH using ^{133}Xe inhalation and emission computerized tomography. J Neurosurg 60: 916–922

Milikan CH (1975) Cerebral vasospasm and ruptured intracranial aneurysm. Arch Neurol 32: 433–449

Miller CA (1980) Biochemistry of vascular smooth muscle: contractile mechanism of human basilar arterry. In: Wilkins RH (ed) Cerebral arterial spasm. Williams & Wilkins, Baltimore, pp 68–75

Miller JD, Stanek AE, Langfitt TW (1973) Cerebral blood flow regulation during experimental brain compression. J Neurosurg 39: 186–196

Miller JR, Myers RE (1972) Neuropathology of systemic circulatory arrest in adult monkeys. Neurology 22: 888–904

Misra HP, Fridovich I (1972) The generation of superoxide radical during the autoxidation of hemoglobin. J Biol Chem 247: 6960–6962

Miyaoka M, Nonaka T, Watanabe H, Ishii S (1976) Etiology and treatment of prolonged vasospasm. Experimental and clinical studies. Neurol Med Chir (Tokyo) 16 (Parr II): 103–114

Mizukami M, Kin H, Araki G, Mihara H. Yoshida Y (1976) Is angiographic spasm real spasm? Acta Neurochir (Wien) 34: 247–259

— Kawase T, Tazawa T, Nagata K, Unoki K, Yoshida Y (1980a) Hypothesis and clinical evidence for the mechanism of chronic cerebral vasospasm after subarachnoid hemorrhage. In: Wilkin RH (ed) Cerebral arterial spasm. Williams & Wilkins, Baltimore, pp 97–106

— Takemae T, Tazawa T, Kawase T, Matsuzaki T (1980b) Value of computed tomography in the prediction of cerebral vasospasm after aneurysm rupture. Neurosurgery 7: 583–586

— Kawase T, Usami T, Tazawa T (1982) Prevention of vasospasm by early operation with removal of subarachnoid blood. Neurosurg 10: 301–307

Modesti LM, Binet EF (1978) Value of computed tomography in the diagnosis and management of subarachnoid hemorrhage. Neurosurg 3: 151–156

Moncada S, Gryglewski RJ, Bunting S, Vane JR (1976) A lipid peroxide inhibits the enzyme in blood vessel microsomes that generates from prostaglandin endoperoxides the substance (prostaglandin X) which prevents platelet aggregation. Prostaglandins 12: 715–737

— Vane JR (1979) Arachidonic acid metabolites and the interactions between platelets and blood-vessel walls. New Engl J Med 300: 1142–1147

Moniz E (1927) L'encéphalographie artérielle, son importance dans la localisation des tumeurs cérébrales. Rev Neurol 34: 72–91

Moran CV, Naidich TB, Gado MH, et al (1978) Leptomeningeal findings in CT of subarachnoid haemorrhage. J Comput Assist Tomogr 2: 520–521

Morisaki N, Lindsey JA, Stitts JM, Zhang H, Gornwell DG (1984) Fatty acid metabolism and cell proliferation. V. Evaluation of pathways for the generation of lipid peroxides. Lipids 19: 381–394

Moskowitz MA, Kiwak KJ, Hekiman K, Levine L (1984) Synthesis of compounds with properties of leukotrienes C_4 and D_4 in gerbil brains after ischemia and reperfusion. Science 224: 886–889

— Puszkin S, Schook W (1983) Characterization of brain synaptic vesicle phospholipase A_2 activity and its modulation by calmodulin, prostaglandin E_2, prostaglandin $F_{2\alpha}$, cyclic AMP, and ATP. J Neurochem 41: 1576–1586

Movat HZ (1979) Tissue injury and inflammation induced by immune complexes: the critical role of the neutrophil leukocyte. Exp Mol Pathol 31: 201–210

Mrsulja BB, Djuricic BM, Cvejic V, Mrsulja BJ, Abe K, Spatz M, Klatzo I (1980) Biochemistry of experimental ischemic brain edema. Adv Neurol 28: 217–230

— — Ueki Y, Cahn R, Cvejic V, Martinez H, Micic DV, Stjanovic T, Spatz M (1985) Cerebral blood flow, energy utilization, serotonin metabolism, (Na, K) ATPase activity and postischemic brain swelling. In: Inaba Y, Klatzo I, Spatz M (eds) Brain edema. Springer, Berlin Heidelberg New York Tokyo, pp 170–177

Myers RE (1979) A unitary theory of causation of anoxic and hypoxic brain pathology. Adv Neurol 26: 195–213

Nagai H, Suzuki Y, Sugiura M, Noda S, Mabe H (1974) Experimental cerebral vasospasm. 1. Factors contributory to early spasm. J Neurosurg 41: 285–292

Naito J, Komatsu H, Ujiie A, Hamano S, Kubota T, Tsuboshima M (1983) Effects of thromboxane synthetase inhibitors on aggregation of rabbit platelets. European J Pharmacology 91: 41–48

Nakamura T, Suzuki N, Imabayashi S, Ishikawa Y, Sasaki T, Asano T (1982) Lipid peroxides in the cerebrospinal fluid after subarachnoid hemorrhage. In: Nozaki M et al. (eds) Oxygenase and oxygen metabolism. Academic Press, New York, pp 611–617

Nakano J, Prancan AV, Moore SE (1972) Metabolism of prostaglandin E_1 in cerebral cortex and cerebellum of the dog and rat. Brain Res 39: 545–548

— Chang ACK, Fisher RG (1973) Effects of prostaglandins E_1, E_2, A_1, A_2, and $F_{2\alpha}$ on canine carotid arterial blood flow, cerebrospinal fluid pressure, and intraocular pressure. J Neurosurg 38: 32–39

Narumiya S, Ogorochi T, Nakao K, Hayaishi O (1982) Prostaglandin D_2 in rat brain, spinal cord and pituitary; basal level and regional distribution. Life Sci 31: 2093–2103

Neely WA, Youmans JR (1963) Anoxia of canine brain without damage. JAMA 183: 1085–1087

Nehls DG, Flom RA, Carter LP, Spetzler RF (1985) Multiple intracranial aneurysms: determining the site of rupture. J Neurosurg 63: 342–348

Neil-Dwyer G, Cruickshank J, Doshi A, Walter P (1980) Systemic effects of subarachnoid hemorrhage. In: Wilkins RH (ed) Cerebral arterial spasm. Williams & Wilkins, Baltimore, pp 256–265

Nelson E, Rennels M (1970) Innervation of intracranial arteries. Brain 93: 475–490

Nelson JR, Goodman SJ (1971) An evaluation of the cerebrospinal fluid infusion test for hydrocephalus. Neurology 21: 1038–1053

Nemoto EM, Hossmann K-A, Cooper HK (1981) Post-ischemic hypermetabolism in cat brain. Stroke 12: 666–676

Nestler EJ, Walaas I, Greengard P (1984) Neuronal phosphoproteins: physiological and clinical implications. Science 225: 1357–1364

Nielsen KC, Owman CH (1967) Adrenergic innervation of pial arteries related to the circle of Willis in the cat. Brain Res 6: 773–776

— — (1971) Contractile response and amine receptor mechanisms in isolated middle cerebral artery of the cat. Brain Res 27: 33–42

Nilsson BW (1977) Cerebral blood flow in patients with subarachnoid haemorrhage studied with an intravenous isotope technique. Its clinical significance in the timing of surgery of cerebral arterial aneurysm. Acta Neurochir (Wien) 37: 33–48

Nishimoto A, Ueta K, Onbe H, et al (1985) Nationwide cooperative study of intracranial aneurysm surgery in Japan. Stroke 16: 48–52

Nishioka H (1966) Report on the cooperative study of intracranial aneurysms and subarachnoid hemorrhage. Section VII. Part I: Evaluation of conservative management of ruptured intracranial aneurysms. J Neurosurg 25: 574–592

Nishizuka Y (1984) Turnover of inositol phospholipids and signal transduction. Science 225: 1365–1370

Noguchi S, Asano T, Takakura K, Usui M, Shimizu T (1986) Production of lipoxygenase metabolites of arachidonic acid (HETEs) in the rat brain: Effect of ischemia and acidosis. Presented in the 45th Meeting of Japanese Neurosurgical Society, Tokyo. Paper in preparation

Nordström CH, Rehncrona S, Siesjö BK (1978) Restitution of cerebral energy state, as well as of glycolytic metabolites, citric acid cycle intermediates and associated amino acids after pronounced incomplete ischemia. J Neurochem 30: 479–486

Norlén G, Olivecrona H (1953) The treatment of aneurysms of the circle of Willis. J Neurosurg 10: 404–415

Nornes H, Magnaes B (1972) Intracranial pressure in patients with ruptured saccular aneurysm. J Neurosurg 36: 537–547

— (1973) The role of intracranial pressure in the arrest of the hemorrhage in patients with ruptured intracranial aneurysm. J Neurosurg 39: 226–234

O'Brien MD, Walz AG, Jordan MM (1970) Ischemic cerebral edema: Distribution of water in brains of cats after occlusion of the middle cerebral artery. Arch Neurol 30: 456–460

Ochiai C, Asano T, Tamura A, Sano K, Fukuda T, Niakamura T (1981) An experimental study on the mechanism of the protective action of pentobarbital and Y-9179 against cerebral ischemia. Neurol Med Chir 21: 303–311

— — Takakura K (1982) Mechanisms of cerebral protection by pentobarbital and nizofenone correlated with the course of local cerebral blood flow changes. Stroke 13: 788–796

Odom GL (1975) Cerebral vasospasm. Clin Neurosurg 22: 29–58

Ogorochi T, Narumiya S, Mizuno N, Yamashita K, Miyazaki H, Hayaishi O (1984) Regional distribution of prostaglandins D_2, E_2, and $F_{2\alpha}$ and related enzymes in postmortem human brain. J Neurochem 43: 71–82

Ohta T, Kajikawa H, Yoshikawa Y, Shimizu K, Funatsu N, Yamamoto M, Toda N (1980) Cerebral vasospasm and hemoglobins: clinical and experimental studies. In: Wilkins RH (ed) Cerebral arterial spasm. Williams & Wilkins, Baltimore, pp 166–172

— Kikuchi H, Hashi K, Kudo Y (1986) Nizofenone administration in the acute stage following subarachnoid hemorrhage. J Neurosurg 64: 420–426

Oldendorf WH (1970) Measurement of brain uptake of radiolabelled substances using a tritiated water internal standard. Brain Res 24: 372–376

Omae T, Takeshita M, Hirota Y (1976) The Hisayama study and joint study on cerebrovascular disease in Japan. In: Scheinberg P (ed) Cerebrovascular diseases. Raven Press, New York, pp 255–265

Ono H, Mizukami M, Kitamura K, Kikuchi H (1984) Ticlopidine: Quo Vadis? Subarachnoid hemorrhage. Agent Actions [Suppl] 15: 259–272

Osaka K (1977) Prolonged vasospasm produced by the breakdown products of erythrocytes. J Neurosurg 47: 403–411

Owman CH, Edvinsson L, Olin T, et al (1979) Pathophysiology of cerebral vasospasm: transmitter changes in perivascular sympathetic nerves and increased pial artery sensitivity to norepinephrine and serotonin. In: Price TR, Nelson E (eds) Cerebrovascular diseases, pp 295–305

Ozaki N, Mullan S (1979) Possible role of the erythrocyte in causing prolonged cerebral vasospasm. J Neurosurg 51: 773–778

Ozawa K, Seta K, Takeda H, Ando K, Handa H, Araki C (1966) On the isolation of mitochondria with high respiratory control from rat brain. J Biochem 59: 501–510

— — Araki H, Handa H (1967) The effect of ischemia on mitochondrial metabolism. J Biochem 61: 512–514

— Kitamura O, Ohsawa T, Murata T, Honjo I (1969) Mitochondrial vulnerability and lipid metabolism. J Biochem 66: 361–367

Palmer GC, Palmer SJ, Christie-Pope BC, Callahan AS, Taylor MD, Eddy LJ (1985) Classification of ischemic-induced damage to Na$^+$, K$^+$-ATPase in gerbil forebrain. Modification by therapeutic agents. Neuropharmacology 24: 509–516

Pappius HM, Wolfe LS (1983) Effects of indomethacin and ibuprofen on cerebral metabolism and blood flow in traumatized brain. J Cereb Blood Flow Metab 4: 448–459

Parker CW (1982) The chemical nature of slow-reacting substances. Adv Inflam Res 4: 1–23

Parkes JD, James IM (1971) Electroencephalographic and cerebral blood flow changes following spontaneous subarachnoid hemorrhage. Brain 94: 69–76

Pasqualin A, Rosta L, Pian RD, Cavazzani P, Scienza R (1984) Role of computed tomography in the management of vasospasm after subarachnoid hemorrhage. Neurosurgery 15: 344–353

Paul KS, Whalley ET, Forster C, Lye R, Dutton J (1982) Prostacyclin and cerebral vessel relaxation. J Neurosurg 57: 334–340

Paxton R, Ambrose J (1974) The EMI scanner: A brief review of the first 650 patients. Br J Radiol 47: 530–565

Peerless SJ, Griffiths JC (1972) Plasma catecholamines following subarachnoid hemorrhage. Ann R Coll Phys Can 5: 48–49

— Kassell NF, Komatsu K, Hunter IG (1980) Cerebral vasospasm: Acute proliferative vasculopathy? II. Morphology. In: Wilkins RH (ed) Cerebral arterial spasm. Williams & Wilkins, Baltimore, pp 88–96

Penfield W (1932) Intravascular vascular nerves. Arch Neurol Psychiat (Chicago) 27: 30–44

Pennink M, White RP, Crockarell JR, et al (1972) Role of prostaglandin F$_{2\alpha}$ in the genesis of experimental cerebral vasospasm. Angiographic study in dogs. J Neurosurg 37: 398–406

Peterson JW (1982) A simple model of smooth muscle myosin phosphorylation and dephosphorylation as rate-limiting mechanism. Biophys J 37: 453–459

Petruk K, Weir BA, Marriot MR, Overton TR (1973) Clinical grade, regional cerebral blood flow and angiographical spasm in the monkey after subarachnoid and subdural hemorrhage. Stroke 4: 431–445

Pia HW (1979) Aneurysms of the internal-carotid-ophthalmic artery junction. In: Pia HW, Langmaid C, Zierski J (ed) Cerebral aneurysms advances in diagnosis and therapy. Springer, Berlin Heidelberg New York, pp 89–93

— (1981) Grading of cerebral aneurysms and the timing of operation. Neurosurg Rev 4: 143–150

Pickard JD (1981) Role of prostaglandins and arachidonic acid derivations in the coupling of cerebral blood flow to cerebral metabolism. J Cereb Blood Flow Metab 1: 361–384

— McKenzie ET (1973) Inhibition of prostaglandin synthesis and the response of baboon cerebral circulation to carbon dioxide. Nature 245: 187–188

— Perry S (1984) Spectrum of altered reactivity of isolated cerebral arteries following subarachnoid haemorrhage-response to potassium, pH, noradrenaline, 5-hydroxytryptamine, and sodium loading. J Cereb Blood Flow Metab 4: 599–609

— Walker V (1984) Problems in the analysis of the role of icosanoids in acute cerebrovascular disease. In: Bes A, Braquet P, Paoletti R, Siesjö BK (eds) Cerebral ischemia. Elsevier, Amsterdam, pp 355–362

— Vinall PE, Simeone FA (1975) Prostaglandins and cerebral vasospasm. A problem of interpretation. Surg Forum 26: 496–498

— MacDonnell LA, MacKenzie ET, Harper AM (1977) Prostaglandin-induced effects in the primate cerebral circulation. Eur J Pharmacol 43: 343–351

— Boisvert DPJ, Graham DI, Fitch W (1979) Late effects of subarachnoid haemorrhage on the response of the primate cerebral circulation to drug-induced changes in arterial blood pressure. J Neurol Neurosurg Psychiatry 42: 899–903

— Matheson M, Paterson J, Wyper DJ (1980) Autoregulation of cerebral blood flow and the predction of late morbidity and mortality after cerebral aneurysm surgery. In: Wilkins RH (ed) Cerebral arterial spasm. Williams & Wilkins, Baltimore, pp 350–355

Piper PJ, Letts LG, Galton SA (1983) Generation of a leukotriene-like substance from porcine vascular and other tissues. Prostaglandins 25: 591–599

Plum F, Posner JB (1963) Edema and necrosis in experimental cerebral infarction. Arch Neurol 9: 563–579

— Cooper AJL, Kraig RP, Petito CK, Pulsinelli WA (1985) Glial cells: The silent partners of the working brain. The Thomas E. Duffy Memorial Lecture. J Cereb Blood Flow Metab 5: S1–S4

Pool JL (1958) Cerebral vasospasm. New Engl J Med 259: 1259–1264

Potter JM (1959) Redistribution of blood to the brain due to localized cerebral arterial spasm. The possible importance of the small peripheral anastomotic cerebral arteries. Brain 82: 367–376

Pressman BD, Gilbert GE, Davis DO (1975) Computerized transverse tomography of vascular lesions of the brain: Part II. Aneurysms. Am J Roentgenol 124: 215–219

Pritz MB (1984) Treatment of cerebral vasospasm.—Usefulness of Sawn-Ganz catheter monitoring of volume expansion. Surg Neurol 21: 239–244

— Giannotta SL, Kindt GW, McGillicuddy JE, Prager RL (1978) Treatment of patients with neurological deficits associated with cerebral vasospasm by intravascular volume expansion. Neurosurgery 3: 364–368

Pryor WA (ed) (1976–1984) Free radicals in biology, vol 1–6. Academic Press, New York

— (1976) The role of free radical reactions in biological systems. In: Pryor WA (ed) Free radicals in biology. Academic Press, New York, pp 1–49

— Stanley JP, Blair E (1976) Autoxidation of polyunsaturated fatty acids: II. A suggested mechanism for the formation of TBA-reactive materials from postaglandin-like endoperoxides. Lipids 11: 370–379

Pulsinelli WA, Brierley JG (1979) A new model of bilateral hemispheric ischemia in the unanesthetized rat. Stroke 10: 267–272

— Waldman S, Rawlinson D, Plum F (1982a) Moderate hyperglycemia augments ischemic brain damage: a neurpathologic study in the rat. Neurology 32: 1239–1246

— Brierley JB, Plum F (1982b) Temporal profile of neuronal damage in a model of transient forebrain ischemia. Ann Neurol 11: 491–498

— Levy DE, Duffy TE (1982c) Regional cerebral blood flow and glucose metabolism following transient forebrain ischemia. Ann Neurol 11: 499–509

Purves MJ (1972) The physiology of the cerebral circulation. Cambridge University Press

— (1978) Control of cerebral blood vessels: present state of art. Ann Neurol 3: 377–383

Ramirez-Lassepas M (1981) Antifibrinolytic therapy in subarchnoid hemorrhage caused by ruptured intracranial aneurysm. Neurology (Ny) 31: 316–322

Rapela CE, Green HD (1967) Baroreceptor reflexes and autoregulation of cerebral blood flow in the dog. Circ Res 21: 559–568

— Martin JB (1975) Reactivity of cerebral extra and intraparenchymal vasculature to serotonin and vasodilator agents. In: Harper M, et al (eds) Flow and metabolism in the brain. Churchill Livingstone, Edinburgh, 4.5–4.9

Rapoport SI (1976) Blood-brain barrier in physiology and medicine. Raven Press, New York

Raynor RB, McMurtry JG, Pool JL (1961) Cerebrovascular effects of topically applied serotonin in the cat. Neurology 11: 190–195

Rehncrona S, Mela L, Siesjö BK (1979) Recovery of brain mitochondrial function in the rat after complete and incomplete cerebral ischemia. Stroke 10: 437–446

— Folbergrová J, Smith DS, Siesjö BK (1980a) Influence of complete and pronounced incomplete cerebral ischemia and subsequent recirculation on cortical concentrations of oxidized and reduced glutathione in the rat. J Neurochem 34: 477–486

— Smith DS, Akesson B, Westerberg E, Siesjö BK (1980b) Peroxidative changes in brain cortical fatty acids and phospholipids, as characterized during Fe^{++}-and ascorbic acid-stimulated lipid peroxidation in vitro. J Neurochem 34: 1630–1638

— Westerberg E, Akesson B, Siesjö BK (1982) Brain cortical fatty acids and phospholipids during and following complete and severe incomplete ischemia. J Neurochem 38: 84–93

— (1984) Indices of free radical production in brain tissue in vitro. In: Bes A, Braquet P, Paoletti R, Siesjö BK (eds) Cerebral ischemia. Excerpta Medica, Amsterdam, pp 285–292

Reivich M (1968) Regulation of the cerebral circulation. Clin Neurosurg 16: 378–418

Rhoton AL Jr (1980) Anatomy of saccular aneurysms. Surg Neurol 14: 59–66

Richardson JC, Hyland HH (1941) Intracranial aneurysms. A clinical and pathological study of subarachnoid and intracerebral haemorrhage caused by berry aneurysms. Medicine (Baltimore) 20: 1–83

Robertson EG (1949) Cerebral lesions due to intracranial aneurysms. Brain 72: 150–185

Robertson JT (1974) Cerebral arterial spasm: current concepts. Clin Neurosurg 21: 100–106

Rodbard S, Saiki H (1952) Mechanism of the pressor response to increased intracranial pressure. Am J Physiol 168: 234–244

Rodrigues AM, Gerritsen ME (1984) Prostaglandin release from isolated rabbit cerebral cortex micro-vessels—Comparison of 6-keto $PGF_{1\alpha}$ and PGE_2 release from micro-vessels incubated in 100% O_2, room air and 95% N_2: 5% CO_2. Stroke 15: 717–722

Rodriguez de Turco EB, Morelli de Liberti S, Bazán NG (1983) Stimulation of free fatty acid and diacylglycerol accumulation in cerebrum and cerebellum during bicuculline-induced status epilepticus. Effect of pretreatment with a-methyl-p-tyrosine and p-chlorophenylamine. J Neurochem 40: 252–259

Ropper AH, Zervas NT (1984) Outcome 1 year after SAH from cerebral aneurysms. J Neurosurg 60: 909–915

Rosenblum WI, Kontos HA (1974) The importance and relevance of studies of the pial microcirculation. Stroke 4: 425–428

— (1975 a) Effects of prostaglandins on cerebral blood vessels. Interaction with vasoactive amines. Neurology 25: 1169–1171

— (1975 b) Constriction of pial arterioles produced by prostaglandin $F_{2\alpha}$. Stroke 6: 293–297

— Giulianti D (1973) Participation of cerebrovascular nerves in generalized sympathetic discherge. Nonspecific release of norepinephrine in the presence or absence of subarachnoid haemorrhage and vasospasm. Arch Neurol 29: 91–94

Ross R, Glomset JA, Kariya B, Harker L (1974) A platelet-dependent serum factor that stimulates the proliferation of arterial smooth muscle cells in vitro. Proc Natl Acad Sci USA 71: 1207–1210

— — (1976) The pathogenesis of atherosclerosis. New Engl J Med 295: 369–377

Rubio R, Berne RM, Winn HR (1978) Production, metabolism and possible functions of adenosine in brain tissue in situ. In: Cerebral vascular smooth muscle and its control, Ciba Foundation Symposium 56, Elsevier, Excerpta Medica, North-Holland, Amsterdam, pp 355–372

Ryan US, Ryan JW (1983) Endothelial cells and inflammation. Clin Lab Med 3 (4): 577–599

Saito I, Sano K (1980) Vasospasm after aneurysm rupture: incidence, onset, and course. In: Wilkins RH (ed) Cerebral arterial spasm. Williams & Wilkins, Baltimore, London, pp 294–301

— Ueda Y, Sano K (1977) Significance of vasospasm in the treatment of ruptured intracranial aneurysms. J Neurosurg 47: 412–429

— Shigeno T, Aritake K, Tanishima T, Sano K (1979) Vasospasm assessed by angiography and computerized tomography. J Neurosurg 51: 466–475

— Asano T, Ochiai C, Takakura K, Tamura A, Sano K (1983) A double-blind clinical evaluation of the effect of nizofenone (Y-9179) on delayed ischemic neurological deficits following aneurysmal rupture. Neurol Res 5: 29–47

Sakurai Y, Ito Z, Uemura K (1975) Pathophysiological aspects of acute stage patients with ruptured intracranial aneurysms. II. Relationships between neurological prognosis and sequential changes of cerebral blood flow dynamics. Brain and Nerve (Tokyo) 27: 1213–1221

Samuelsson B (1983) Leukotrienes: Mediators of immediate hypersensitivity reactions and inflammation. Science 220: 568–575

— Goldyne M, Granstrom E, Hamberg M, Hammarstrom S, Malmsten C (1978) Prostaglandins and thromboxanes. Ann Rev Biochem 47: 997–1029

— Hammarstrom S, Murphy RC, Borgeat P (1980) Leukotrienes and slow reacting substance of anaphylaxis (SRS-A). Allergy 35: 375–381

Sano K (1979) Basilar artery aneurysms—transoral transclival approach. In: Pia HW, Langmaid C, Zierski J (ed) Cerebral aneurysms. Springer, Berlin Heidelberg New York, pp 326–328

Sano K (1980a) A multipurpose all-angle clip applier for aneurysm surgery—Technical note. J Neurosurg 53: 260–261

— (1980b) Temporo-polar approach to aneurysms of the basilar artery at and around the distal bifurcation—Technical note. Neurol Res 2: 253–272

— (1983) Cerebral vasospasm and aneurysm surgery. Clin Neurosurg 30: 13–58

— (1985) Timing of surgery, operative mortality, and follow-up results in cases with subarachnoid haemorrhage due to aneurysm rupture. In: Auer LM (eds) Timing of aneurysm surgery. Walter de Gruyter, Berlin, pp 227–236

— Malamud N (1953) Clinical significance of sclerosis of the cornu ammonis. Ictal "psychic phenomena". Arch Neurol Psychiat 70: 40–53

— Saito I (1978) Timing and indication of surgery for ruptured intracranial aneurysms with regard to cerebral vasospasm. Acta Neurochir (Wien) 41: 49–60

— Saito I (1980) Early operation and washout of blood clots for prevention of cerebral vasospasm. In: Wilkins RH (ed) Cerebral arterial spasm. Williams and Wilkins, Baltimore, London, pp 510–513

— Tamura A (1985) A proposal for grading of subarachnoid haemorrhage due to aneurysm rupture. In: Auer LM (ed) Timing of aneurysm surgery. Walter de Gruyter, Berlin, pp 3–7

— Asano T, Tanishima T, Sasaki T (1980) Lipid peroxidation as a cause of cerebral vasospasm. Neurol Res 2: 253–272

— Handa H, Suzuki S, Asano T, Tamura A, Yonekawa Y, Ono H, Tachibana N (1986) The utility of OKY-046, a TxA$_2$ synthetase inhibitor, for angiographic and symptomatic vasospasm following subarachnoid hemorrhage due to rupture of intracranial aneurysms: Clinical assessment by a multi-institutional, double blind study. Igaku no Ayumi (Tokyo) 138 (6–7): in press

Sasaki T, Wakai S, Asano T, Watanabe T, Kirino T, Sano K (1981a) The effect of a lipid hydroperoxide of arachidonic acid on the canine basilar artery. J Neurosurg 54: 357–365

— Murota S, Wakai S, Asano T, Sano K (1981b) Evaluation of prostacyclin biosynthetic activity in canine basilar artery following subarachnoid hemorrhage. J Neurosurg 55: 771–778

— Wakai S, Asano T, Takakura K, Sano K (1982a) Prevention of cerebral vasospasm after SAH with a thromboxane synthetase inhibitor OKY-1581. J Neurosurg 57: 74–82

— Asano T, Takakura K (1982b) Inhibitory effects of drugs on the vasoconstriction induced by the cerebrospinal fluid of SAH patients. In: Iwabuchi T (ed) Cerebral vasospasm. Proceedings of the 11th Conference of Surgical Treatment of Stroke. Neuron-Sha, Tokyo, pp 183–188

— — — Sano K, Nakamura T, Suzuki N, Imabayashi S, Ishikawa Y (1982c) Cerebral vasospasm and lipid peroxidation: Lipid peroxides in the cerebrospinal fluid after subarachnoid hemorrhage. Brain Nerve (Tokyo) 34: 1191–1196

— — — — Kassell NF (1984) Nature of the vasoactive substance in CSF from patients with subarachnoid hemorrhage. J Neurosurg 60: 1186–1191

— Kassell NF, Yamashita M, Fujiwara S, Zuccarello M (1985a) Barrier disruption in the major cerebral arteries following experimental subarachnoid hemorrhage. J Neurosurg 63: 433–440

— — Colohan ART, Nazar GB (1985b) Cerebral vasospasm following subarachnoid hemorrhage. In: McDowell F, Caplan LR (eds) Cerebrovascular survey report. For the National Institute of Neurological and Communicative Disorders and Stroke, NIH Public Health Service

Sato K, Yamaguchi M, Mullan S, Evans JP, Ishii S (1969) Brain edema. A study of biochemical and structural alteration. Arch Neurol 21: 413–424

Sautebin L, Spagnuolo C, Galli C, Galli G (1978) A mass fragmentographic procedure for the simultaneous determination of HETE and PGF$_{2a}$ in the central nervous system. Prostaglandins 16: 985–988

Schneck SA (1964) On the relationship between ruptured intracranial aneurysm and cerebral infarction. Neurology 14: 691–702

— Kricheff II (1964) Intracranial aneurysm rupture, vasospasm, and infarction. Arch Neurol 11: 668–680

Schorstein J (1940) Carotid ligation in saccular intracranial aneurysms. Br J Surg 28: 50–70

Schuier FJ, Hossmann K-A (1980) Experimental brain infarcts in cats. II. Ischemic brain edema. Stroke 11: 593–601

Schutz H, Silverstein PR, Vapalahti M, Bruce DA, Mela L, Langfitt TW (1973) Brain mitochondrial function after ischemia and hypoxia. I. Ischemia induced by increased intracranial pressure. Arch Neurol 29: 408–416

Schwartz JP, Mrsulja BB, Mrsulja BJ, Passonneau JV, Klatzo I (1976) Alterations of cyclic nucleotide-related enzymes and ATPase during unilateral ischemia and recirculation in gerbil cerebral cortex. J Neurochem 27: 101–107

Serbinenko FA (1974) Balloon catheterization and occlusion of major cerebral vessels. J Neurosurg 41: 125–145

Serhan CN, Hamberg M, Samuelsson B (1984) Lipoxins: Novel series of biologically active compounds formed from arachidonic acid in human leukocytes. Proc Natl Acad Sci USA 81: 5335–5339

Shapiro HM, Stromberg DD, Lee DR, Wiederhielm CA (1971) Dynamic pressures in the pial arterial microcirculation. Am J Physiol 221: 279–283

Shibata S, Hodge CP, Pappius HM (1974) Effect of experimental ischemia on cerebral water and electrolytes. J Neurosurg 41: 146–159

Shigeno S, Fritschka E, Shigeno T, Brock M (1985a) Effects of indomethacin on rCBF during and after focal cerebral ischemia in the cat. Stroke 16: 235–240

Shigeno T (1981) Norepinephrine (NE) in cerebral vasospasm: Assay of NE in CSF from patients with a ruptured intracranial aneurysm and its role in experimental vasospasm. Brain and Nerve 33: 537–545

— Asano T, Watanabe E, Johshita H, Takakura K (1985b) Does capillary Na$^+$, K$^+$-ATPase play a role in the development of ischemic brain edema? In: Inaba Y, Klatzo I, Spatz M (eds) Brain edema. Springer, Berlin Heidelberg New York Tokyo, pp 461–464

— — Hanamura T, Mima T, Takakura K (1985) Enhanced activity of capillary Na$^+$, K$^+$-ATPase promotes sodium flux across the blood-brain barrier and causes ischemic brain edema. Presented in Symposium on the Blood-Brain Barrier, Copenhagen. Paper in submission

Shiguma M (1982) Change in the ionic environment of cerebral arteries after subarachnoid haemorrhage especially of potassium ion concentration in subarachnoid haematoma and its role in cerebral vasospasm. Neurol Med Chir (Tokyo) 22: 805–812

Shikinami A, Yamada H, Sakai N, Ando T, Kagawa Y (1985) Clinical study on effect of thromboxane A synthetase inhibitor (OKY-046) administered prophylactically to prevent cerebral vasospasm after early surgery of ruptured aneurysm. Jpn J Stroke 7: 200–209

Shimizu T, Kondo K, Hayaishi O (1981) Role of prostaglandin endoperoxides in the serum thiobarbituric acid reaction. Arch Biochem Biophys 206: 271–276

— Kito K, Hoshi T, Yamazaki N, Takahashi K, Takahashi M, Yamane K, Sim C, Kitamura K, Sendo S (1982) Immunological study of late cerebral vasospasm in subarachnoid hemorrhage. Neurol Med Chir (Tokyo) 22: 613–619

— Isumi T, Seyama Y, Tadokoro K, Radmark O, Samuelsson B (1986) Characterization of leukotriene A4 synthase from murine mast cells. Evidence for its identity to arachidonate 5-lipoxygenase. Proc Natl Acad Sci USA 83: 4175–4179

— Tagusagawa Y, Watanabe T, Asano T: Activation of the arachidonate 5-lipoxygenase pathway in the canine basilar artery after experimental subarachnoid hemorrhage. Paper in preparation

Shrago E (1978) The effect of long chain fatty acyl CoA esters on the adenine nucleotide translocase and myocardial metabolism. Life Sci 22: 1–6

Siegel BA, Studer RK, Potchen EJ (1973) Brain ^{22}Na uptake in experimental cerebral microembolism. J Neurosurg 38: 739–742

Siesjö BK, Zwetnow NN (1970) Effects of increased cerebrospinal fluid pressure upon adenine nucleotides and upon lactate and pyruvate in rat brain tissue. Acta Neurol Scand 46: 187–202

Siesjö BK (1978) Brain Energy Metabolism. John Wiley and Sons, New York

— (1981) Cell damage in the brain: A speculative synthesis. J Cereb Blood Flow Metab 1: 155–185

— Bendek G, Koide T, Westerberg E, Wieloch T (1985 a) Influence of acidosis on lipid peroxidation in brain tissues *in vitro*. J Cereb Blood Flow Metab 5: 253–258

— Wieloch T (1985 b) Brain injury: Neurochemical aspects. In: Becker DP, Povlishock JT (eds) Central nervous system trauma status report. William Byrd Press Inc, Richmond, VA, pp 513–532

— (1985 c) Acid-base homeostasis in the brain: physiology, chemistry, and neurochemical pathology. Prog Brain Res 63: 121–154

Simeone FA, Trepper PJ, Brown DJ (1972) Cerebral blood flow evaluation of prolonged experimental vasospasm. J Neurosurg 37: 302–311

Sinet PM, Heikkila RE, Cohen G (1980) Hydrogen peroxide production by rat brain *in vivo*. J Neurochem 34: 1421–1428

Slater TF (1972) Free radical mechanisms in tissue injury. Pion Ltd, London

Smith B (1963) Cerebral pathology in subarachnoid haemorrhage. J Neurol Neurosurg Psychiat 26: 535–539

Smith RR, Clower BR, Peeler DF, Yoshioka J (1983) The angiopathy of subarachnoid hemorrhage: angiographic and morphologic correlates. Stroke 14: 240–245

— — Grotendorst GM, Yabuno N, Cruse JM (1985) Arterial wall changes in early human vasospasm. Neurosurgery 16: 171–176

Sobieszek A, Smoll JV (1977) Regulation of actin-myosin Interaction in vertebrate smooth muscle: Activation via a myosin light-chain kinase and the effect of tropomyosin. J Mol Biol 112: 559–576

Solomon RA, Post KD, McMurtry III JG (1984) Depression of circulating blood volume in patients after subarachnoid hemorrhage: Implications for the management of symptomatic vasospasm. Neurosurgery 15: 354–360

Sommer W (1880) Erkrankung des Ammonshorns als aetiologisches Moment der Epilepsie. Arch Psychiat 10: 631–675

Sonobe M, Suzuki J (1978) Vasospasmogenic substance produced following subarachnoid haemorrhage, and its fate. Acta Neurochir (Wien) 44: 97–106

Spagnuolo C, Sautebain L, Galli G, Racagni G, Galli C, Mazzari S, Finesso M (1979) $PGF_{2\alpha}$, thromboxane B_2 and HETE levels in gerbil brain cortex after ligation of common carotid arteries and decapitation. Prostaglandins 18: 53–61

Spector AA (1975) Fatty acid binding to plasma albumin. J Lipid Res 16: 165–176

Spielmeyer W (1925) Zur Pathogenese oertlich elektiver Gehirnveraenderungen. Z ges Neurol Psychiatr 99: 756–776

Steiner L, Löfgren J, Zwetnow NN (1975) Lethal mechanism in repeated subarachnoid hemorrhage in dogs. Acta Neurol Scand 52: 268–293

Stornelli SA, French JD (1964) Subarachnoid hemorrhage-factors in prognosis and management. J Neurosurg 21: 769–780

Sugita K, Hirota T, Tsugane R (1975) Application of nasopharyngeal mirror for aneurysm operation—Technical note. J Neurosurg 43: 244–246

Sun AY (1972) The effect of lipoxidation on synaptosomal Na^+, K^+ ATPase isolated from the cerebral cortex of squirrel monkey. Biochim Biophys Acta 266: 350–360

Sun FF, Chapman JP, McGuire JC (1977) Metabolism of prostaglandin endoperoxide in animal tissues. Prostaglandins 14: 1055–1074

Sun GY, Horrocks LA (1969) The metabolism of palmitic acid in the phospholipids, neutral glycerides and galactolipids of mouse brain. J Neurochem 16: 181–189

— Su KL (1979) Metabolism of arachidonyl phosphoglycerides in mouse brain subcellular fractions. J Neurochem 32: 1053–1059

— — Der OM, Tang W (1979) Enzymic regulation of arachidonate metabolism in brain membrane phosphoglycerides. Lipids 14: 229–235

Sundt TM Jr, Whisnant JP (1978) Subarachnoid hemorrhage from intracranial aneurysms: Surgical management and natural history of disease. N Engl J Med 299: 116–122

Suzuki J (1979) Grading and timing of the operation on cerebral aneurysms. In: Pia

HW, *et al.* (eds) Cerebral aneurysms. Advances in diagnosis and therapy. Springer, Berlin Heidelberg New York, pp 201–208

— Yoshimoto T (1979) Distribution of cerebral aneurysms. In: Pia HW, Langmaid C, Zierski J (eds) Cerebral aneurysms. Springer, Berlin Heidelberg New York, pp 127–133

— Onuma T, Yoshimoto T (1979) Results of early operations on cerebral aneurysms. Surg Neurol 11: 407–412

— Komatsu S, Sato T, Sakurai Y (1980) Correlation between CT findings and subsequent development of cerebral infarction due to vasospasm in subarachnoid hemorrhage. Acta Neurochir (Wien) 55: 63–70

Suzuki K (1981) Chemistry and metabolism of brain lipids. In: Siegel GJ, Albers RW, Agranoff BW, Katzman R (eds) Basic Neurochemistry, 3rd edition. Little, Brown and Company, Boston, pp 355–370

Suzuki N, Nakamura T, Imabayashi S, Ishikawa Y, Sasaki T, Asano T (1983) Identification of 5-hydroxy eicosatetraenoic acid in cerebrospinal fluid after subarachnoid hemorrhage. J Neurochem 41: 1186–1189

Suzuki R, Yamaguchi T, Kirino T, Orzi F, Klatzo I (1983 a) The effects of 5-minute ischemia in Mongolian gerbils: I. Blood-brain barrier, cerebral blood flow, and local cerebral glucose utilization changes. Acta Neuropathol (Berl) 60: 207–216

— — Li C-L, Klatzo I (1983 b) The effects of 5-minute ischemia in Mongolian gerbils. II. Changes of spontaneous neuronal activity in cerebral cortex and CA 1 sector of hippocampus. Acta Neuropathol (Berl) 60: 217–222

Suzuki S, Iwabushi T, Tanaka T, Kanayama S, Ottomo M, Hatanaka M, Aihara H (1985) Prevention of cerebral vasospasm with OKY-046 an imidazole derivative and a thromboxane synthetase inhibitor. A preliminary co-operative clinical study. Acta Neurochir (Wien) 77: 133–141

Svendgaard NA, Edvinsson L, Owman C, *et al* (1977) Increased sensitivity of the basilar artery to norepinephrine and 5-hydroxytryptamine following experimental subarachnoid hemorrhage. Surg Neurol 8: 191–195

Symonds CP (1923) Contributions to the clinical study of intracranial aneurysms. Guy's Hosp Rec 73: 139–158

Tagari P, Boulay GH du, Aitken V, Boullin DJ (1983) Leukotriene D_4 and the cerebral vasculature *in vivo* and *in vitro*. Prostaglandins Leukotrienes Med 11: 281–297

Takayasu M, Shintani A, Negoro M, Asai T (1985) The value of contrast-enhanced computed tomography for the diagnosis of intracranial aneurysms. Neurol Med Chir (Tokyo) 25: 27–31

Takemae T, Mizukami M, Kin H, Kawase T, Araki G (1978) Computed tomography of ruptured intracranial aneurysm in acute stage.—Relationship between vasospasm and high density on CT scan. Brain Nerve (Tokyo) 30: 861–866

Tamura A, Asano T, Sano K, Tsumagari T, Nakajima A (1979) Protection from cerebral ischemia by a new imidazole derivative (Y-9179) and pentobarbital. A comparative study in chronic middle cerebral artery occlusion in cats. Stroke 10: 126–134

— Graham DE, McCulloch J, Teasdale GM (1981 a) Focal cerebral ischaemia in the rat. I. Description of technique and early neuropathological consequences following middle cerebral artery occlusion. J Cereb Blood Flow Metab 1: 53–60

— — — — (1981 b) Focal cerebral ischaemia in the rat. 2: Regional cerebral blood flow determined by [^{14}C]-iodoantipyrine autography follwing middle cerebral artery occlusion. J Cereb Blood Flow Metab 1: 61–69

Tanabe T, Saitoh T, Tachibana S, Takagi H, Yada K (1982) Effect of hyperdynamic therapy on cerebral ischaemia caused by vasospasm associated with subarachnoid haemorrhage. Acta Neurochir (Wien) 63: 291–296

— Sakata K, Yamada H, Ito T, Takado M (1978) Cerebral vasospasm and ultrastructural changes in cerebral arterial wall. J Neurosurg 49: 229–238

Tanaka R, Tanimura K, Ueki K (1977) Ultrastructural and biochemical studies on ouabain-induced oedematous brain. Acta Neuropathol (Berl) 37: 95–100

Taneda M (1982) Effect of early operation for ruptured aneurysms on prevention of delayed ischemic symptoms. J Neurosurg 57: 622–628

— Sakamoto T, Shimada M, Hiraga S, Kim A (1985) Risk of rebleeding during angiography in patients with subarachnoid haemorrhage at the peracute stage. In: Auer LM (ed) Timing of aneurysm surgery. Walter de Gruyter, Berlin, pp 653–656

Tani E, Maeda Y, Fukumori T, Nakano M, Kochi N, Morimura T, Yokota M, Matsumoto T (1984) Effect of selective inhibitor of thromboxane A synthetase on cerebral vasospasm after early surgery. J Neurosurg 61: 24–29

Tanishima T (1980) Cerebral vasospasm: contractile activity of hemoglobin in isolated canine basilar arteries. J Neurosurg 53: 787–793

— Asano T, Sasaki T, Sano K (1979) Role of peroxidation in the genesis of cerebral arterial spasm. Acta Neurol Scand [Suppl] 60: 484–485

Tappel AL (1953) The mechanism of the oxidation of unsaturated fatty acids catalyzed by hematin compounds. Arch Biochem Biophys 44: 378–395

— (1975) Lipid peroxidation and fluorescent molecular damage to membranes. In: Trump BF, Arstila AV (eds) Pathobiology of cell membranes, vol 1. Academic Press, New York, pp 145–170

— (1980) Measurement of and protection from in vivo lipid peroxidation. In: Pryor WA (ed) Free radicals in biology, vol IV. Academic Press, New York, pp 1–47

Tazawa T, Mizukami M, Kawase T, Usami T, Togashi O, Hyodo A, Eguchi T (1983) Relationship between contrast enhancement on computed tomography and cerebral vasospasm in patients with subarachnoid hemorrhage. Neurosurgery 12: 643–648

Teasdale GM, Murray G, Parker L, Jennett B (1979) Adding up the Glasgow coma score. Acta Neurochir [Suppl] 28: 13–16

— Knill-Jones RP, Lindsay KW (1983) Clinical assessment of SAH. J Neurosurg 59: 550–551

— Lindsay KW, Allardyce G, Dharker S, Ward P (1985) Standardized clinical grading of patients with subarachnoid haemorrhage: A

uniform internationl system? In: Auer LM (ed) Timing of aneurysm surgery. Walter de Gruyter, Berlin, pp 9–14

Terada T, Kikuchi H, Karasawa J, Kuriyama Y (1985) Sequential changes in the autoregulation of cerebral blood flow in patients with vasospasm. Neurol Med Chir (Tokyo) 25: 89–94

Thomas MJ, Mehl KS, Pryor WA (1982) The role of superoxide in xanthine oxidase-induced autooxidation of linoleic acid. J Biol Chem 257: 8343–8347

Thompson RK, Malina S (1959) Dynamic axial brain-stem distortion as a mechanism explaining the cardiorespiratory changes in increased intracranial pressure. J Neurosurg 16: 664–675

Toda N (1980) Responses to prostaglandin H_2 and I_2 of isolated dog cerebral and peripheral arteries. Am J Physiol 238: H 111–H 117

— Fujita Y (1973) Responsiveness of isolated cerebral and peripheral arteries to serotonin, norepinephrine, and transmural electrical stimulation. Circ Res 33: 98–104

— Miyazaki M (1978) Responses of isolated dog cerebral and peripheral arteries to prostaglandins after application of aspirin and polyphloretin phosphate. Stroke 9: 490–498

— Ozaki T, Ohta T (1977) Cerebrovascular sensitivity to vasoconstricting agents induced by subarachnoid hemorrhage and vasospasm in dogs. J Neurosurg 46: 296–303

Tomlinson BE (1959) Brain changes in ruptured intracranial aneurysm. J Clin Pathol 12: 391–399

Tönnis W (1934) Traumatischer Aneurysma der linken Art. carotis int., etc. Zbl Chir 61: 844–848

Tourtellotte WW, Metz LN, Bryan ER, Dejong RN (1964) Spontaneous subarachnoid hemorrhage: Factors affecting the rate of clearing of the cerebrospinal fluid. Neurology 14: 301–306

Tsementzis SA, Kennett RP, Hitchcock ER (1984) Rupture of intracranial vascular lesions during arteriography. J Neurol Neurosurg Psychiatry 47: 795–798

Uchiwa H, Kato T, Onishi H, Isobe T, Okuyama T, Watanabe S (1982) Purification of chicken gizzard myosin light-chain kinase, and its calcium and strontium sensitivities as compared with those of superprecipetation and ATPase activities of actomyosin. J Biochem 91: 273–282

Uski T, Andersson KE, Brandt L, Devinsson L, Ljunggren B (1983) Responses of isolated feline and human cerebral arteries to prostacyclin and some of its metabolites. J Cereb Blood Flow Metab 3: 238–245

Usui M, Asano T, Terao S, Takakura K (1985) Identification of a lipoxygenase product (11-HETE) in rat brain microvessel, and its relevance to ischemic brain edema. In: Inaba Y, Klatzo I, Spatz M (eds) Brain edema. Springer, Berlin Heidelberg New York Tokyo, pp 396–402

— — Takakura K: Identification and quantitative analysis of hydroxyeicosatetraenoic acids in the rat brain exposed to regional ischemia. Stroke, in press

Uyama O, Nagatsuka K, Nakabayashi S, Isaka Y, Yoneda S, Kimura K, Abe H (1985) The effect of a thromboxane synthetase inhibitor, OKY-046, on urinary excretion of immunoreactive thromboxane B and 6-keto prostaglandin $F_{2\alpha}$ in patients with ischemic cerebrovascular disease. Stroke 16: 241–244

Varsos VG, Liszczak TM, Han DH, Kistler JP, Vielma J, Black PM, Heros RC, Zerva NT (1983) Delayed cerebral vasospasm is not reversible by aminophylline, nifedipine, or papaverine in a "two-hemorrhage" canine model. J Neurosurg 58: 11–17

Vermeulen MJ, Lindsay KW, Murray GD, Cheah F, Hijdra A, Muizelaar JP, Schannong M, Teasdale GM van, Crebel H van, Gijn J (1984) Antifibrinolytic treatment in subarachnoid hemorrhage. N Engl J Med 311: 432–437

Vladimirov YUA, Olenev VI, Suslova TB, Cheremisina ZP (1980) Lipid peroxidation in mitochondrial membrane. Adv Lipid Res 17: 173–249

Vogt C, Vogt O (1922) Erkrankungen der Gehirnrinde in Lichte der Topitik, Pathoklise und Pathoarchitektonik. J Psychol Neurol 28: 1–171

Wade JG, Amtorp O, Sorensen SC (1975) No-flow state following cerebral ischemia. Arch Neurol 32: 381–384

Wakai S, Aritake K, Asano T, Takakura K (1982) Selective destruction of the outer leaflet of the capillary endothelial membrane

after intracerebral injection of arachidonic acid. Acta Neuropathol (Berl) 58: 303–306

Walker V, Pickard JD, Smythe P, Eastwood S, Perry S (1983) Effects of subarachnoid haemorrhage on intracranial prostaglandins. J Neurol Neurosurg Psychiatry 46: 119–125

— — (1985) Prostaglandins, thromboxane, leukotrienes and the cerebral circulation in health and disease. In: Symon L, et al (eds) Advances and technical standards in neurosurgery, vol 12. Springer, Wien New York, pp 3–90

Watanabe K, Shimizu T, Iguchi S, Wakatsuka H, Hayashi M, Hayaishi O (1980) An NADP-linked prostaglandin D dehydrogenase in swine brain. J Biol Chem 255: 1779–1782

Watanabe T, Asano T, Takakura K, Watanabe K (1985) Measurement of HETEs and effect of glutathione, AVS, and steroid on vasospasm in double hemorrhage SAH model. Presented in the 44th Meeting of Japan Neurosurgical Society, Nagasaki, Japan: Paper submitted to J. Neurosurg.

— — — Shimizu T (1986) Participation of arachidonic acid lipoxygenase products in the pathogenesis of cerebral vasospasm. Presented in 6th International Conference of Prostaglandins and Related Compounds, in Florence, Italy. Paper in preparation

Watson BD, Busto R, Goldberg WJ, Santiso M, Yoshida S, Ginsberg MD (1984) Lipid peroxidation in vivo induced by reversible global ischemia in rat brain. J Neurochem 42: 268–274

Weed LH (1935) Certain anatomical and physiological aspects on the meninges and cerebrospinal fluid. Brain 58: 383–397

Wei EP, Kontos HA, Dietrich WD, Povlishock JT, Ellis EF (1981) Inhibition by free radical scavengers and by cyclooxygenase inhibitors of pial arteriolar abnormalities from concussive brain injury in cats. Circ Res 48: 95–103

Weinberger LM, Gibbon MA, Gibbon JH (1940) Temporary arrest of the circulation to the central nervous system. II. Pathologic effects. Arch Neurol Psychiat 43: 961–986

Weinstein JD, Langfitt TW, Kassell NF (1964) Vasopressor response to increased intracranial pressure. Neurology 141: 1118–1131

Weir B, Grace M, Hansen J, Rothberg C (1978) Time course of vasospasm in man. J Neurosurg 48: 173–178

— Okwuasaba FK, Cook DA, Krueger CA (1980) Pharmacology of vasospasm-effects of various agents including blood on isolated cerebral arteries. In: Wilkins RH (ed) Cerebral arterial spasm. Williams & Wilkins, Baltimore, pp 237–243

— Myles T, Kahn M, Maroun F, Malloy D, Benoit B, McDermott M, Cochrane D, Mohr G, Ferguson G, Durity F (1984) Management of acute subdural hematomas from aneurysmal rupture. Can J Neurol Sci 11: 371–376

Wellum GR, Irvine TW Jr, Zervas NT (1980) Dose responses of cerebral arteries of the dog, rabbit, and man to human hemoglobin *in vitro*. J Neurosurg 53: 486–490

— — — (1982) Cerebral vasoactivity of heme proteins *in vitro*. Some mechanistic considerations. J Neurosurg 56: 777–783

— Peterson JW, Zervas NT (1985) The relevance of *in vitro* smooth muscle experiments to cerebral vasospasm. Stroke 16: 573–581

Welsh FA, Greenberg JH, Jones SC, Ginsberg MD, Reivich M (1980) Correlation between glucose utilization and metabolite levels during focal ischemia in cat brain. Stroke 11: 79–84

Werf AJM van der, Muizelaar JP, Hageman LM, Albrecht KW (1985) The use of the Ca-antagonist nimodipine in the prevention and the treatment of clinical ischaemia after subarachnoidal haemorrhage and the value of early aneurysm surgery. In: Voth D, Glees P (eds) Cerebral vascular spasm. Walter de Gruyter, Berlin, pp 399–404

Wermuth B (1981) Purification and properties of an NADPH-dependent carbonyl reductase from human brain. Relationship to prostaglandin 9-keto-reductase and senobiotic ketone reductase. J Biol Chem 256: 1206–1213

Westergaard E, Go G, Klatzo I, Spatz M (1976) Increased permeability of cerebral vessels to horseradish peroxidase induced by ischemia in Mongolian gerbils. Acta Neuropathol (Berl) 35: 307–325

White RP, Hagen AA, Morgan H, Dawson WN, Robertson JT (1975) Experimental study on the genesis of cerebral vasospasm. Stroke 6: 52–57

— (1979 a) Multiplex origins of cerebral vasospasm. In: Price TR, Nelson E (eds) Cerebrovascular diseases. Raven Press, New York, pp 307–319

— Hagen AA, Robertson JT (1979 b) Effect of nonsteroid anti-inflammatory drugs on subarachnoid hemorrhage in dogs. J Neurosurg 51: 164–171

— Chapleau CE, Dugdale M, Robertson JT (1980) Cerebral arterial contractions induced by human and bovine thrombin. Stroke 11: 363–368

— Hagen AA (1982) Cerebrovascular actions of prostaglandins. Pharmacol Ther 18: 303–331

— (1983) Vasospasm I. Experimental Findings. In: Fox JL (ed) Intracranial aneurysms. Springer, New York, pp 218–249

Wieloch T, Harris RJ, Symon L, Siesjö BK (1984) Influence of severe hypoglycemia on brain extracellular calcium and potassium activities, energy, and phospholipid metabolism. J Neurochem 43: 160–168

— Lindvall O, Blomqvist P, Gage FH (1985) Evidence for amelioration of ischaemic neuronal damage in the hippocampal formation by lesions of the perforant path. Neurol Res 7: 24–26

— (1985) Neurochemical correlates to selective neuronal vulnerability. Prog Brain Res 63: 69–85

Wilkins RH, Alexander JA, Odom GL (1968) Intracranial arterial spasm: a clinical analysis. J Neurosurg 29: 121–134

— Levitt P (1970) Intracranial arterial spasm in the dog. A chronic experimental model. J Neurosurg 33: 260–269

— — (1971) Potassium and the pathogenesis of cerebral arterial spasm in dog and man. J Neurosurg 35: 45–50

— (1976) Aneurysm rupture during angiography: does acute vasospasm occur? Surg Neurol 5: 299–303

— (ed) (1980) Cerebral arterial spasm. Proceedings of the Second International Workshop, Amsterdam, The Netherlands, 1979. Williams & Wilkins, Baltimore

Willis ED (1966) Mechanisms of lipid peroxide formation in animal tissues. Biochem J 99: 667–676

Winn HR, Rubio GR, Berne RM (1981) Editorial. The role of adenosine in the regulation of cerebral blood flow. J Cereb Blood Flow Metab 1: 239–244

— Almaani WS, Berga SL, Jane JA, Richardson AE (1983) The long-term outcome in patients with multiple aneurysms. J Neurosurg 59: 642–651

Winterbourn CC, McGrath BM, Carrell RW (1976) Reactions involving superoxide and normal and unstable haemoglobins. Biochem J 155: 493–502

Witting LA (1980) Vitamin E and lipid antioxidants in free-radical-initiated reactions. In: Pryor WA (ed) Free radicals in biology, vol IV. Academic Press, New York, pp 295–319

Wölbling RH, Aehringhaus U, Peskar BM, Peskar BA (1983) Release of slow-reacting substance of anaphylaxis from layers of guinea pig aorta. Prostaglandins 25: 823–828

Wojtczak L, Lehninger AL (1961) Formation and disappearance of an endogenous uncoupling factor during swelling and contraction of mitochondria. Biochim Biophys Acta 51: 442–456

— (1976) Effect of long-chain fatty acids and acyl-CoA on mitochondrial permeability, transport, and energy-coupling processes. J Bioenerget Biomembr 8: 293–311

Wolfe LS (1982) Eicosanoids: Prostaglandins, thromboxanes, leukotrienes, and other derivatives of carbon-20 unsaturated fatty acids. J Neurochem 38: 1–14

— Pappius HM (1984) Arachidonic acid metabolites in cerebral ischemia and brain injury. In: Bes A, Braquet P, Paoletti R, Siesjö BK (eds) Cerebral ischemia. Elsevier, Amsterdam, pp 223–231

— Rostworowski K, Marion J (1976a) Endogenous formation of the prostaglandin endoperoxide metabolite, thromboxane B_2 by brain tissue. Biochem Biophys Res Commun 70: 907–913

— Pappius HM, Marion J (1976b) The biosynthesis of prostaglandins by brain tissue *in vitro*. Adv Prostaglandin Thromboxane Res 1: 345–355

Wolff HG (1963) Experimentally induced elevation of intracranial pressure and headache observations. In: Headache and other head pain. Oxford University Press, New York, p 120

— Forbes HS (1928) The cerebral circulation. V. Observations of the pial circulation during changes in intracranial pressure. Arch Neurol Psychiat 20: 1035–1047

Wright EM (1972) Mechanisms of ion transport across the choroid plexus. J Physiol (Lond) 226: 545–571

Yamamoto I, Hara M, Ogura K, Suzuki Y, Nakane T, Kageyama N (1983) Early operation for ruptured intracranial aneurysms: Comparative study with computed tomography. Neurosurgery 12: 169–174

Yamamoto YL, Feindel W, Wolfe LS, Katoh H, Hodge CP (1972) Experimental vasoconstriction of cerebral arteries by prostaglandins. J Neurosurg 37: 385–397

Yaşargil MG, Fox JL (1975) The microsurgical approach to intracranial aneurysms. Surg Neurol 3: 7–14

— Antic J, Laciga R, Jain KK, Hodosh RM, Smith RD (1976) Microsurgical pterional approach to aneurysms of the basilar bifurcation. Surg Neurol 6: 83–91

Yashon D, Brown RJ, Hunt WE (1977) Vasoactive properties of prostaglandin compounds on the *in vitro* human basilar artery. Surg Neurol 8: 111–115

Yasuda H, Shimada O, Nakajima A, Asano T (1981) Cerebral protective effect and radical scavenging action. J Neurochem 37: 934–938

— Kishiro K, Izumi N, Nakanishi M (1985) Biphasic liberation of arachidonic and stearic acids during cerebral ischemia. J Neurochem 45: 168–172

Yasui N, Kawamura S, Ohta H, Suzuki A, Kamiyama H, Sayama I, Kubota S (1985a) Clinical grading for subarachnoid hemorrhage caused by ruptured intracranial aneurysm. Neurol Med Chir (Tokyo) 25: 448–454

— Suzuki A, Ohta H, Kamiyama H, Kawamura S (1985b) Rebleeding attack of cerebral aneurysms—clinical significance of early aneurysmal rebleeding. In: Auer LM (ed) Timing of aneurysm surgery. Walter de Gruyter, Berlin, pp 663–672

Yazawa M, Yagi K (1977) A calcium-binding subunit of myosin light chain kinase. J Biochem 82: 287–290

Yoshida S, Inoh S, Asano T, Sano K, Kubota M, Shimazaki H, Ueta N (1980) Effect of transient ischemia on free fatty acids and phospholipids in the gerbil brain. Lipid peroxidation as possible cause of postischemic injury. J Neurosurg 53: 323–331

— Abe K, Busto R, Watson BD, Kogure K, Ginsberg MD (1982) Influence of transient ischemia on lipid-soluble antioxidants, free fatty acids and energy metabolites in rat brain. Brain Res 245: 307–316

— Harik SI, Santiso M, Martinez E, Ginsberg MD (1984) Free fatty acids and energy metabolites in ischemic cerebral cortex with noradrenaline depletion. J Neurochem 42: 711–717

— Ikeda M, Busto R, Santiso M, Martinez E, Ginsberg MD (1986) Cerebral phosphoinositide, triacylglycerol and energy metabolism in reversible ischemia: origin and fate of free fatty acids. J Neurochem, in press

Zervas NT, Lavyne MH, Negoro M (1975) Neurotransmitters and the normal and ischemic cerebral circulation. N Engl J Med 293: 812–816

Zingesser LH, Schechter MM, Dexter J, Katzman R, Scheinberg LC (1968) On the significance of spasm associated with rupture of a cerebral aneurysm. The relationship between spasm as noted angiographically and regional blood flow determinations. Arch Neurol 18: 520–528

— Schechter MM, Dexter J, Katzman R, Scheinberg LC (1969) Regional cerebral blood flow in patients with subarachnoid hemorrhage. Acta Radiol 9: 573–588

Zwetnow NN (1968) CBF autoregulation to blood pressure and intracranial pressure variations. Scand J Clin Lab Invest 22, [Suppl] 102, V: A

— (1970) Effects of increased cerebrospinal fluid pressure on the blood flow and on the energy metabolism of the brain. An experimental study. Acta Physiol Scand [Suppl] 339: 1–31

— Kjällquist A, Siesjö BK (1968) Cerebral blood flow during intracranial hypertension related to tissue hypoxia and to acidosis in cerebral extracellular fluids. Prog Brain Res 30: 87–92

SUBJECT INDEX

B. George / C. Laurian

The Vertebral Artery

Pathology and Surgery

1987. 97 figures. VIII, 258 pages.
Cloth DM 185,–, öS 1295,–
ISBN 3-211-81968-1
Prices are subject to change without notice

In considering vascular surgery in the neck, the carotid artery has usually been regarded as the only vascular axis amenable to surgical approach; surgery on the vertebral artery was contemplated with the idea of mastering a challenge.

The authors demonstrate that the surgical approach, exposure and control of the VA on any part of its cervical course has become a reliable technique at whatever level. Precise technique permits exposure of the VA with a sufficiently large field to allow any surgical procedure identical to that on any other vessel to be performed.

The authors have now performed more than 150 VA operations for various indications with excellent results, no mortality and a very limited morbidity. But the choice of the procedure remains a difficult and controversial problem. Profound knowledge of the anatomy and the frequent variations and anomalies of each VA are of utmost importance.

Therefore the book includes: anatomy of the VA and its variations as far as relevant for the surgeon; extensive description of the pathologies involving the VA and consequences upon the cerebral blood supply; detailed description of surgical approaches to any part of the cervical VA including an original approach with as much safety as carotid artery surgery; report on the authors' personal experience with successes, pitfalls and failures.

These possibilities should be known by every specialist having to cope with deep lesions in the neck or managing cerebral ischemia.

Springer-Verlag Wien New York

Moelkerbastei 5, P.O. Box 367, A-1011 Wien
175 Fifth Avenue, New York, NY 10010, USA
Heidelberger Platz 3, D-1000 Berlin 33
37-3, Hongo 3-chome, Bunkyo-ku, Tokyo 113, Japan

J. Suzuki

Treatment of Cerebral Infarction

Experimental and Clinical Study

1987. 240 partly colored figures. XIV, 380 pages.
Cloth DM 165,–, öS 1160,–
ISBN 3-211-81933-9
Prices are subject to change without notice

Contents: Introduction. – Experimental Study: Experimental Models – Histological Study – Cerebral Blood Flow – Ischemic Brain Edema – Cerebral Metabolism and Free Radical Pathology – The Development of New Brain Protective Agents. – Clinical Study: Epidemiology and Symptomatology – Diagnostic Techniques – Medical and Surgical Treatments of Cerebral Infarction – Revascularization in Acute Stage. – Appendixes: Temporary Occlusion of Trunk Arteries of the Brain During Surgery – The Pathology of Cerebral Vasospasms and Its Treatment – Surgical Therapy for Moyamoya Disease. – References. – Subject Index.

Among the developed nations, Cerebrovascular Disease (CVD) ranks among the top three causes of death and must therefore be regarded as a major health hazard. Although a gradual decrease in the incidence of hemorrhagic CVD has been observed, the number of ischemic CVDs still increases steadily, partly due to higher longevity. Therefore the development of techniques for the prevention and treatment of ischemic diseases of the brain is of utmost importance.

The causes of ischemic CVD are well known, but there remain considerable uncertainties concerning the nature of the gradual intracerebral changes which occur after the ischemic attack. The volume introduces the most recent developments concerning cerebral infarction, reviewing the present state-of-knowledge and the author's own extensive research which resulted in a significant advance in the treatment of the acute stage cerebral infarct, the "Sendai Cocktail", a combination of Mannitol, Vitamin E, Steroids and Phenytoin.

Springer-Verlag Wien New York

Moelkerbastei 5, P.O. Box 367, A-1011 Wien
175 Fifth Avenue, New York, NY 10010, U.S.A.
Heidelberger Platz 3, D-1000 Berlin 33
37-3, Hongo 3-chome, Bunkyo-ku, Tokyo 113, Japan